Advances in Neurobiology

Volume 21

Series Editor
Arne Schousboe

More information about this series at http://www.springer.com/series/8787

Albert Cheung-Hoi Yu • Lina Li
Editors

Systems Neuroscience

Springer

Editors
Albert Cheung-Hoi Yu
Neuroscience Research Institute
Peking University
Beijing, China

Hai Kang Life Corporation Ltd.
China

Lina Li
Neuroscience Research Institute
Peking University
Beijing, China

Hai Kang Life Corporation Ltd.
China

ISSN 2190-5215 ISSN 2190-5223 (electronic)
Advances in Neurobiology
ISBN 978-3-030-06880-6 ISBN 978-3-319-94593-4 (eBook)
https://doi.org/10.1007/978-3-319-94593-4

This Springer imprint is published by the registered company Springer Nature Switzerland AG
The registered company address is: Gewerbestrasse 11, 6330 Cham, Switzerland

Preface

Systems neuroscience is essential for the understanding of higher cognitive functions such as language, memory, emotion, reasoning, pain, neural networks, and self-awareness and is also vital to the study of pathologies and diseases in brain systems. Brain is the most important and complex organ in the body. We believe that a full understanding of its mechanisms is only possible by addressing the properties of single cell types and how they interact with other groups of cells and function as a system through the lens of systems neuroscience.

Over the years, the definition of "Systems Neuroscience" has often been misinterpreted and confused by people who are not working in the field. For example, in an open content online encyclopedia, "Systems Neuroscience" has been misinterpreted as a subdiscipline of systems biology and neuroscience. Systems neuroscience and systems biology, in many aspects, are two entirely different disciplines. In particular, "Systems Neuroscience" is not the "Systems Biology" of neuroscience. Even throughout our invitation process for manuscripts, we have encountered a lot of difficulties due to the chaos in the interpretation of the subjects. With the recent advancement in the field of neuroscience and the initiation of "Brain Initiatives" in many countries, clarification of the terms systems neuroscience, neuroscience and systems biology, and their interconnections becomes urgent and necessary for all of us to have a better understanding of the field. It is particularly worth mentioning that compiling this book has been acclaimed as "an excellent initiative" by some experts in the field which is indeed encouraging!

This timely volume of *Systems Neuroscience* contains chapters from renowned scientists in this emerging field. Topics range from the basic study of systems neuroscience to its applications in disease therapy. The book covers, in more detail, how different neural circuits analyze sensory information, form perceptions of the external world, make decisions, and execute movements; how nerve cells behave when connected together to form the neural networks; and how the relationships between molecular and cellular approaches assist in understanding brain structure and function, high-level mental functions, brain pathologies, and therapy. Both traditional systems neuroscience approaches and systems biology analysis methods were included. A hierarchy of neurobiological complexity, therefore, arises from the genome, transcriptome, proteome, peptidome, metabolome, cells, synapses, circuits, coding, plasticity, brain regions, whole brain, behavior, and other higher brain functions.

It took us 2 years to complete this book. We would like to sincerely thank all the authors for their great contribution, support, and patience, without which this work can never be completed. We would also like to acknowledge Matthew E. R. Butchbach, Tunahan Cakir, Jianguo Chen, Alexander C. Jackson, Veronika Koren, Ying Li, Grace Y. Sun, Miklos Vegh, Yuguo Yu, and Li Zhang, who enthusiastically contributed their knowledge and time in reviewing manuscripts for this special volume. Last but not least, we would like to thank our students Jiangshan Zhan, Tarokh Mollafarajzadeh, and Sara Rashidi for their assistance in the peer review process. We hope that this book would fill some gaps in knowledge and provide tantalizing samples of recent progress in research to promote the continuous development in the field of systems neuroscience and provide great insight in allowing effective integration of systems biology into neuroscience.

Beijing, China

Albert Cheung-Hoi Yu
Lina Li

Contents

Contributors

Yuri I. Alexandrov Department of Psychology, National Research University Higher School of Economics, Moscow, Russia

Shvyrkov's Lab, Neural Bases of Mind, Institute of Psychology, Russian Academy of Sciences, Moscow, Russia

Paulo E. Andrade Department of Psychology, Goldsmiths, University of London, London, UK

Vishwanath T. Anekonda VA Puget Sound Health Care System, Office of Research and Development Medical Research Service, Department of Veterans Affairs Medical Center, Seattle, WA, USA

Joydeep Bhattacharya Department of Psychology, Goldsmiths, University of London, London, UK

James E. Blevins VA Puget Sound Health Care System, Office of Research and Development Medical Research Service, Department of Veterans Affairs Medical Center, Seattle, WA, USA

Division of Metabolism, Endocrinology, and Nutrition, Department of Medicine, University of Washington School of Medicine, Seattle, WA, USA

Mark J. Buckley Department of Experimental Psychology, Oxford University, Oxford, UK

Gully A. P. C. Burns Information Sciences Institute, Viterbi School of Engineering, University of Southern California, Marina del Rey, CA, USA

Matthew E. R. Butchbach Center for Applied Clinical Genomics, Nemours Biomedical Research, Nemours/Alfred I. duPont Hospital for Children, Wilmington, DE, USA

Center for Pediatric Research, Nemours Biomedical Research, Nemours/Alfred I. duPont Hospital for Children, Wilmington, DE, USA

Department of Pediatrics, Thomas Jefferson University, Philadelphia, PA, USA

Department of Biological Sciences, University of Delaware, Newark, DE, USA

Tunahan Çakır Department of Bioengineering, Gebze Technical University, Gebze, Kocaeli, Turkey

Kai Gao Department of Pediatrics, Peking University First Hospital, Peking University Health Science Center, Beijing, China

Vladimir V. Gavrilov Shvyrkov's Lab, Neural Bases of Mind, Institute of Psychology, Russian Academy of Sciences, Moscow, Russia

Alexander G. Gorkin Shvyrkov's Lab, Neural Bases of Mind, Institute of Psychology, Russian Academy of Sciences, Moscow, Russia

Alice H. Grant UTEP Systems Neuroscience Laboratory, University of Texas at El Paso, El Paso, TX, USA

Department of Biological Sciences, University of Texas at El Paso, El Paso, TX, USA

Graduate Program in Pathobiology, University of Texas at El Paso, El Paso, TX, USA

Yuwu Jiang Department of Pediatrics, Peking University First Hospital, Peking University Health Science Center, Beijing, China

Arshad M. Khan Sr UTEP Systems Neuroscience Laboratory, University of Texas at El Paso, El Paso, TX, USA

Department of Biological Sciences, University of Texas at El Paso, El Paso, TX, USA

Border Biomedical Research Center, University of Texas at El Paso, El Paso, TX, USA

Weijing Kong Department of Pediatrics, Peking University First Hospital, Peking University Health Science Center, Beijing, China

Evgeniya A. Kuzina Shvyrkov's Lab, Neural Bases of Mind, Institute of Psychology, Russian Academy of Sciences, Moscow, Russia

Ying Li Department of Biomedical Sciences, City University of Hong Kong, Kowloon, Hong Kong

Centre for Biosystems, Neuroscience, and Nanotechnology, City University of Hong Kong, Kowloon, Hong Kong

School of Veterinary Medicine, City University of Hong Kong, Kowloon, Hong Kong

Xiaoyan Liu Department of Pediatrics, Peking University First Hospital, Peking University Health Science Center, Beijing, China

Farshad A. Mansouri Cognitive Neuroscience Laboratory, Department of Physiology, Monash Biomedicine Discovery Institute, Monash University, Clayton, VIC, Australia

Roger Marek Queensland Brain Institute, The University of Queensland, Brisbane, QLD, Australia

Anais Martinez UTEP Systems Neuroscience Laboratory, University of Texas at El Paso, El Paso, TX, USA

Department of Biological Sciences, University of Texas at El Paso, El Paso, TX, USA

Graduate Program in Pathobiology, University of Texas at El Paso, El Paso, TX, USA

Daniel H. Moralejo Division of Neonatology, Department of Pediatrics, University of Washington School of Medicine, Seattle, WA, USA

Emrah Özcan Department of Bioengineering, Gebze Technical University, Gebze, Kocaeli, Turkey

Zachary S. Roberts VA Puget Sound Health Care System, Office of Research and Development Medical Research Service, Department of Veterans Affairs Medical Center, Seattle, WA, USA

Pankaj Sah Queensland Brain Institute, The University of Queensland, Brisbane, QLD, Australia

Alexey A. Sozinov Shvyrkov's Lab, Neural Bases of Mind, Institute of Psychology, Russian Academy of Sciences, Moscow, Russia

Faculty of Psychology, National Academic University of Humanities, Moscow, Russia

Olga E. Svarnik Shvyrkov's Lab, Neural Bases of Mind, Institute of Psychology, Russian Academy of Sciences, Moscow, Russia

Susumu Takahashi Graduate School of Brain Science, Doshisha University, Kyoto, Japan

Brendan S. Thatcher VA Puget Sound Health Care System, Office of Research and Development Medical Research Service, Department of Veterans Affairs Medical Center, Seattle, WA, USA

Benjamin W. Thompson VA Puget Sound Health Care System, Office of Research and Development Medical Research Service, Department of Veterans Affairs Medical Center, Seattle, WA, USA

Jingmin Wang Department of Pediatrics, Peking University First Hospital, Peking University Health Science Center, Beijing, China

Xiru Wu Department of Pediatrics, Peking University First Hospital, Peking University Health Science Center, Beijing, China

Ye Wu Department of Pediatrics, Peking University First Hospital, Peking University Health Science Center, Beijing, China

Han Xie Department of Pediatrics, Peking University First Hospital, Peking University Health Science Center, Beijing, China

Albert Cheung-Hoi Yu Laboratory for Functional Study of Astrocytes, Department of Neurobiology, School of Basic Medical Sciences, Key Laboratory for Neuroscience, Ministry of Education, National Health and Family Planning Commission, Neuroscience Research Institute, Peking University, Beijing, China

Hai Kang Life Corporation Ltd., China

Yuguo Yu State Key Laboratory of Medical Neurobiology, School of Life Science and Human Phenome Institute, Institutes of Brain Science, Institute of Science and Technology for Brain-Inspired Intelligence, Fudan University, Shanghai, China

Feng Zhang Institute of Reproduction and Development, Obstetrics and Gynecology Hospital of Fudan University, Shanghai, China

Ling Zhang Institute of Reproduction and Development, Obstetrics and Gynecology Hospital of Fudan University, Shanghai, China

Yuehua Zhang Department of Pediatrics, Peking University First Hospital, Peking University Health Science Center, Beijing, China

Yujia Zhang Department of Pediatrics, Peking University First Hospital, Peking University Health Science Center, Beijing, China

Shanglin Zhou State Key Laboratory of Medical Neurobiology, School of Life Science and Human Phenome Institute, Institutes of Brain Science, Institute of Science and Technology for Brain-Inspired Intelligence, Fudan University, Shanghai, China

Chapter 1
Neuronal Bases of Systemic Organization of Behavior

Yuri I. Alexandrov, Alexey A. Sozinov, Olga E. Svarnik, Alexander G. Gorkin, Evgeniya A. Kuzina, and Vladimir V. Gavrilov

Abbreviations

IEG	Immediate early gene
RSC	Retrosplenial cortex
TFS	Theory of Functional Systems
LTP	Long-term potentiation

1.1 The Systems View of Neuroscience

1.1.1 Goal-Directed Behavior and the Result

In contemporary neuroscience there is a problem of isolation of meaningful segments of behavior and related to them brain activity. In many cases brain activity is averaged over the period right after the presented stimulus, despite the fact that in

Y. I. Alexandrov (✉)
Department of Psychology, National Research University Higher School of Economics, Moscow, Russia

Shvyrkov's Lab, Neural Bases of Mind, Institute of Psychology, Russian Academy of Sciences, Moscow, Russia

A. A. Sozinov
Shvyrkov's Lab, Neural Bases of Mind, Institute of Psychology, Russian Academy of Sciences, Moscow, Russia

Faculty of Psychology, National Academic University of Humanities, Moscow, Russia

O. E. Svarnik · A. G. Gorkin · E. A. Kuzina · V. V. Gavrilov
Shvyrkov's Lab, Neural Bases of Mind, Institute of Psychology, Russian Academy of Sciences, Moscow, Russia

A. Cheung-Hoi Yu, L. Li (eds.), *Systems Neuroscience*, Advances in Neurobiology 21, https://doi.org/10.1007/978-3-319-94593-4_1

this case all predictive activity is completely lost. Considering all behavior as goal-directed entails the necessity of viewing brain activity as related to the future events. Such perspective was suggested in the framework of the functional systems theory by P.K. Anokhin (1974) and has been developed further within the system-evolutionary theory (Shvyrkov 1986).

An important point of the Theory of Functional Systems (TFS) is the definition of a factor that unites sparse brain and body elements into a system ("systems-creating factor"). The factor is a *result* of the system—an adaptive effect in the organism-environment interaction achieved upon realization of that system. The result is isomorphic in relation to any system. Therefore, the systems approach can be applied to various research objects and behavioral situations. Notably, in relation to performed behavior the result is a future event, not a past one, like stimulus. The system is understood as a dynamic organization of activity of components with different anatomical localization, where interaction becomes a mutual facilitation in the process of ensuring an adaptive for an organism result (see also Alexandrov et al. 2000).

How can the result, a future event, be a reason of current activity? In the TFS this "time paradox" is explained via the concept of goal—a model of the result which contains its predicted parameters and is provided by the "acceptor of the result of an action". Thus P. Anokhin (1974) had resolved the disconnection between causality and teleology in the description of behavior in the form acceptable for those dedicated to causality as a necessary principle of scientific analysis (Bunge 1963).

The TFS enables a holistic view of behavior through studying the result-driven organization of entire organism-environment interactions (Anokhin 1974), unlike the traditional view of functions as direct effects of a certain substrate, including the nerve tissue, e.g. motor functions, sensory, emotional, motivational etc. A similar view expressed by other authors also suggests considering the activity of any brain area with respect to behavioral performance as explained by implementation and selection of systems (Cisek and Kalaska 2010). The "task space" representation, proposed by Weible et al. (2009) for the anterior cingulate cortex, is an idea to a certain extent alike, if expanded to the whole organism. In the TFS the function is defined as achievement of a result. Such systemic function can not be localized, it is applied to the whole organism that interacts with the environment.

Multiple research efforts within the TFS framework have led to creation of the system-evolutionary approach (Shvyrkov 1986; see also Alexandrov 2015). One of the most important steps of this approach was the solution of the psycho-physiological (mind-body) problem. The psychic processes describe the organism and its behavior as a whole. Physiological processes are considered on the level of elements. The *organization* of physiological processes into a system is based on neither psychic, nor physiological, but specific (informational) systemic processes. Their substrate is physiological activity, whereas their informational content is psychic. Thus, behavioral acts are not only based on localized physiological processes, but also on the processes of their organization. In other words, psychical and physiological processes are different aspects of the same systemic processes.

It is important to note that the described solution of the psychophysiological problem excludes "theoretical reductionism" (see Dudai 2002 for types of reductionism), i.e. degrading the psychic processes to physiological ones. This aspect seems especially important within the problem of consciousness (Alexandrov and Sams 2005): since consciousness cannot be simplified by analytical means, it is often discarded from the scope of scientific investigation (Kandel 2006).

The specific issues under investigation of systemic psychophysiology are formation and realization of systems, their taxonomy, and the dynamics of intersystem relations in behavior. The systemic psychophysiology rejects the reactivity paradigm and employs goal-directed principles in the analysis of activity of individuals and, more importantly, neurons. Therefore, it is free from the eclectic explanations of goal-directed behavior by reflex-based mechanisms (see Alexandrov 2015 for more details).

We acknowledge that the presented view shares certain aspects of cognitive structures. For example, U. Neisser's cognitive schema concept also includes prediction of incoming information, guidance of the exploration, and modification during execution (Neisser 1976; see also Moscovitch et al. 2016). Moreover, it presumes simultaneous activation of schemas on different levels of hierarchy (see related assertions in paragraph 4). Other resemblant views are presented in the brain activity interpretations by Engel et al. (2001) and von Stein et al. (2000). The cognitive maps (Tolman 1948), reconstructed via place-cell activity analysis etc. (Burgess and O'Keefe 2011; Hartley et al. 2013; O'Keefe 1976), and cognitive schemas (Bartlett 1995), modeled in consolidation research (Hennies et al. 2016; Tse et al. 2007, 2011) also reveal some of these properties (see also Dudai et al. 2015). However, a clear formulation of what makes a system (the result) is of critical importance for considering cognitive units in terms of the individual and its interaction with the environment, rather than in terms of environment proper. This leads to interpretation of neuronal firing and other physiological measures from living organisms as a manifestation of their *activity*.

1.1.2 Activity Paradigm

The view of behavior as aimed at future results assumes that the principal feature of the living matter is its activity. The concrete form of this activity is defined by the level of the matter organization (Anokhin 1974). The activity principle presumes that behavior is driven by a model of its result.

The classical TFS includes the concept of the "starting stimulus". All the processes of system organization are goal-directed, whereas the starting stimulus solely triggers the execution of integrated elements. And even this role of the stimulus, that seems necessary, disappears when the behavioral act is considered not as an isolated entity, but as a component of a behavioral continuum, that is, the succession of behavioral acts performed by an individual during lifetime. The given behavioral act within a continuum is deployed after the result of the previous act has been achieved and evaluated. The evaluation of the results of the given act is a necessary part of the next act initiation. These processes serve as a transition from the execution of one act to a subsequent one. There is no room for a stimulus in the continuum. The environmental changes described as stimuli are contained in the model of a preceding result. They are conditions, but not causes of behavior. Unexpected changes will either have no effect on the continuum (i.e. "ignored"), or serve a condition for behavior that interrupts the succession: either repetition of the interrupted behavioral act, or building a new one via systemogenesis (see Sect. 1.2.1). In any case, both are aimed at the future, and their cause is a mismatch (see Sect. 1.3.1).

Provided that the whole organism is active on the level of behavior, the neuronal firing would also be considered as manifestation of activity.

1.1.3 Active Neuron

Within the reactivity paradigm behavior is a reaction based on the transmission of excitation in a circuit (or a net). The function of a neuron is therefore forwarding of excitation. Events recorded in the neuron are considered as a response to a stimulus, that had affected some part of it and may travel further along the cell to be a stimulus to other nerve cells. Thus, a neuron, just like an organism, responds to stimuli.

The activity paradigm also dictates coherent understanding of both the whole organism functioning, and that of a single cell in a multicellular organism. This correspondence has been achieved by treating events in a neuron or any other living cell as execution of a genetic program. The execution requires receiving metabolites from other cells (Shvyrkov 1986; Alexandrov 2015). Consequently, neuronal activity, alike behavior of an organism, is not a response, but a way of changing its relation to environment. Events in the neuron are "actions" that change its microenvironment with respect to its "needs", causing modifications in blood flow, metabolic inflow from glial cells, and activity of other neurons. Therefore, a neuron is not a conductor or a calculator—it's an organism inside organism.

A neuron can satisfy its metabolic "needs" only by co-action with other elements of an organism to form a functional system. Their cooperative, joint activity leads to a new relation between the whole organism and its environment, as well as (at the cellular level) to satisfying metabolic "needs" of the cells. As soon as the result is achieved by the organism (and metabolites are received by a neuron), the firing of the neuron ceases.

This view of neuronal activity corresponds to the evolutionary perspectives that show similarities between survival principles of single-cell organisms and neurons within nervous system. It has been shown that the colonies of single-cell organisms and cells in a multicellular organism provide for breath, nutrition, and other group functions via cooperation—they synchronize their metabolic processes (e.g. Weber et al. 2012). The satisfaction of all various metabolic requirements of an organism is achieved by diverse behavioral acts. It can also be argued that besides regular functioning a neuron is active during apoptosis, or "altruistic suicide" (see Sect. 1.3.3).

The systemic view of the neuronal activity requires a corresponding approach to investigation of learning and memory (see also Alexandrov 2008).

1.2 The Formation of Memory during Learning and Systemic Structure of Behavior

1.2.1 Systemogenesis

The key notion in the TFS besides the system is that of development, revealed in the concept of systemogenesis. Systemogenesis refers to the idea that maturation of organs is not homogeneous—the first elements to mature during early ontogenesis are those parts of organs and tissues that are essential for achieving the results of the systems to ensure the survival of the organism at these stages of individual development (Anokhin 1974).

It has been argued that the systemogenesis also occurs during learning in adults—the emergence of a new system provides a new behavioral act. It was proposed that due to successive emergence of new systems during lifetime the role of different neurons in the organization of behavior should be considered on the basis of individual history (Alexandrov and Alexandrov 1982). This hypothesis has led to the systems-evolutionary theory and the system-selection concept of learning (Shvyrkov 1986). The latter construct is in accordance with G. Edelman's (1987) view of neuronal ensemble formation during learning as a selection process (excerpt of cells on the basis of their features), as opposed to learning as instruction (modification of neuronal features caused by stimulation). The principle of selection underlies the immune and evolutionary processes in a similar way, although at different time scales.

According to G. Edelman, the selection process starts in the early ontogeny, when lots of neurons die during brain maturation. The cells that survive this selection were termed a "primary assortment". The "secondary assortment" is formed upon learning via behavioral interaction with the environment at the second stage of the selection.

Changeux had also defined two stages of the selection: the "neural Darwinism" in the early (including prenatal) ontogeny, proposed to be a selection of effective synapses, and the "mental Darwinism" in adults considered as modification of the existing synaptic connections. However, the units assumed to be subject for the selection are not the connections, but groups of neurons that were selected at the first stage and had cooperative activity (Changeux and Connes 1999).

The essence of the selection process in both individual development and evolution is the resulting outcome. As Wright (1995) puts it, it's not the "truth" that the brain needs, but a success. Similarly, the evolution "supports" those who have survived, not those who were right (Cacioppo and Gardner 1999). The selection during evolution applies not to independent features, but to the holistic organisms, the phenotypes. The basis for the selection is the achievement of results beneficial for the given phenotypic variation. These phenotypes are the only objects of the selection (Shvyrkov 1986; Fodor 2007). Success in the selection process, defined by the quality of the achieved results, includes formation of "pre-specialized" (see below in this section) and "specialized" neurons (see example in Fig. 1.1).

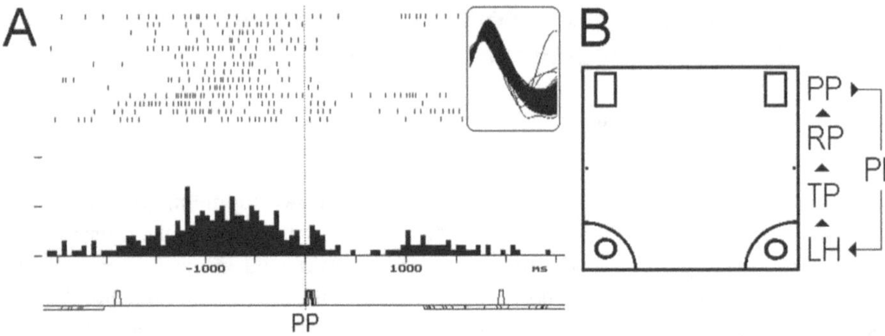

Fig. 1.1 Activity of a representative specialized neuron #1938 recorded in rabbit's anterior cingulate cortex. (**a**) The raster plot shows spikes during successive turns of the animal towards a pedal aligned to the pedal pressing (PP). The inset shows spike waveforms, selected after sorting. Below is the histogram of these spikes (50 ms in a bin, the ticks of the ordinate show tens of spikes). The bottom panel shows superimposed behavioral markers (up-deflection for pedal-pressing and down-deflection for lowering the head into the feeder). In this experiment (Sozinov et al. 2012) the activity of neurons was recorded with glass electrodes (3–6 MOhms @ 1 KHz) after reaching the learning criterion—see text and panel (**b**). For definition of specialization, see text. (**b**) The schematic view of the experimental chamber from the top. The behavioral markers (beam crossing) were used to identify the following behavioral acts (that corresponded to the stages of learning): lowering head and taking food from the feeder (LH); lifting head from feeder and turning it toward pedal (TP); moving to pedal corner (RP); pedal pressing (PP); running from pedal to feeder (PF). This sequence of acts was looped (arrows) during neuronal recordings (10+ cycles on each side)

In the frames of the systemic approach, the individual development is a sequence of systemogeneses that provide new interactions with environment. Formation of a new system during systemogenesis is considered as emergence of a new element of individual (subjective) experience during learning. The new system consists of the neurons that were selected from the "reserve" cells—presumably, those are low active or silent cells. These cells correspond to the primary assortment (according to Edelman 1987) and termed here "pre-specialized". These are the neurons subjected to selection during learning, when some of them become specialized in relation to a system of a new behavioral act. This selection process is defined by specific metabolic features of these cells. Accordingly, the group of selected units can be termed the secondary assortment. Therefore, the new system is an addition to the existing ones, it is 'superimposed' on them.

These evolutionary considerations presume that the specialization of neurons remains for their whole lifetime. The learning process would then be provided by recruiting new cells, rather than by retraining the ones previously trained. This is in accordance with experimental data (Schmidt et al. 1976; Thompson and Best 1990; Wilson and McNaughton 1993; Swadlow and Hicks 1997; Williams et al. 1999; Greenberg and Wilson 2004; Brecht et al. 2005; Chestek et al. 2007; Jackson et al. 2007; Fraser and Schwartz 2012) that show "silent" cells in the brain of different species, increase of the number of active cells during learning, and constancy of new neuronal specializations (during the whole registration period—weeks or even months and years, see also McMahon et al. 2014).

The adult neurogenesis in birds and mammals (Paton and Nottebohm 1984; Carleton et al. 2003) has been shown to be related to learning. The learning process facilitates the survival of newborn neurons (the "use it or loose it" principle—Kempermann et al. 1998), as well as proliferation (Prickaerts et al. 2004), whereas inhibition of neurogenesis disrupts memory formation (Shors et al. 2001). On the basis of these and other related data (Anacker and Hen 2017; Frankland et al. 2013) we propose that the adult neurogenesis may support the formation of new systems (see general scheme on Fig. 1.7). Therefore, both the "reserve" cells and newborn cells can be specialized in relation to new systems during learning. The adult neurogenesis may also contribute to the reinstatement of the primary and secondary assortments of neurons in pathology (Xue 1998). This compensation of loss of neurons, including the pre-specialized neurons can possibly also occur in a healthy organism. Since the latter assumption is less grounded, the corresponding relation is marked with a question on Fig. 1.7.

1.2.2 The Formation of Neuronal Specializations during Individual Development Continues Phylogenesis

The emergence of the nervous system is a "revolutionary" event in the evolution, because it had provided radical increase of complexity and variability of behavior. The complexity of organisms and the genome size do not seem to correlate (Gregory 2001). However, the number of cell types does correspond to the phylogenetic complexity (Bonner 1988). Importantly, it is the nervous system that had contributed the most to this increase. The cell types in the nervous systems are of great, evidently innumerable, variety (DeFelipe 2011). Moreover, the combinations of different cell specializations are individual, because the specialization is formed in relation to the elements of individual experience—the functional systems. Therefore, number of unique sets of specializations equals the number of individuals. In other words, every individual has a unique (although culture-specific) composition of systems. The scope of all possible specialization types depends on the species and the subset of neurons pre-specialized during early ontogeny (the primary assortment). Accordingly, the composition of neuronal specializations (the secondary assortment) is individual. Within the presented view of development as a formation of new specializations the ontogeny appears as phylogeny continued through the increase of the number of cell specialization types.

1.2.3 The Patterns of Neuronal Specializations in Different Species

The research in our laboratory includes recording of neuronal spikes from brains of animals during cyclic operant appetitive behavior (see below in this section). The experimental protocols are in accordance with the Council of the European

Communities Directive of November 24, 1986 (86/609 EEC) and the National Institutes of Health "Guidelines for the Care and Use of Animals for Experimental Procedures", and were approved by the ethics committee of the Institute of Psychology, Russian Academy of Sciences. The specialization of a neuron in relation to a system is assessed via probability of activation in behavioral acts. If this probability reaches 100% in one or more acts, then the neuron is considered specialized, and the activations of a given act are called "specific" activations (see Alexandrov et al. 2013 for more details). An example of specific activations of a specialized neuron is presented in Fig. 1.1a.

According to the framework presented above, the individual reflects interaction with the physical world, rather than the world itself. This reflection depends on the individual goals and history and can be described on the basis of the individual structure of memory (see paragraph 4). Any individual is essentially a composition of both the phylogenic and ontogenic memory. Thus, we have proposed that different species and even different individuals, who acquire new behavior in the same "resultative milieu" (operant food-acquisition behavior), would have memory structure that has similarities and differences, revealed by comparing patterns of neuronal specializations in various brain regions. The similarities would be explained by the identity of achieved results, whereas the differences would reveal the peculiarities of species and the history of learning. The patterns of neuronal specializations are relative numbers of neurons specialized in relation to different systems. Therefore, the pattern reveals a particular" "set" of systems.

We have compared the patterns of neuronal specializations in the homological areas of cingulate and motor cortices of rats and rabbits (retrosplenial cortex, RSC, according to Paxinos and Watson 1997 in rats, and according to Vogt et al. 1986 in rabbits). The animals acquired operant appetitive behavior in a chamber with two pedals that activated corresponding feeders. The chamber viewed from the top was axially symmetric with a pedal and a feeder in adjacent corners of each of the two sides of a square—Fig. 1.1b). The rats' chamber was 1/3 of the size of the rabbits' chamber.

The correct performance was a looped movement from pressing a pedal through turning to corresponding feeder facing a wall to eating in the feeder, and turning back to the same pedal (10–15 loops until switch to the opposite side). The behavioral cycle was divided into several acts (Fig. 1.1b): lowering head and taking food from the feeder; lifting head from feeder and turning it toward pedal; moving to pedal corner; pedal pressing; running from pedal to feeder. This division was based on the stages of learning that had been introduced daily during training. The significant increase of the firing rate above background frequency was termed activation. Details of the experimental setup and procedures have been described in more detail elsewhere (e.g. Alexandrov et al. 1990, 2013).

The symmetric arrangement of the chamber allows for classification of neurons according to how discriminatory its firing is in relation to the behavioral acts. For example, the neuron on Fig. 1.1a has activations in each of the consecutive runs to

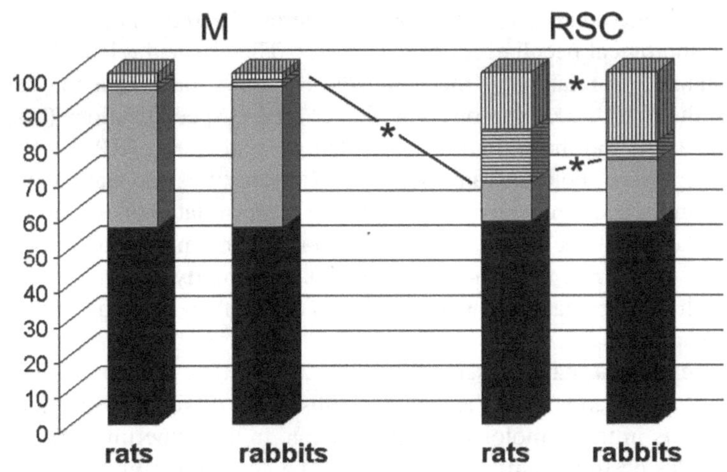

Fig. 1.2 Percentage of neurons with different specialization in motor and retrosplenial cortical areas in rats and rabbits. Black—unidentified, grey—"old", white—"new" (horizontal—feeder-related, vertical—pedal-related). Asterisk: significant differences between percentages of neurons. See text for details

the pedal on the right side of the chamber. Then its classification would depend on specific activations on the left side. If there's none, the neuron is specialized in relation to a system that subserves approaching the right pedal, which is peculiar for the new behavior (the case for this neuron). However, if there's activation during approaching the right feeder, the firing might be explained by a left turn of the animal's head and/or body (verified in additional tests). Then the neuron might belong to a system that subserves (and presumably have subserved) acts of behavior beyond the experimental chamber. Consequently, as opposed to the neurons of the "new" systems, the latter were termed neurons of the "old" systems.

In one of the experiments within this paradigm, activity of single neurons was recorded from RSC and motor (M) cortices in rats and rabbits (see Gavrilov et al. 2002 for more details). In the RSC most of the specialized units were classified as belonging to the systems of "new" behavioral acts, whereas most of the M neurons were of "old" systems (e.g. context-independent activations during any particular movement, or during taking any food or non-eatable objects from the feeder or anywhere in the chamber). Thus, the percentage of units with "new" specializations in RSC was (at least seven times) higher than in the M (Fig. 1.2). The difference between the numbers of neurons with old and new specializations between the two cortical areas was significant and equi-directional in rats and rabbits. The relative old/new systems content in the two cortices shows general similarity of the specialization patterns in the two species.

Meanwhile, the details of specialization patterns differed—presumably, according to the ethological peculiarities of the species. The rats had relatively more neurons specialized in relation to the new acts of taking food from the feeder. We contrasted the numbers of neurons with feeder-related specializations (acts PF and LH on Fig. 1.1b) and those with pedal-related specializations (RP and PP). While these numbers were about the same in rats' RSC, the rabbits had significantly (some four times) more pedal-related neurons than the feeder-related ones. This difference may be explained by the peculiarities of food taking and manipulation, which is of great variety in the rats (Whishaw et al. 1998). Apparently, the greater number of neurons with new specializations in rats is also due to their more differentiated and complex behavior.

Consequently, the data presented reveal both the task-related similarities and species-derived differences of the neuronal subserving of similar behavior between rats and rabbits in the homological cortical areas. In this experiment the recording of neurons was performed after acquisition of the pedal-pressing behavior, and the specializations were revealed during asymptotic performance. Meanwhile, the investigation of cognitive components necessitates consideration of their emergence. Within our framework, the systems that underlie behaviors appear via systemogenesis, whereas in the conventional terms, new memory undergoes a process of consolidation.

1.2.4 The Traditional View of Memory Consolidation

The processes of the acquisition and consolidation of memory attract the best modern expertise in both the methods and conceptual schemes (e.g. Feld and Born 2017; Kitamura et al. 2017; Moscovitch et al. 2016). However, most of the schemes and investigations are based on the old Descartes' concept of memory traces: the traces are made of the pores that become more permeable as the spirit repeatedly passes through them during the behavior execution.

The issues that follow this idea are those of the mechanisms and limitations of pore enlargement, the brain structures with different amount of pores, of pore permeability duration, etc. These issues, translated from the seventeenth century to the modern terms (from pores to synapses, from spirits to neuronal firing), maintain their essence under the concept of engram. Unfortunately, the approaches to consolidation, albeit very distinguished (see Dudai 2012 for review), rely mainly on long-term increases of conduction effectiveness in circuits, networks, etc. "The current central dogma of synaptic consolidation is that it involves stimulus ("teacher")-induced activation of intracellular signaling cascades, resulting in posttranslational modifications, modulation of gene expression, and synthesis of gene products that alter synaptic efficacy" (Dudai 2012, p. 228).

1.2.5 The Systems View of Memory Consolidation

From the systems point of view, the neuron is a result achiever, rather than a conductor of excitation. Therefore, the issue of the conduction efficiency increase is out of the scope of the systems approach. The learning process is considered as formation of a new system of co-active cells (including neurons) of different localization, not necessarily directly connected. This view excludes the concept of "trace" left solely by the instructive input due to plasticity of the nervous system.

The systems view of consolidation was formed on the basis of the systems approach (above). However, the experimental evidence leads other authors to similar conclusions. For example, G. Horn claims that the cross-correlational analysis of neuronal activity in IMHV of domestic chicks does not confirm that the connectivity of "imprint-responsive" neurons is increased during learning, as predicted by the Hebbian rule. "Rather, – the author concludes, – the neurons might form a set of parallel, largely uncoupled elements that are likely to provide a larger storage capacity than a system with tightly coupled elements" (Horn 2004, p. 121). Although functional connectivity may indeed increase after learning (Abdou et al., 2018), we believe that the connectivity affords synchronization of metabolic activity between structurally connected neurons (see Sect. 1.1.3), and G. Horn's conclusion remains accurate for the rest of neurons of the same specialization with no direct connections, and even more so for the somatic cells. Different kinds of network approach in the analysis of synaptic (Hoshiba et al. 2017), cellular (Adams et al. 2017) and whole-brain (Lohmann et al. 2016) processes share some aspects of the systems approach, albeit they largely retain instruction-based view on learning, and hence the issue of a unit-of-analysis (Korhonen et al. 2017).

The systems description of the consolidation process necessarily includes two groups of interdependent processes: the systemic specialization, and accommodative reconsolidation. The former applies to the morphological and functional modifications of a neuron that provide its involvement into a new system (described above). The definition of the latter process requires several preliminary considerations.

Of importance is that a new memory is dynamic and adaptive, rather than a stable entity (Bartlett 1995). The recent progress of memory reconsolidation research shows the modification of memory after post-consolidation retrieval at the molecular level (Nader 2015; Sara 2000). Memory formation and reactivation require protein synthesis, although the consolidation and reconsolidation processes are not identical (Anokhin et al. 2002; Dudai and Eisenberg 2004). Therefore, the protein synthesis-dependent consolidation reveals a wide range of "active" memory processes (Nader 2003, 2015), rather than just those of "new" memory.

The idea of reconsolidation does not contradict to the above notion of permanent specialization. The reconsolidation does not rule out the changes that had underlied the long term memory formation (Nader et al. 2000). However, it does constitute another, supposively less influential, step of differentiation process for a neuron.

We consider learning as specialization of a group of neurons in relation to a new system. The new system is not a substitution, but an addition to the previously formed systems. It follows that this addition would necessitate the coordination between new and prior elements. Current scope of evidence on reconsolidation shows that reconsolidation may indeed be the general mechanism of prior memory reorganization after new learning (see Dudai et al. 2015; Hupbach et al. 2008; McKenzie and Eichenbaum 2011).

We have suggested earlier that the neurons that are specialized in relation to a system of one behavioral act may modify their activity and be involved in another behavior without changing the specialization (see Alexandrov 2008). Later, the acute (Alexandrov et al. 2001) and chronic tetrode (see Alexandrov 2008) recordings provided more evidence for reorganization of an existing system upon acquisition of a new behavioral act. Namely, the chronic recording of neuronal activity was made during acquisition of the appetitive operant behavior, described in Sect. 1.2.3. When the animals reached an asymptote level of pedal-pressing on the first side of the chamber, the pedal was turned off to start training on the second side. Upon reaching the same criterion there, the animals were returned to the first side. Consequently, the sides were alternated 10–20 cycles of pedal-pressing each. Thus, activity of several neurons was tracked on the first side of the chamber before and after training on the second one.

Three of these neurons with activations specific to acts on the first side changed their activity patterns after initial training series on the second side. Activity of one of these cells is shown in Fig. 1.3 (the firing frequency of this cell changed significantly in several acts, including the specific act LH (see Sect. 1.2.3); see also panel A on Fig. 1.4 for learning-induced activation changes in a neuron with activations specific to preceding behavioral acts). Notably, these changes remained significant in all subsequent series unlike temporary changes of activity of specialized neurons in the first trials of specific acts after alternation or rest periods. The modifications of this kind were termed by us "accommodative" reconsolidation (Alexandrov et al. 2001).

Results that point to reorganization of previously formed system after acquisition of a new one were also received by us via immediate early gene (IEG) expression analysis (Svarnik et al. 2013). This study was designed to control for learning prior to operant food-acquisition by pedal-pressing to reveal activation of the first-skill-specific neurons during acquisition of the second one. In the experimental group of rats the first skill was a "whisking task"—that of using left or right whiskers to receive a water drop. These animals were overtrained for 5 days before the second skill of pedal pressing for food had been introduced. The control group acquired the same food-acquisition behavior, but the first task was a non-instrumental drinking instead of the "whisking". Albeit the second skill did not involve the whiskers, we have found c-Fos expression in significantly greater number of barrel-field neurons in animals of the experimental group compared to the control. These data may suggest that c-Fos induction during the second training took place in neurons that were specialized in relation to the first, "whisking" task, which is a sign of accommodative reconsolidation. Therefore, besides the specialization of neurons in relation to new systems, we consider morphological and functional modifications of previously specialized neurons. These modifications do not change the specialization and provide inclusion of a new system into the existing structure of individual experience.

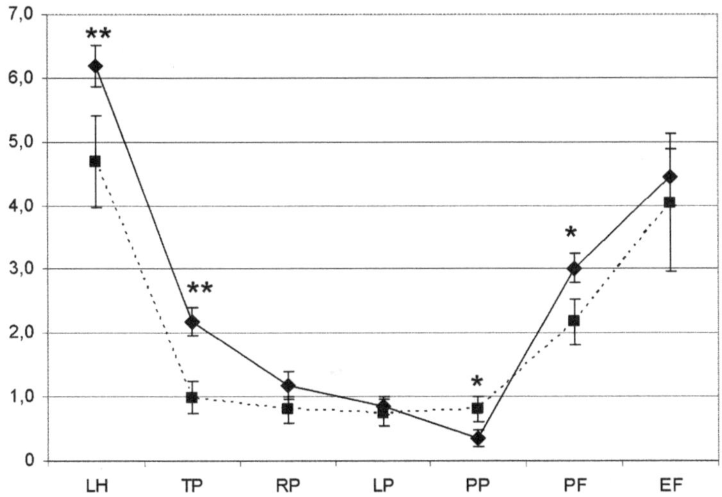

Fig. 1.3 Firing frequency of neuron #261204-cl6 in consecutive series of behavioral acts on the first side of the chamber before (solid line) and after (dashed) training on the second side. Ordinate: mean ± SEM spike frequency (spikes per second). Abscissa: acts of food-acquisition behavior on first side of the cage (see Fig. 1.1b for definition of behavioral acts; LP—locating in pedal corner before pedal pressing; EF—visiting empty feeder). Significant difference between the two series: Mann-Whitney $*p < 0.05$; $**p < 0.01$

It had been previously proposed within the cognitive theories that the memory reorganization may be either routine (reordering the interactions of existing schemas), or heuristic (emergence of new components along with the modification of the prior ones) (Piaget 1951). The specialization and accommodative reconsolidation processes refer to the second type of the reorganization. As far as the first type, the modifications of neurons that belong to existing systems without emergence of a new system may be referred to as "reorganizational" reconsolidation. Presumably, a gradual increment of efficacy may be one of the manifestations of the latter, whereas the former would be signified by curt transition to good performance—like the one we see during our food-acquisition training.

We consider the difference between the specialization and accommodative reconsolidation processes as essential for investigation of underpinnings of learning. Ignored in most of the studies, these processes may be indistinguishable in the data on molecular and cellular learning-related processes. The differentiation of the processes that manifest emergence of new experience from those of prior experience modification is necessary in the contemporary research of memory principles.

In the systems perspective that we develop, learning is the key process under investigation, as it covers the most essential changes of individual experience—the emergence of new systems and modifications within existing ones—and presumes that memory is active and dynamic. Therefore, we next present our view of fundamental

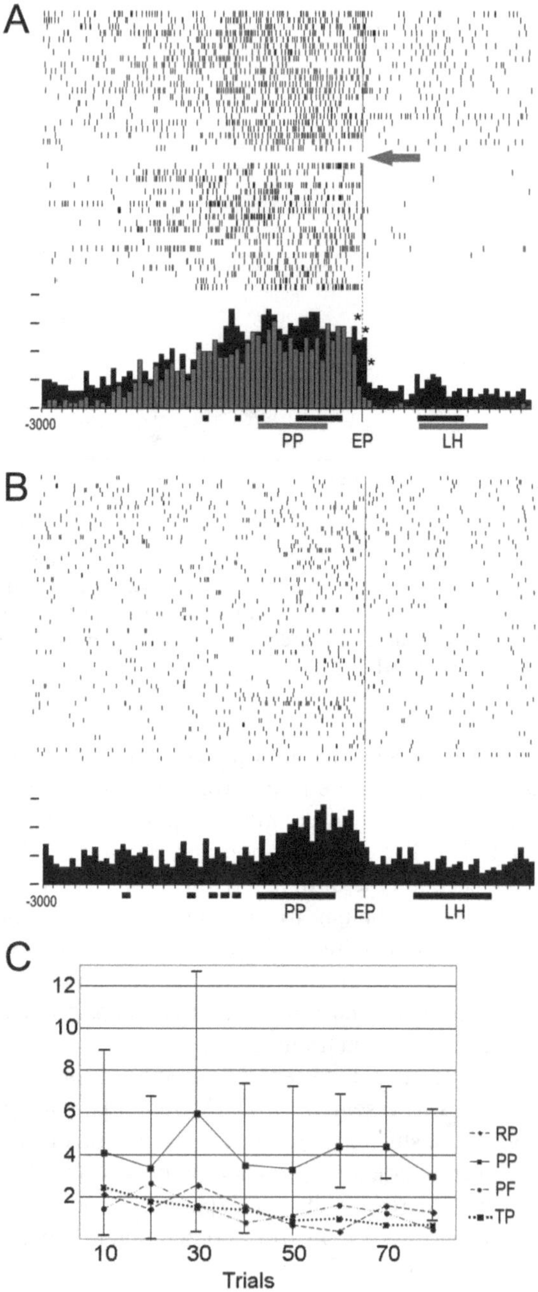

Fig. 1.4 Activity of a rat RSC single neuron specialized in relation to approaching and pressing the first pedal during stable performance of the previously acquired behavior and learning to press the second pedal on the opposite side of the experimental chamber. Recording time: 27 min. See Fig. 1.1b for designation of behavioral acts. (**a**) Spike raster plot (top) and histogram (bottom)

processes that underlie or accompany learning: the mismatch between "needs" and recent environmental "input" in the whole organism as well as in neurons, "altruistic suicide", and long-term potentiation.

1.3 Fundamentals of Learning within the Systems Perspective

1.3.1 Memory Formation Starts with Mismatch

As we mentioned earlier the initial step of the cascade of subcellular molecular events that determine the morphological modifications of neurons, both in the process of morphogenesis (early ontogenesis), and in memory consolidation in adults, is the expression of immediate early genes, followed by the expression of "late "genes that might be directly related to the structural modifications of a neuron. These days the relations between IEG expression and learning, noted a while ago (Maleeva et al. 1989; Tischmeyer et al. 1990; Anokhin and Rose 1991), have become widely accepted (e.g., Horn 2004; Kubik et al. 2007; Barry and Commins 2011; Minatohara et al. 2016).

In the framework of systems neuroscience it seems a logical assumption that the expression of IEGs and the formation of neuronal specializations are related. Indeed, we showed earlier that those structures that contained a lot of neurons specialized in relation to operant behavior also demonstrated a higher number of Fos-positive neurons after learning (Svarnik et al. 2005).

Induction of IEG expression in the adult takes place not only during learning, but also during stress, intoxication, lesions of the nervous system, brain ischemia and other conditions (Herrera and Robertson 1996; Meyer 2015). It was also shown that an artificial change in the microenvironment of neurons causes the appearance of activity in previously silent cells and the expression of IEGs (Stone et al. 1993). Given that activity at neuron is determined by mismatch between neuron's needs and the current influx of metabolites (as discussed above), IEG expression—a specific manifestation of cellular activity (Clayton 2000) arising in a situation of

Fig. 1.4 (continued) aligned to the end of pedal-pressing (EP) on the side trained first. Neuronal activity during repeated trials before (22 trials, above arrow, black histogram) and after (below arrow, grey histogram) onset of learning to press the second pedal (20 trials, 10 effective trials in a row). Ordinate: ticks of the histogram—tens of spikes in 50-ms bins Abscissa: hundreds of milliseconds Horizontal bars represent spans of act onsets (PP and LH). * significant differences between numbers of spikes within 100 ms bins. (**b**) Spike raster plot and histogram aligned to the end of pedal-pressing (EP) on the second side. (67 trials). All markers as in panel (**a**). (**c**) Dynamics of activity of the same neuron in different acts during acquisition of the pedal-pressing on the second side: mean ± SD spike frequencies in tens of consecutive trials

novelty (Anokhin and Sudakov 2003; Aggleton et al. 2012)—is suggested to be evident in a general bottom line of all these situations i.e.—during the mismatch.

The mismatch arises due to the fact that the previous possibilities of meeting the metabolic neuronal "needs" within the existing memory structure turned out to be ineffective in the condition of a stable change in microenvironment of neurons. The latter occurs upon change in the circumstances of corresponding behavior. Neuron, as noted above, may provide "needs" of its metabolism by combining with other elements of the organism and forming the functional system. Achieving the result of the system simultaneously eliminates the mismatch between "needs" and the state of the microenvironment of neurons, and provides the desired result for the organism on the behavioral level. This may happen only when the corresponding behavior has already been formed. However, learning in normal conditions and recovery in pathology (for example, after a stroke, traumatic brain injury) occur when the "needs" cannot be conformed with existing matching methods of the individual (i.e., within the available individual experience). The mismatch in this situation is different from that in the definitive behavior: it is eliminated not by reactivating existing memory, but by systemogenesis, i.e. selection and fixation of new elements and variants of combining them (see Sect. 1.2.1).

1.3.2 From Mismatch through Match to Consolidation

The emergence of a new system (systemogenesis) may lead both to achievement of the desired result for the organism, and to satisfaction of metabolic "needs" of neurons. However, the new integration is not constant. It was shown that the activity of the human brain changes not only in the process of learning, but also during hours (and days) after learning criteria achievement (e.g. Karni et al. 1995). It was also shown in animal experiments that the parameters of neuronal activations, as well as number of activated cells change within hours and days from the first successful behavioral trial (Erickson and Desimone 1999; Kuzina et al. 2016; Horn 2004; McKenzie and Eichenbaum 2011; Smith et al. 2012; etc).

Our results (Svarnik et al. 2005) show that the number of neurons in which the IEG expression is detected exceeds in many times the number of neurons in this area specialized in relation to the system of the formed behavior. We believe that part of these genetically activated cells are neurons specialized in the relation to the earlier formed systems, and the IEG expression in those cells reflects the beginning of a process of accommodative reconsolidation (see above). Others are pre-specialized neurons, and their gene expression induction is a prerequisite for the transition of cells into a state of readiness for selection during the trials.

As we hypothesized, it is in the trial-and-error process that certain neurons are selected from activated ones (activated both genetically and, presumably, electrophysiologically) and become specialized in relation to the formed system. Decrease in the number of activations as well as in heterogeneity of activity of specialized neurons in the course of memory consolidation demonstrated by us earlier (Kuzina et al. 2016) reflects this selection process and changes in neuronal subserving of new behavior. We compared neuronal activity in rat RSC recorded during the pedal-pressing (see Sect. 1.2.3) within either first five days (group 1), or from days 7 to 15 after its acquisition (group 2), which corresponds to the "early" and "later" stages of consolidation in rodents (e.g. Buitrago et al. 2004). The second group of animals was kept in the homecage during the first week after acquisition. There were significantly more neurons specialized in relation to new behavioral acts in group 1 that had specific activations during both approaching and pressing the pedal, i.e. acts that were acquired just before the start of recordings. In contrast, most neurons of the "pedal" category of group 2 were specifically active only in one of these acts: either approaching, or pressing the pedal. Within the first 5 days (in group 1) there were significantly more pedal-specific neurons with highly differentiated activity in other acts than in group 2. On the other hand, the enhanced selectivity of individual neurons in group 1 was accompanied by more variable activity in acts associated with a "feeder" part of the behavioral cycle. Activity of "pedal" neurons in group 2 was consistently reduced in "feeder" acts. Apparently, the reduction in the variability of activity may be associated with the completion of the selection process and stabilization of neuronal population involved in the newly formed behavior. It is possible that such stabilization requires not only time, but also a certain number of repetitions of experience reactivation (Weible et al. 2009, 2012; McKenzie et al. 2013).

It has been shown that some cells are activated only during the initial stages of learning, and when behavior is stabilized, their activations decrease and disappear (Shima et al. 1996; Wirth et al. 2003). In our view, some of these cells are likely to be pre-specialized neurons activated during trials. In the case of training for behavioral acts similar to previously formed ones (e.g., pedal-pressing on the second side in our setup) activity of neurons during pressing the first pedal may look like variable nonspecific activity, presumably reflecting the process of specialization (ref. to Fig. 13 in Alexandrov 2008). In addition, as our data show, neurons specialized in relation to previously formed behavior may be active during formation of a new one. For example, activity of a neuron on Fig. 1.4 was recorded during the following periods of the experiment: pressing pedal 1 before pedal 2 training (Fig. 1.4a, above arrow, black histogram), acquisition of pedal 2 pressing (Fig. 1.4b), and pedal 1 pressing after acquisition of both (Fig. 1.4a, below arrow, grey histogram). Spike frequencies in certain behavioral acts (TP, PF, and LH, see Fig. 1.1b) had significantly decreased after acquisition of the pedal-pressing on the second side. The activity of this neuron during the specific acts (RP and PP) on the first side had also changed: activations started and ended earlier after acquisition of the pedal-pressing on the second side. Additionally, activity of neurons during pedal 2 training and the following pedal 1 pressing might be considered as a neuronal basis of learning transfer: it accompanies speeded up learning after previous similar experience. Accordingly, mean spike frequencies were significantly higher along the whole period of learning in PP, than in any other acts including RP (Fig. 1.4b, c). Also, there was a significant decrease of activity in all acts, except PP, from the beginning to the end of learning to press pedal on the second side (Fig. 1.4c).

It might be also suggested that the first trials during learning in organisms with highly developed nervous system are subserved by co-activation of not only changing sets of specialized and pre-specialized neurons but also so called "novelty" neurons possibly specialized in relation to orienting behavior (for further details see Ranganath and Rainer 2003; Aleksandrov 2006). This co-activation may provide trial performance as well as achievements of the first positive results during learning. After stabilization of the behavior "novelty" neurons as well as a number of other previously specialized neurons cease their activity. Therefore, the decline of the activity of previously specialized and "novelty" neurons corresponds to the consolidation process and signifies serious reorganization of neuronal supply of memory reactivation. Meanwhile, the specific activation of specialized neurons may be evident from the very first implementation of the new behavior.

The chronic tetrode recordings described above (Sect. 1.2.5) have also revealed neurons specialized in relation to the behavior on the second side. Activation of these neurons that satisfied the criterion of specific activity (see Sect. 1.2.3) was indeed evident since the first relevant behavioral act acquired during learning. An example of such neuronal activity is shown in Fig. 1.5. Smith et al. (2012) also found the phenomenon of emergence of supposedly specific activity in neurons during the first trials of new behavior. However, in this study, animals were not pre-trained (except for familiarization with new environment), as in our experiments, where they were pre-trained to press the first pedal before they learned to press the second one. Perhaps this difference is one of the reasons why the activation (in the place field of a neuron) during the first implementations of behavior were evident in the hippocampus, but not in the RSC.

Earlier (Alexandrov et al. 1991) we showed that under acute ethanol treatment as compared to control the percentages of neurons of different specializations were not changed in the motor cortex. Meanwhile, the neuronal population is different under ethanol: the upper layer neurons are mostly excluded, and the lower layer neurons become more included into the neuronal population that subserves the behavior. Thus, at the early stages of learning the processes of neuronal specialization may proceed differently in different brain structures.

In studies of brain activity reorganization during learning changing roles of brain structures have been repeatedly demonstrated at various stages of training (Rose 1993; Kelly and Garavan 2005). It is known that learning scores, memory, and "cognitive control" depend on the intact cingulate cortical regions of the human brain (Hayden et al. 2010). Furthermore, outwardly the same behavior is accompanied by activation of different areas or layers of the cingulate cortex as learning progresses. On the one hand, by both methods of functional anatomy in humans (Tracy et al. 2003) and multiunit activity recording in animals (Freeman and Gabriel 1999) it was shown that activation of the anterior regions of the cingulate cortex declines and activation of the posterior regions increases in the process of learning. On the other hand, the posterior cingulate cortex is activated during aversive behavior (and also needed for its implementation) both at early and late stages of learning (Gabriel et al. 1991; Katche et al. 2013). It was also shown that the anterior cingulate cortex is involved in the "context-freezing" task at both early and late stages of learning

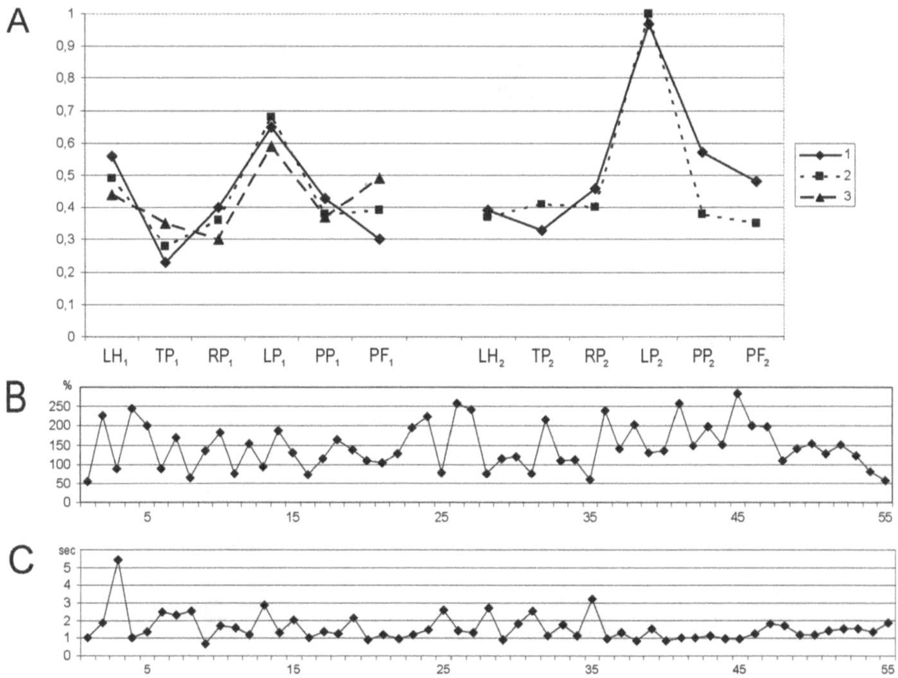

Fig. 1.5 Activity of neuron RAT27904–1 in consecutive series of trials after onset of training on the second side of experimental chamber. (**a**) Mean frequencies in different acts of behavior on the side trained first (left panel) and second (right panel). LP—locating in pedal corner before pedal pressing; see Fig. 1.1b for designation of behavioral acts. The frequencies are normalized to maximum. Lines 1, 2, and 3 represent series of behavior, separated by switches to alternative side. (**b**) Frequency excess over background in consecutive trials (along abscissa) for specific act LP2 on the second side. The series 2 starts from trial 41. (**c**) Duration of correponding trials of act LP_2 seconds

and necessary for reconsolidation of this memory (Vetere et al. 2011; Einarsson and Nader 2012). If the specialization of the neuron, as we noted above, is constant (i.e. neuronal differentiation is irreversible—Sect. 1.2.1), and evident from the first implementations of the newly formed acts (above), it might be assumed that the most significant contribution into the described reorganization is made not by the dynamics of the activity of the specialized neurons but other ("unidentified") neurons.

In our experiment we recorded activity of the neurons in anterior and posterior areas of rabbit cingulate cortices at "early" (the first week of learning) and "late" (the second week of learning) stages of training of pedal-pressing. We analyzed average frequency of spikes, percentage of specialized and unidentified neurons, and the number of behavioral acts with non-specific activations (i.e. acts with probability of activation between 40% and 100% in unidentified neurons, see Sozinov et al. 2012). We found that all these variables for specialized neurons did not differ between the first and the second week of training (Fig. 1.6, top), whereas the aver-

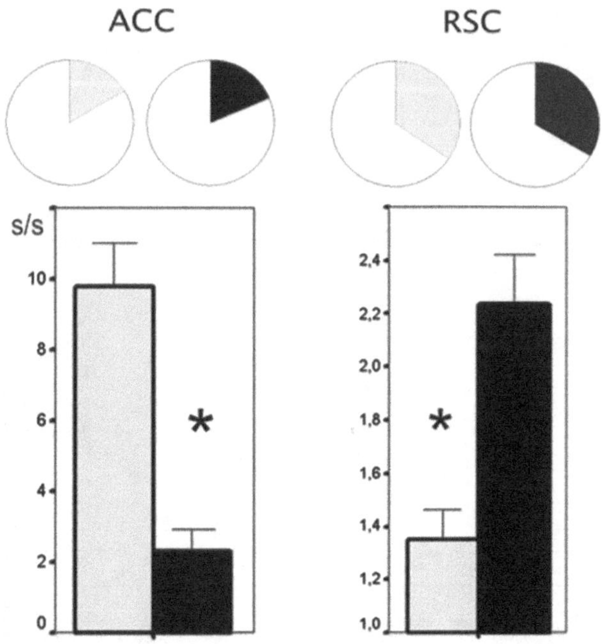

Fig. 1.6 Activity of neurons in the rabbits' anterior cingulate (ACC) and retrosplenial (RSC) cortices in the course of the first (light grey) and the second (black) weeks of pedal-pressing training. Top: Percentage of specialized neurons recorded during the two weeks. Bottom: Spike frequencies of unidentified neurons in ACC and average number of non-specific acts in RSC. The asterisk shows significant differences

age frequencies and the number of acts with non-specific activations were different for neurons with unidentified specializations (Fig. 1.6, bottom). Therefore, neuronal activity turned out to differ between the first and the second week of training of food-acquisition skill. These differences were primarily indicators of activity of unidentified neurons, rather than specialized neurons.

On the basis of these results it might be possible to propose that in unidentified neurons of the anterior cingulate cortex the frequency decreases from the first to the second week of training, but in such neurons of the posterior cingulate cortex the number of activations in new behavioral acts increases during the same period. In other words, the dynamics of brain activations is less accounted for newly specialized neurons, but is due to activity of those neurons whose specializations are not identified. As we argued earlier (Alexandrov et al. 1993) unidentified neurons are probably specialized in relation to systems of other behavioral acts than those formed in our training. That is why the established differences in characteristics of neuronal activity at successive learning stages might be connected to the processes of reorganization of that experience which served as a basis for newly formed behavior, rather than to changes in cohort of specialized neurons.

These data support mentioned above reasoning that it is necessary to differentiate characteristics of new experience formation and old experience reorganization (Alexandrov et al. 2001; Grosmark and Buzsáki 2016; McKenzie and Eichenbaum 2011). However, saying about higher manifestation of reconsolidative changes we should take into account possibility of maintenance of the stable percentage of differently specialized neurons and, at the same time, changes in neurons of other specializations.

Treating a neuron as an active living organism has additional consequences for several well-known phenomena. Among them are "altruistic suicide" and long-term potentiation covered in the remainder of this paragraph.

1.3.3 "Altruistic Suicide"

As it was mentioned above IEG expression is induced when the organism does not have experience of satisfaction of metabolic "needs" of its cells in some situation, or when repetitive impulses of co-activated neurons do not lead to the result achievement (goal achievement). The IEG expression might be considered not only as the first step of consolidation process, but also as induction of other transcriptional factors underlying cell's "decision to live or die" (Lee et al. 1998, p. 2736). If the mismatch between "needs" of neurons and their microenvironment is prolonged, neurons become hyperactive, and waves of IEG expression repeat. In such cases "death" gene expression might be induced, which will result in neuronal death—apoptosis (see Fig. 1.7). Thus when the mismatch between "needs" of the neuron and its microenvironment cannot be eliminated in the conditions of existing experience the neuron has two alternatives: to be changed during systemogenesis (new system formation) or to die (Fig. 1.7). These two alternatives exist both in normal conditions (during early ontogenesis and adulthood) and in pathological conditions. The involvement into systemogenesis might be either system specialization process or the process of accommodative (reorganizational) reconsolidation. Cell death is often observed during early development and under pathology, when existing experience of the organism is inapplicable for agreement among metabolisms of different cells of the organism. But this is true not only for such cases. There are data showing that apoptosis is evident in brains of healthy adults and is necessary for normal functioning of animals (e.g. Leist and Jäättelä 2001). Since systemogenesis is a general principle for early development and learning at any age, adaptation and recovery, the discussed data allow concluding that "change or die" options exist in normal conditions. It was shown (Sherstnev et al. 2013) that elimination of neurons (observed as neuronal selection in early ontogenesis important for behavioral repertoire formation at that stage) also contributes to the process of systemogenesis during adulthood (Fig. 1.7). Thus the formulated position states that there are no two alternatives ("systemogenesis or death") but two interconnected roads to

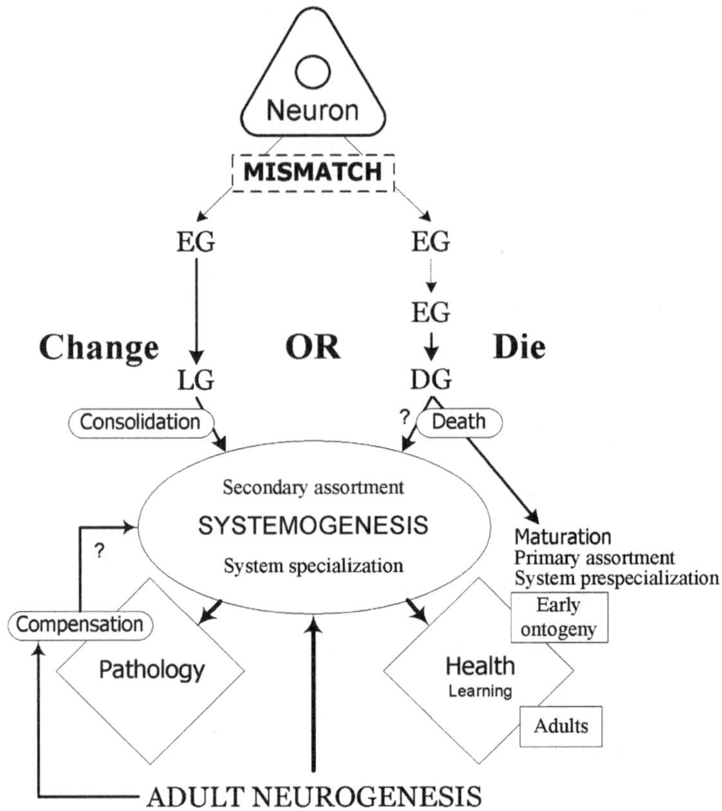

Fig. 1.7 General framework of the systemic organization of behavior: Theoretical schema of the "Change or Die" principle. EG—"early" genes; LG—"late" genes; DG—"death" genes. See Sects. 1.2.1 and 1.3.3 for explanations

systemogenesis: modification of a neuron or its death. It might be suggested that death of some cells is a necessary payment for a possibility of successful systemogenesis during individual ontogenesis in all those situations when metabolic needs of some cells are in an unavoidable conflict with new means of agreement among cellular needs. The activity principle is applicable for all periods and aspects of existence of a neuron including the processes connected to the "change or die" alternative. Each stage of cellular elimination is an active process (Raoul et al. 2000), and thus neuronal elimination is a suicide (Leist and Jäättelä 2001). This suicide is altruistic in a sense that the neuron turns on the program of self-elimination in order to abolish metabolic conflict and provide survival of other neurons that belong to the same cellular clone. Earlier other authors have already argued for the existence of "altruistic cellular suicide" in the nervous system (Allsopp and Fazakerley 2000) and in unicellular organisms (Strassmann et al. 2000).

1.3.4 Long-Term Potentiation: Traditional and Systems Approach

Above we described the system approach to learning and memory processes. If someone wanted to argue for alternative traditional framework of learning mechanisms, she (he) would probably refer to the phenomenon of long-term potentiation (LTP), which is considered to be a physiological mechanism of long-term memory and regarded as an experimental model of activity-dependent plasticity. Studies of LTP have for many years been seen as the most important and urgent approach mainly because this phenomenon is well demonstrated in the framework describing the formation of memory as an increase in synaptic efficiency of impulse conductance in neuronal networks. Within the systems approach, LTP can be regarded as an electrophysiological description of the mismatch (see Sect. 1.3.1). If we consider neuronal activity as determined by the mismatch we may conclude that an artificial electrical (or chemical) stimulation used to elicit an influx not accordant with the neuron's preceding activity and not caused by it serves as a powerful mismatch factor. And the increased cell excitability persistently found during testing is a reflection of this mismatch. Not only theoretical framework but also experimental data argue for the link between LTP and the mismatch process, among them the data that show similarity between LTP and the processes that take place during pathological conditions, when metabolic cellular environment is strongly changed (McEachern and Shaw 1996; Vikman et al. 2003). Thus, although experimenters using tetanization do not intend to induce the mismatch, they do. And the mismatch is, as argued above, the initial stage of learning and the formation of a new memory. Therefore, we do consider LTP as a phenomenon that may be related to mechanisms of learning and memory but for different reasons: because it models the initial stage of learning—the mismatch. However, it is not known whether the mismatch obtained during the experimental induction of LTP has the properties of characteristics of natural mismatch during learning. Note that the discrepancy between the traditional concept of LTP and data accumulated from studies of this phenomenon requires an alternative explanation even for those authors who have no doubt that the increase in synaptic efficiency between neurons provides the basis for the formation of memories. McEachern and Shaw (1996) believe that the mechanisms of receptor regulation allow neurons an attempt to prevent long-term changes in their synaptic excitability, which is harmful for neurons. LTP (like depression), acting against this regulation, is not the basis of learning but is an initial manifestation of the cascade of processes leading to the reorganization of the activity of a neuronal group, which "strives" to maintain homeostasis. Shors and Matzel (1997) also came to the conclusion that there is a non-correspondence between the properties of LTP, particularly its duration, and those required if LTP is to support the retention of long-term memory. They believe that LTP is a mechanism related not to maintenance of long-term memory but to the initial stage of its formation. As we see the presented conclusions are in accordance with the suggestion that the mismatch is the initial stage of systemogenesis, and that LTP is an electrophysiological description of the mismatch process. If we consider this suggestion about LTP, we may conclude that although the duration of LTP is insufficient for it to be regarded as the basis of long-term memory, it may be adequate for it to be regarded as an electrophysiological manifestation of prolonged mismatch leading to cell death. Put more simply, the logic of the ideas proposed here suggests a link between LTP and neuronal death. There are some data showing this connection (McEachern and Shaw 1996; Manahan-Vaughan et al. 1999; Ambrogini et al. 2004). Thus, within the systems approach the phenomenon of LTP might be related to the mechanisms of learning and memory, but not because it models increased effectiveness of impulse transmission in neuronal networks, but because it models the mismatch process, which is a characteristic of initial stages of learning.

1.4 The History of Memory Formation and the Memory Structure Are Interrelated

We showed earlier that any behavior is subserved by activation of not only new systems formed during learning but also older systems activation formed at earlier stages of individual ontogenesis (see Alexandrov 2008; Alexandrov et al. 2000). Thus, behavior is reflection of history of its formation (phylogenetic history as well as ontogenetic), i.e. realization of multiple systems, each of which fixates a stage of behavior acquisition. It might be suggested from this statement that system organization of overt similar actions differs if the history of their formation differs. We showed earlier for complex operant behavior in rabbits that neuronal activity in the cingulate cortex differs significantly when rabbits learned the same behavioral acts but in different order (Gorkin and Shevchenko 1996).

In other experiments we checked the hypothesis about connection between activity of recent-task-related system-specialized neurons and the number of stages of learning. The following logic underlied this hypothesis. We showed earlier that in different brain structures of rabbits there were neurons activated during different acts of the acquired (on a daily basis) instrumental behavior: approach to the feeder, turn from the feeder to the pedal, approach to the pedal, the pedal pressing. Since all these acts constituted the sequence during performance, this behavior is accomplished due to reactivation of all systems of these acts, and, hence, due to activation of the system-specialized neurons. Thus, on the basis of the transformation of learning stages into systems of the learned behavioral acts we could propose that if the number of learning stages differ, then organization of neuronal activity during overtly identical behavior differ also. In order to check this assumption we compared neuronal activity in the RSC after rats learned the instrumental behavior within one stage (only pedal pressing was reinforced) with the neuronal activity during the same behavior but when rats learned it in several stages (each stage of instrumental behavior described above was reinforced). We found that the number of newly specialized neurons did not differ between the two cases. However, aggregated activations of all new system-specialized neurons in RP and PP were higher in the case of multiple stages learning, than in the case of one-stage learning. In the latter group we found more neurons with specific activations during turning to the pedal and higher spike frequency of specialized neurons in all acts. These data suggest that there is a connection between activities of the neurons specialized in relation to newly acquired behavior and the number of stages used to acquire this behavior. Thus, in different species we find consistent general principle—the organization of neuronal activity during behavior depends on the history of its acquisition.

This principle is also manifested at the molecular level. We showed that the number of neurons that changed their gene expression (detected by Fos expression) during new behavior learning depended on the number of acquisition stages of the previous training (Svarnik et al. 2013). In these experiments we trained animals to press a pedal on one side of the experimental cage in one or several stages, and then compared the number of Fos-positive neurons after the acquisition of the same skill on the other side of the cage. Despite the fact that the overt behavior during the second acquisition was similar in these two cases, the number of neurons with changes in gene expression was significantly different. It might be suggested that such changes depended on the processes of accommodative reconsolidation described above. Thus, learning involves not only acquisition of new information but also assimilation of this information into earlier established experience structure or schemas (see also Tse et al. 2011).

1.5 Conclusion

We put forward the following sequence of memory formation and functioning, brought together on Fig. 1.8. Learning starts from the mismatch between individual needs and possibilities to meet them using acquired memory. It is manifested on a cellular level as a mismatch between metabolic cellular "needs" and recent metabolic input. In a familiar situation this mismatch might be cleared up by performance of previously formed behavior: Fig. 1.8a shows memory reactivation during behavior acquired earlier. Memory reactivation might be connected to changes of experience structure due to "reactivational reconsolidation". This type of reconsolidation may appear as continuous memory formation.

In many experimental models of reconsolidation the acquisition is followed by presentation of a reminder—namely, by additional training (see, e.g. Davis et al. 2010). Therefore, Fig. 1.8b shows modification of individual experience structure in a new situation limited by reorganization of previously formed systems. In this situation new element of experience (a new system) is not formed. This type of modification occurs due to "reorganizational reconsolidation".

The type of modification on Fig. 1.8c is the one where the mismatch can neither be eliminated by reactivation of existing memory (as in A), nor by reorganization of earlier formed systems and intersystems relations (as in B). Then the mismatch is

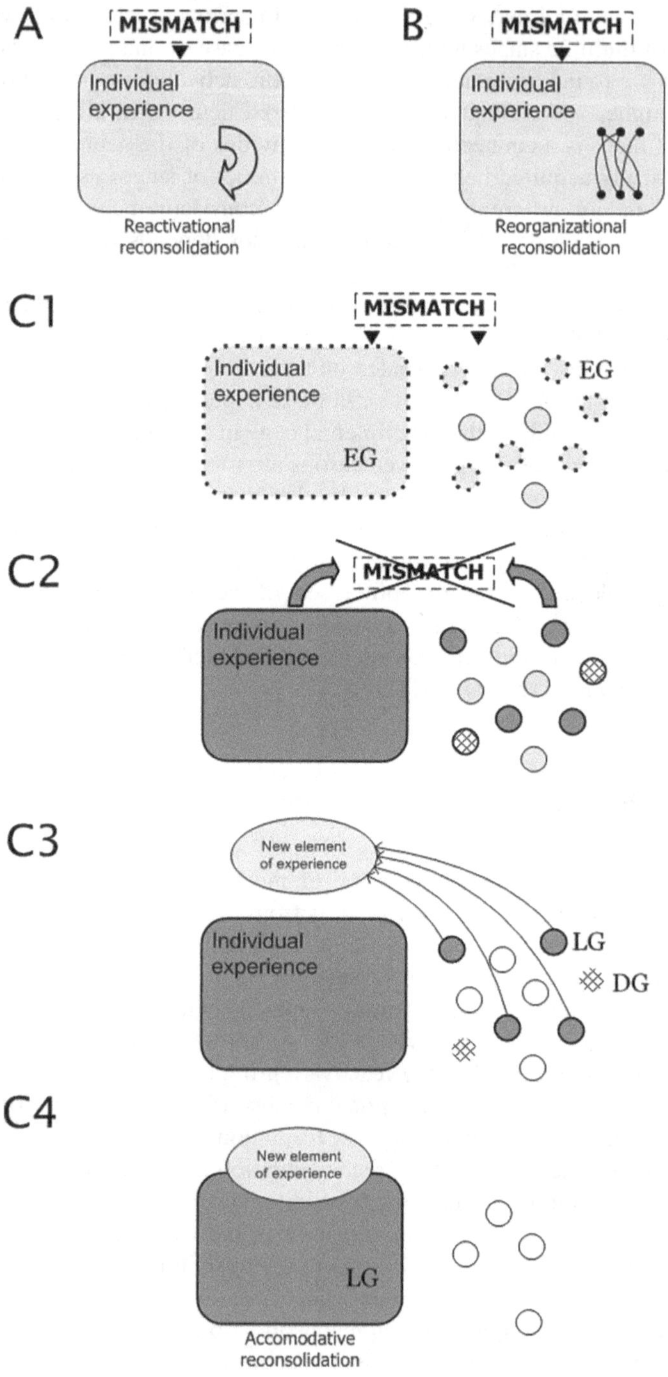

Fig. 1.8 General framework of the systemic organization of behavior: Types and stages of modification of the individual experience structure. See Fig. 1.7 for abbreviations and Conclusion for explanations

eliminated by formation of a new system. This process involves several stages: C1—early gene expression (EG), which is manifested at early stages of acquisition in pre-specialized neurons (circles), as well as in neurons that belong to the systems of prior individual experience; C2—selection during trial behavior: among activated pre-specialized neurons (that appeared also during adult neurogenesis) a necessary combination (darker circles) is selected; C3—during the process of selection a neuron has a choice of being changed and involved into a new system consolidated later through late gene expression (LG), or die (crossed circles; "death" gene expression, DG); C4—accommodative modification of neurons, specialized in relation to earlier formed systems (the rectangle), determined by inclusion of a new system into the structure of individual experience.

The formation of new integration, preceded by "internal" testing and hypothesis selection, is also manifested in trial behavior. At the cellular level this trial behavior is based on testing combinations of activated neurons; successful combinations provide result achievement and elimination of mismatch (Fig. 1.8, C2). The success is accomplished by modifications of some cells and elimination of others (Fig. 1.8, C3). As the first results are achieved, the cells presumably pre-specialized in relation to searching behavior, as well as other cells that belonged to earlier formed systems, gradually decrease and eliminate their activity. It is probably manifested in temporal changes in overt behavior (that seems as already formed) and in changes of the set of activated cells. Gradual stabilization of the set of activated cells may be manifested in more stable relations between neuronal activations and behavior. Late gene expression provides reorganization of selected neurons and their transition to being constantly specialized in relation to a newly formed system. This system modifies earlier specialized neurons during the process of accommodative reconsolidation (Fig. 1.8, C4). Thus, stability of neuronal specialization in a sense of belonging to certain system does not mean that formed memory is unchangeable. Some of the processes proposed by the schema on Fig. 1.8 remain hypothetical (including the modification of intersystem relations), but we consider this as a consistent framework that provides testable propositions.

Acknowledgements This work was supported by Russian Science Foundation grant #14-28-00229 for Institute of Psychology, Russian Academy of Sciences.

Conflict of Interest The authors declare that they have no conflict of interest.

References

Abdou K, Shehata M, Choko K, Nishizono H, Matsuo M, Muramatsu SI, Inokuchi K. Synapse-specific representation of the identity of overlapping memory engrams. Science. 2018;360(6394):1227–31

Adams NE, Sherfey JS, Kopell NJ, Whittington MA, Lebeau FE. Hetereogeneity in neuronal intrinsic properties: a possible mechanism for hub-like properties of the rat anterior cingulate cortex during network activity. eNeuro. 2017;4(1):ENEURO-0313.

Aggleton JP, Brown MW, Albasser MM. Contrasting brain activity patterns for item recognition memory and associative recognition memory: insights from immediate-early gene functional imaging. Neuropsychologia. 2012;50(13):3141–55. https://doi.org/10.1016/j.neuropsychologia.2012.05.018.

Aleksandrov YI. Learning and memory: traditional and systems approaches. Neurosci Behav Physiol. 2006;36(9):969–85.

Alexandrov YI. How we fragment the world: the view from inside versus the view from outside. Soc Sci Inf. 2008;47:419–57.

Alexandrov YI. Cognition as systemogenesis. In: Nadin M, editor. Anticipation: learning from the past. Cognitive systems monographs, vol. 25. Cham: Springer; 2015. p. 193–220.

Alexandrov YI, Alexandrov IO. Specificity of visual and motor cortex neurons activity in behavior. Acta Neurobiol Exp. 1982;42:457–68.

Alexandrov YI, Sams M. Emotion and consciousness: ends of a continuity. Cogn Brain Res. 2005;25(2):387–405.

Alexandrov YI, Grinchenko YV, Laukka S, Jarvilehto T, Maz VN, Svetlaev IA. Acute effect of ethanol on the pattern of behavioral specialization of neurons in the limbic cortex of the freely moving rabbit. Acta Physiol Scand. 1990;140:257–68.

Alexandrov YI, Grinchenko YV, Laukka S, Jarvilehto T, Matz VN. Acute effects of alcohol on unit activity in the motor cortex of freely moving rabbits: comparison with the limbic cortex. Acta Physiol Scand. 1991;142:429–35.

Alexandrov YI, Grinchenko YV, Laukka S, Jarvilehto T, Maz VN, Korpusova AV. Effect of ethanol on hippocampal neurons depends on their behavioral specialization. Acta Physiol Scand. 1993;149:429–35.

Alexandrov YI, Grechenko TN, Gavrilov VV, et al. Formation and realization of individual experience: a psychophysiological approach. In: Miller R, Ivanitsky AM, Balaban PM, editors. Conceptual advances in brain research. Vol. 2. Conceptual advances in Russian neuroscience: complex brain functions. Amsterdam: Harwood Academic Publishers; 2000. p. 181–200.

Alexandrov YI, Grinchenko YV, Shevchenko DG, Averkin RG, Matz VN, Laukka S, Korpusova AV. A subset of cingulate cortical neurons is specifically activated during alcohol-acquisition behavior. Acta Physiol Scand. 2001;171:87–97.

Alexandrov YI, Grinchenko YV, Shevchenko DG, Averkin RG, Matz VN, Laukka S, Sams M. The effect of ethanol on the neuronal subserving of behavior in the hippocampus. J Behav Brain Sci. 2013;3:107–30.

Allsopp TE, Fazakerley JK. Altruistic cell suicide and the specialized case of the virus-infected nervous system. Trends Neurosci. 2000;23:284–90.

Ambrogini P, Orsini L, Mancini C, Ferri P, Ciaroni S, Cuppini R. Learning may reduce neurogenesis in adult rat dentate gyrus. Neurosci Lett. 2004;359:13–6.

Anacker C, Hen R. Adult hippocampal neurogenesis and cognitive flexibility—linking memory and mood. Nat Rev Neurosci. 2017;18(6):335–46.

Anokhin KV, Rose SP. Learning-induced increase of immediate early gene messenger RNA in the chick forebrain. Eur J Neurosci. 1991;3(2):162–7.

Anokhin KV, Sudakov KV. Genome of brain neurons in organization of systemic mechanisms of behavior. Bull Exp Biol Med. 2003;135(2):107–13.

Anokhin KV, Tiunova AA, Rose SPR. Reminder effects -reconsolidation or retrieval deficit? Pharmacological dissection with protein synthesis inhibitors following reminder for a passive-avoidance task in young chicks. Eur J Neurosci. 2002;15:1759–65.

Anokhin PK. Biology and neurophysiology of the conditioned reflex and its role in adaptive behavior. New York: Pergamon Press; 1974.

Barry DN, Commins S. Imaging spatial learning in the brain using immediate early genes: insights, opportunities and limitations. Rev Neurosci. 2011;22:131–42. https://doi.org/10.1515/RNS.2011.019.

Bartlett FC. Remembering: a study in experimental and social psychology. New York: Cambridge University Press; 1995.

Bonner JT. The evolution of complexity by means of natural selection. Princeton, NJ: Princeton University Press; 1988.
Brecht M, Sceneider M, Manns ID. Silent neurons in sensorimotor cortices: implication for cortical plasticity. In: Ebner FF, editor. Neural plasticity in adult somatic sensory-motor systems. Boca Raton: Taylor & Francis Group, LLC; 2005. p. 1–19.
Buitrago MM, Ringer T, Schulz JB, Dichgans J, Luft AR. Characterization of motor skill and instrumental learning time scales in a skilled reaching task in rat. Behav Brain Res. 2004;155:249–56.
Bunge MA. Causality: the place of the causal principle in modern science. Cambridge: Harvard University Press; 1963.
Burgess N, O'Keefe J. Models of place and grid cell firing and theta rhythmicity. Curr Opin Neurobiol. 2011;21(5):734–44. https://doi.org/10.1016/j.conb.2011.07.002.
Cacioppo JT, Gardner WL. Emotion. Annu Rev Psychol. 1999;50:191–214.
Carleton A, Petreanu LT, Lansford L, Lledo P-M. Becoming a new neuron in the adult olfactory bulb. Nat Neurosci. 2003;6:507–18.
Changeux JP, Connes A. Conversations on mind, matter, and mathematics. Princeton, NJ: Princeton University Press; 1999.
Chestek CA, Batista AP, Santhanam G, Yu BM, Afshar A, Cunningham JP, Gilja V, Ryu SI, Churchland MM, Shenoy KV. Single-neuron stability during repeated reaching in macaque premotor cortex. J Neurosci. 2007;27:10742–50.
Cisek PI, Kalaska JF. Neural mechanisms for interacting with a world full of action choices. Annu Rev Neurosci. 2010;33:269–98. https://doi.org/10.1146/annurev.neuro.051508.135409.
Clayton DF. The genomic action potential. Neurobiol Learn Mem. 2000;74:185–216.
Davis S, Renaudineau S, Poirier R, Poucet B, Save E, Laroche S. The formation and stability of recognition memory: what happens upon recall? Front Behav Neurosci. 2010;4:177. https://doi.org/10.3389/fnbeh.2010.00177.
Defelipe J. The evolution of the brain, the human nature of cortical circuits, and intellectual creativity. Front Neuroanat. 2011;5:29. https://doi.org/10.3389/fnana.2011.00029.
Dudai Y. Memory from a to Z. Keywords, concepts and beyond. Oxford: Oxford University Press; 2002.
Dudai Y. The restless engram consolidations never end. Annu Rev Neurosci. 2012;35:227–47. https://doi.org/10.1146/annurev-neuro-062111-150500.
Dudai Y, Eisenberg M. Rites of passage of the engram: reconsolidation and the lingering consolidation hypothesis. Neuron. 2004;44:93–100.
Dudai Y, Karni A, Born J. The consolidation and transformation of memory. Neuron. 2015;88(1):20–32. https://doi.org/10.1016/j.neuron.2015.09.004.
Edelman GM. Neural Darwinism: the theory of neural group selection. New York: Basic Books; 1987.
Einarsson EO, Nader K. Involvement of the anterior cingulate cortex in formation, consolidation, and reconsolidation of recent and remote contextual fear memory. Learn Mem. 2012;19:449–52. https://doi.org/10.1101/lm.027227.112.
Engel AK, Fries P, Singer W. Dynamic predictions: oscillations and synchrony in top–down processing. Nat Rev Neurosci. 2001;2(10):704–16. https://doi.org/10.1038/35094565.
Erickson CA, Desimone R. Responses of macaque perirhinal neurons during and after visual stimulus association learning. J Neurosci. 1999;19:10404–16.
Feld GB, Born J. Sculpting memory during sleep: concurrent consolidation and forgetting. Curr Opin Neurobiol. 2017;44:20–7. https://doi.org/10.1016/j.conb.2017.02.012.
Fodor J. Against darwinism. In: Vosniadou S, Kayser D, Protopapas A, editors. Proceedings of EuroCogSci07. Hillsdale, NJ: Lawrence Erlbaum Associates; 2007. p. 23–8.
Frankland PW, Kohler S, Josselyn SA. Hippocampal neurogenesis and forgetting. Trends Neurosci. 2013;36:497–503. https://doi.org/10.1016/j.tins.2013.05.002.
Fraser GW, Schwartz AB. Recording from the same neurons chronically in motor cortex. J Neurophysiol. 2012;107:1970–8. https://doi.org/10.1152/jn.01012.2010.

Freeman JH Jr, Gabriel M. Changes of cingulothalamic topographic excitation patterns and avoidance response incubation over time following initial discriminative conditioning in rabbits. Neurobiol Learn Mem. 1999;72:259–72.
Gabriel M, Vogt BA, Kubota Y, Poremba A, Kang E. Training-stage related neuronal plasticity in limbic thalamus and cingulate cortex during learning: a possible key to mnemonic retrieval. Behav Brain Res. 1991;46:175–85.
Gavrilov V, Grinchenko YV, Alexandrov YI. Do neurons in homologous cortical areas of rabbits and rats have similar behavioral specialization? FENS Abstr. 2002;1:A040.8.
Gorkin AG, Shevchenko DG. Distinctions in the neuronal activity of the rabbit limbic cortex under different training strategies. Neurosci Behav Physiol. 1996;26(2):103–12.
Greenberg PA, Wilson FAW. Functional stability of dorsolateral prefrontal neurons. J Neurophysiol. 2004;92:1042–55.
Gregory TR. Coincidence, coevolution, or causation? DNA content, cell size, and the C-value enigma. Biol Rev. 2001;76:65–101.
Grosmark AD, Buzsáki G. Diversity in neural firing dynamics supports both rigid and learned hippocampal sequences. Science. 2016;351(6280):1440–3. https://doi.org/10.1126/science.aad1935.
Hartley T, Lever C, Burgess N, O'Keefe J. Space in the brain: how the hippocampal formation supports spatial cognition. Philos Trans R Soc Lond Ser B Biol Sci. 2013;369(1635):20120510. https://doi.org/10.1098/rstb.2012.0510.
Hayden BY, Smith DV, Platt ML. Cognitive control signals in posterior cingulate cortex. Front Hum Neurosci. 2010;4:1–8. https://doi.org/10.3389/fnhum.2010.00223.
Hennies N, Ralph MA, Kempkes M, Cousins JN, Lewis PA. Sleep spindle density predicts the effect of prior knowledge on memory consolidation. J Neurosci. 2016;36(13):3799–810. https://doi.org/10.1523/JNEUROSCI.3162-15.2016.
Herrera DG, Robertson HA. Activation of c-fos in the brain. Prog Neurobiol. 1996;50(2–3):83–107.
Horn G. Pathways of the past: the imprint of memory. Nat Rev Neurosci. 2004;5:108–21.
Hoshiba Y, Wada T, Hayashi-Takagi A. Synaptic ensemble underlying the selection and consolidation of neuronal circuits during learning. Front Neural Circuits. 2017;11:12. https://doi.org/10.3389/fncir.2017.00012.
Hupbach A, Gomez R, Hardt O, Nadel L. The dynamics of memory: context-dependent updating. Learn Mem. 2008;15:574579. https://doi.org/10.1101/lm.1022308.
Jackson A, Mavoori J, Fetz EE. Correlations between the same motor cortex cells and arm muscles during a trained task, free behavior, and natural sleep in the macaque monkey. J Neurophysiol. 2007;97:360–74.
Kandel ER. In search of memory: the emergence of a new science of mind. New York: WW Norton & Company; 2006.
Karni A, Meyer G, Jezzard P, Adams MM, Turner R, Ungerleider LG. Functional MRI evidences for adult motor cortex plasticity during motor skill learning. Nature. 1995;377:155–8.
Katche C, Dorman G, Slipczuk L, Cammarota M, Medina JH. Functional integrity of the retrosplenial cortex is essential for rapid consolidation and recall of fear memory. Learn Mem. 2013;20:170–3. https://doi.org/10.1101/lm.030080.112.
Kelly AMC, Garavan H. Human functional neuroimaging of brain changes associated with practice. Cereb Cortex. 2005;15:1089–102.
Kempermann G, Kuhn GH, Gage FH. Experience-induced neurogenesis in the senescent dentate gyrus. J Neurosci. 1998;18:3206–12.
Kitamura T, Ogawa SK, Roy DS, Okuyama T, Morrissey MD, Smith LM, Redondo RL, Tonegawa S. Engrams and circuits crucial for systems consolidation of a memory. Science. 2017;356(6333):73–8. https://doi.org/10.1126/science.aam6808.
Korhonen O, Saarimäki H, Glerean E, Sams M, Saramäki J. Consistency of regions of interest as nodes of fMRI functional brain networks. Netw Neurosci. 2017;1(3):254–74. https://doi.org/10.1162/NETN_a_00013.

Kubik S, Miyashita T, Guzowski JF. Using immediate-early genes to map hippocampal subregional functions. Learn Mem. 2007;14:758–70.
Kuzina EA, Gorkin AG, AlexandrovYu.I. Neuron activity in the retrosplenial cortex of the rat at the early and late stages of memory consolidation. Neuroscience and Behavioral Physiology. 2016;46(7):789–793. https://doi.org/10.1007/s11055-016-0312-z.
Lee Y, Park KH, Baik SH, Chi C. Attenuation of c-Fos basal expression in the cerebral cortex of aged rat. Neuroreport. 1998;9:2733–6.
Leist M, Jäättelä M. Four deaths and a funeral: from caspases to alternative mechanisms. Nat Rev. 2001;2:1–10.
Lohmann G, Stelzer J, Zuber V, Buschmann T, Margulies D, Bartels A, Scheffler K. Task-related edge density (TED)—a new method for revealing dynamic network formation in fMRI data of the human brain. PLoS One. 2016;11(6):e0158185. https://doi.org/10.1371/journal.pone.0158185.
Maleeva NE, Ivolgina GL, Anokhin KV, Limborskaia SA. Analysis of the expression of the c-fos proto-oncogene in the rat cerebral cortex during learning. Genetika. 1989;25(6):1119–21. (in Russian).
Manahan-Vaughan D, Behnish T, Reymann KG. ACPD-mediated slow-onset potentiation is associated with cell death in the rat CA1 region in vivo. Neuropharmacology. 1999;38:487–94.
Mceachern JC, Shaw CA. An alternative to the LTP orthodoxy: a plasticity-pathology continuum model. Brain Res Rev. 1996;22:51–92.
Mckenzie S, Eichenbaum H. Consolidation and reconsolidation: two lives of memories? Neuron. 2011;71:224–33. https://doi.org/10.1016/j.neuron.2011.06.037.
Mckenzie S, Robinson NT, Herrera L, Churchill JC, Eichenbaum H. Learning causes reorganization of neuronal firing patterns to represent related experiences within a hippocampal schema. J Neurosci. 2013;33(25):10243–56. https://doi.org/10.1523/JNEUROSCI.0879-13.2013.
Mcmahon DB, Jones AP, Bondar IV, Leopold DA. Face-selective neurons maintain consistent visual responses across months. Proc Natl Acad Sci U S A. 2014;111(22):8251–6. https://doi.org/10.1073/pnas.1318331111.
Meyer RE. Physiologic measures of animal stress during transitional states of consciousness. Animals (Basel). 2015;5(3):702–16. https://doi.org/10.3390/ani5030380.
Minatohara K, Akiyoshi M, Okuno H. Role of immediate-early genes in synaptic plasticity and neuronal ensembles underlying the memory trace. Front Mol Neurosci. 2016;8:78. https://doi.org/10.3389/fnmol.2015.00078.
Moscovitch M, Cabeza R, Winocur G, Nadel L. Episodic memory and beyond: the hippocampus and neocortex in transformation. Annu Rev Psychol. 2016;67:105–34. https://doi.org/10.1146/annurev-psych-113011-143733.
Nader K. Response to Arshavsky: challenging the old views. Trends Neurosci. 2003;26:466–8.
Nader K. Reconsolidation and the dynamic nature of memory. Cold Spring Harb Perspect Biol. 2015;7:1–16. https://doi.org/10.1101/cshperspect.a021782.
Nader K, Schafe GE, Le Doux JE. Fear memories require protein synthesis in the amygdala for reconsolidation after retrieval. Nature. 2000;406:722–6.
Neisser U. Cognition and reality: principles and implications of cognitive psychology. New York: Freeman; 1976.
O'Keefe J. Place units in the hippocampus of the freely moving rat. Exp Neurol. 1976;51:78–109.
Paton JA, Nottebohm FN. Neurons generated in the adult brain are recruited into functional circuits. Science. 1984;225:1046–8.
Paxinos G, Watson C. The rat brain in stereotaxic co-ordinates. New York: Academic; 1997.
Piaget J. Play, dreams, and imitation in childhood. New York: Norton; 1951.
Prickaerts J, Koopmans G, Blokland A, Scheepens A. Learning and adult neurogenesis: survival with or without proliferation? Neurobiol Learn Mem. 2004;81:1–11.
Ranganath C, Rainer G. Neural mechanisms for detecting and remembering novel events. Nat Rev Neurosci. 2003;4:193–202.

Raoul C, Pettmann B, Henderson CE. Active killing of neurons during development and following stress: a role for p75NTR and Fas? Curr Opin Neurobiol. 2000;10:111–7.

Rose S. The making of memory: from molecules to mind. London: Bantam Books; 1993.

Sara SJ. Retrieval and reconsolidation: toward a neurobiology of remembering. Learn Mem. 2000;7:73–84.

Schmidt EM, Bak MJ, Mcintosh JS. Long-term chronic recordings from cortical neurons. Exp Neurol. 1976;52:496–506.

Sherstnev VV, Gruden MA, Alexandrov YI, Storozheva ZI, Golubeva ON, Proshin AT. Different populations of neurons in relevant brain structures are selectively engaged in the functioning of long-term spatial memory. Neurochem J. 2013;7(4):278–83.

Shima K, Mushiake H, Saito N, Tanji J. Role for cells in the presupplementary motor area in updating motor plans. Proc Natl Acad Sci U S A. 1996;93:8694–8.

Shors TJ, Matzel LD. Long-term potentiation [LTP]: what's learning got to do with it? Behav Brain Sci. 1997;20:597–655.

Shors TJ, Miesegaes G, Beylin A, Zhao M, Rydel T, Gould E. Neurogenesis in the adult is involved in the formation of trace memories. Nature. 2001;410:372–6.

Shvyrkov VB. Behavioral specialization of neurons and the system-selection hypothesis of learning. In: Klix F, Hagendorf H, editors. Human memory and cognitive capabilities. Amsterdam: Elsevier; 1986. p. 599–611.

Smith DM, Barredo J, Mizumori SJY. Complimentary roles of the hippocampus and retrosplenial cortex in behavioral context discrimination. Hippocampus. 2012;22:1121–33. https://doi.org/10.1002/hipo.20958.

Sozinov AA, Kazymaev SA, Grinchenko YV, Alexandrov YI. Percent of task-specialized cingulate cortex neurons does not change during training. FENS Abstr. 2012;6:114.08.

Stone EA, Zhang Y, John S, Filer D, Bing G. Effect of locus coeruleus lesion on c-fos expression in the cerebral cortex caused by yohimbine injection or stress. Brain Res. 1993;603:181–5.

Strassmann JE, Zhu Y, Queller DC. Altruism and social cheating in the social amoeba Dictyostelium discoideum. Nature. 2000;408:965–7.

Svarnik OE, Alexandrov YI, Gavrilov VV, Grinchenko YV, Anokhin KV. Fos expression and task-related neuronal activity in rat cerebral cortex after instrumental learning. Neuroscience. 2005;136:33–42.

Svarnik OE, Bulava AI, Alexandrov YI. Expression of c-Fos in the rat retrosplenial cortex during instrumental re-learning of appetitive bar-pressing depends on the number of stages of previous training. Front Behav Neurosci. 2013;7:78. https://doi.org/10.3389/fnbeh.2013.00078.

Swadlow HA, Hicks TP. Subthreshold receptive ields and baseline excitability of "silent" S1 callosal neurons in awake rabbits: contributions of AMPA/kainate and NMDA receptors. Exp Brain Res. 1997;115:403–9.

Thompson LT, Best PJ. Long-term stability of the place-field activity of single units recorded from the dorsal hippocampus of freely behaving rats. Brain Res. 1990;509:299–308.

Tischmeyer W, Kaczmarek L, Strauss M, Jork R, Matthies H. Accumulation of c-fos mRNA in rat hippocampus during acquisition of a brightness discrimination. Behav Neural Biol. 1990; 54(2):165–71.

Tolman EC. Cognitive maps in rats and men. Psychol Rev. 1948;55(4):189–208.

Tracy J, Flanders A, Madi S, Laskas J, Stoddard E, Pyrros A, Natale P, Delvecchio N. Regional brain activation associated with different performance patterns during learning of a complex motor skill. Cereb Cortex. 2003;13:904–10.

Tse D, Langston RF, Kakeyama M, Bethus I, Spooner PA, Wood ER, Witter MP, Morris RGM. Schemas and memory consolidation. Science. 2007;316:76–82.

Tse D, Takeuchi T, Kakeyama M, Kajii Y, Okuno H, Tohyama C, Bito H, Morris RGM. Schema-dependent gene activation and memory encoding in neocortex. Science. 2011;333:891–5. https://doi.org/10.1126/science.1205274.

Vetere G, Restivo L, Novembre G, Aceti M, Lumaca M, Ammassari-Teule M. Extinction partially reverts structural changes associated with remote fear memory. Learn Mem. 2011;18:554–7. https://doi.org/10.1101/lm.2246711.

Vikman KS, Duggan AW, Siddall PJ. Increased ability to induce long-term potentiation of spinal dorsal horn neurons in monoarthritic rats. Brain Res. 2003;990:51–7.

Vogt BA, Sikers RW, Swaldow HA, Weyand TG. Rabbit cingulate cortex: cytoarchitecture, physiological border with visual cortex, and different cortical connections of visual, motor, postsubicular and intracingulate origin. J Comp Neurol. 1986;248:74–94.

von Stein A, Chiang C, König P. Top-down processing mediated by interareal synchronization. Proc Natl Acad Sci. 2000;97(26):14748–53. https://doi.org/10.1073/pnas.97.26.14748.

Weber A, Prokazov Y, Zuschratter W, Hauser MJB. Desynchronisation of glycolytic oscillations in yeast cell populations. PLoS One. 2012;7(9):e43276.

Weible AP, Rowland DC, Pang R, Kentros C. Neural correlates of novel object and novel location recognition behavior in the mouse anterior cingulate cortex. J Neurophysiol. 2009;102:2055–68. https://doi.org/10.1152/jn.00214.2009.

Weible AP, Rowland DC, Monaghan CK, Wolfgang NT, Kentros CG. Neural correlates of long-term object memory in the mouse anterior cingulate cortex. J Neurosci. 2012;32:5598–608. https://doi.org/10.1523/JNEUROSCI.5265-11.2012.

Whishaw IQ, Sarna JR, Pellis SM. Evidence for rodent-common and species-typical limb and digit use in eating, derived from a comparative analysis of ten rodent species. Behav Brain Res. 1998;96:79–91.

Williams JC, Rennaker RL, Kipke DR. Stability of chronic multichannel neural recordings: implications for a long-term neural interface. Neurocomputing. 1999;26:1069–76.

Wilson MA, Mcnaughton BL. Dynamics of the hippocampal ensemble code for space. Science. 1993;261:1055–8.

Wirth S, Yanike M, Frank LM, Smith AC, Brown EN, Wendy AS. Single neurons in the monkey hippocampus and learning new associations. Science. 2003;300:1578–81.

Wright R. The moral animal: evolutionary psychology and everyday life. New York: Vintage Books; 1995.

Xue ZM. The studies on neurogenesis induced by brain injury in adult ring dove. Cell Res. 1998;8:151–62.

Chapter 2
Neural Circuits Mediating Fear Learning and Extinction

Roger Marek and Pankaj Sah

2.1 Introduction

The brain is a complex organ, and disorders of brain function result in a host of disorders, that together make up over 50% of the burden of disease in most societies today. Understanding how neural activity results in thought and behaviour is not only an intrinsically interesting question, but is a crucial step toward finding reliable and specific treatments for neurological disorders. As a result, interest in encoding the neural circuits that underlie specific behaviour across many species has risen immensely. The development of tools such as optogenetics, multi-unit recordings of neurons in behaving animals, and the use of designer drugs that bind to engineered receptors have greatly accelerated this endeavour. One circuit that has evoked intense interest is the neural circuit that triggers fear responses. This circuit is evolutionary preserved and allows animals, including humans, to react rapidly and appropriately to adverse events, and is an essential survival mechanism. Physiological responses during fear include changes in the activity of the limbic system, with activation of the sympathetic nervous system that triggers a fight-or-flight response. As a result, such a response leads to an increase in heart rate, blood pressure and skin conductance, as well as a change in posture and mobility. This circuit and the biochemical mechanisms that underpin its activity have been studied for many years, however, recent findings have started to reveal the precise neural correlates and circuits involved.

R. Marek · P. Sah (✉)
Queensland Brain Institute, The University of Queensland, Brisbane, QLD, Australia
e-mail: pankaj.sah@uq.edu.au

A. Cheung-Hoi Yu, L. Li (eds.), *Systems Neuroscience*, Advances in Neurobiology 21, https://doi.org/10.1007/978-3-319-94593-4_2

2.2 Fear Conditioning and Extinction

Fear is a normal physiological response triggered by specific events such as a natural threats and disasters or predators, and evokes a transient physiological and behavioural state that returns to baseline after some time. However, repetitive or single traumatic exposure can also lead to abnormal fear processing that result in anxiety related disorders such as post-traumatic stress disorder (PTSD). Due to the complexity of human fear processes, and the limitations in human imaging techniques, the neural circuits that underpin fear have been studied using a classical Pavlovian conditioning procedure called fear conditioning. In this procedure, an emotionally neutral stimulus, the conditioned stimulus (CS), such as a light or tone, is temporally paired with an aversive stimulus, the unconditioned stimulus (US), typically a mild shock. Following a small number of pairings, subjects form an association between the CS and US such that the CS predicts the subsequent US, and subjects begin to respond to the CS with an avoidance response, called the conditioned response (CR). The CR is rapidly acquired, long lasting, and results from the formation and storage of a long-term memory associating the CS with the US. However, subsequent presentations of the CS, not paired with the US, break this association, and lead to a gradual reduction of the CR through a process known as extinction. Since the first studies of Pavlov, it has been appreciated that extinction does not results from an erasure of previous memory associated with the CS but is due, at least in part, to new learning (Pavlov 1927). This idea rests on three key observations in extinction. First, the learnt fear response to the CS can reappear with the passage of time (spontaneous recovery). Secondly, the CR returns when the CS is presented in a context different from the one in which extinction training originally took place (renewal). Third, unexpected delivery of the US following extinction can restore the response to the CS (reinstatement). Both renewal and reinstatement show that the CS retains its ability to drive the CR following extinction. Some of these features of extinction are illustrated in Fig. 2.1. Thus, although the original memory is still present, extinction training results in a new memory trace that inhibits the response to the original CS. In effect, the subject has learnt that a previously aversive situation is no longer dangerous.

Fear conditioning and extinction are evolutionarily conserved, and can be demonstrated in all species from insects to humans. Indeed, one of the most famous experiments is the classical fear conditioning trial with the baby called "Albert". Albert was initially allowed to play with a rat, which he enjoyed, before the experimenter played an unpleasant auditory sound every time Albert touched the rat. After some time, Albert got very distressed just by seeing the rat, which in this case is the conditioned response. Dysfunction in the circuits that mediate fear conditioning and extinction is widely thought to be responsible for a range of anxiety related disorders including phobias and post-traumatic stress disorder. Moreover, treatments for some of these disorders using exposure therapy in which there is controlled repetitive exposure to the fearful stimulus, are based on extinction. Stimulation and lesion studies in animals have identified three key brain regions that contribute to fear

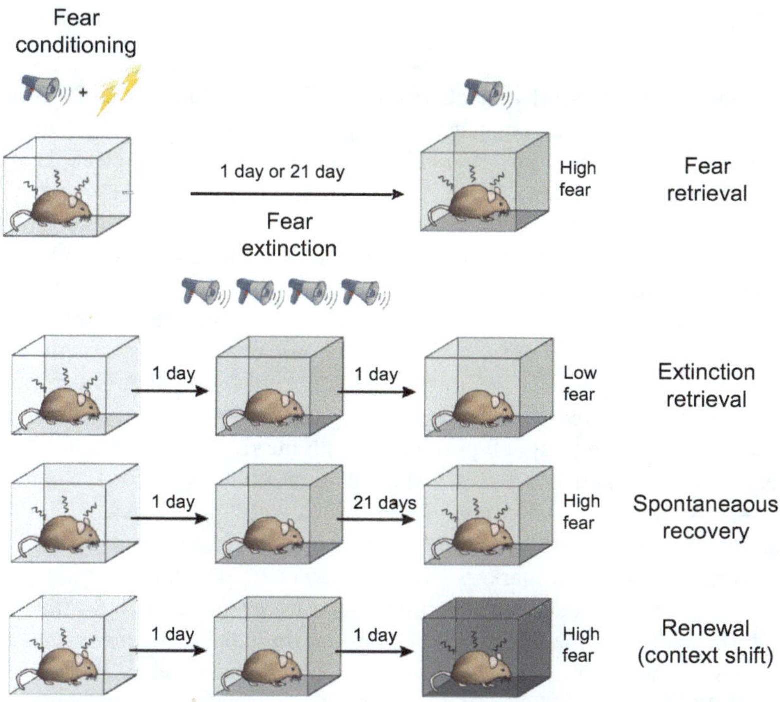

Fig. 2.1 Fear learning and fear extinction. In fear conditioning subjects are presented with a neutral stimulus, the conditioned stimulus (CS), such as a tone, that is contingently paired with an aversive stimulus, the unconditioned stimulus (US) typically a foot shock. Following one or several pairings, subjects respond to the initially neutral stimulus with a conditioned response (high fear). Following fear conditioning, subjects are repeatedly presented with the neutral stimuli, but not paired with footshock, and result in a reduction of the fear response (extinction learning). Retrieval of extinction in the same context as extinction learning results in low fear due to extinction of the previously learnt fear. Spontaneous recovery of fear can occur with passage of time (days or years) and subjects again respond with a fearful response. Fear can also be renewed by exposing the animal to the conditioned stimulus outside the extinction context

learning and extinction: these are the amygdala, medial prefrontal cortex and hippocampus (Marek et al. 2013). Studies in humans using functional magnetic resonance imaging (fMRI), have found that the same three regions are also engaged in humans, suggesting the underlying neural circuits are shared.

2.3 Neural Circuits of Fear and Extinction

Fear responses are not the result of neural activity of single brain structures, but rather result from the orchestrated activity of multiple nuclei, mediated by synaptic connections between them to allow the fear response to occur.

2.4 Anatomy

The amygdala is a temporal lobe structure that is divided into over 20 subnuclei with extensive internuclear connections (Pape and Pare 2010; Sah et al. 2003). These subnuclei are commonly divided into three groups (Price et al. 1987; Sah et al. 2003): a deep basolateral (BLA) group that includes the lateral nucleus, basal nucleus and accessory basal nucleus; a more superficial or cortical-like group that includes the cortical nuclei and nucleus of the lateral olfactory tract; and a centromedial group composed of the medial and central nuclei (CeA). Of these, the BLA and CeA are the most widely studied nuclei of the amygdala, and both structures are highly involved in fear and extinction learning, and are generally thought to form the input and output regions of the amygdala respectively. The BLA is a cortical like structure and contains two types of neurons: glutamatergic principal neurons that form nearly 80% of the total cell population with the remaining being GABAergic interneurons (McDonald 1982, 1992; McDonald and Mascagni 2001; Spampanato et al. 2011). As with cortical regions, interneurons within the BLA are divided into different populations that show distinct electrophysiological properties and expression of particular cytosolic markers (Spampanato et al. 2011). Recent studies have begun to reveal the intrinsic organization and roles of some of these interneuron families, showing that specific types of interneuron make different types of local connections (Jasnow et al. 2013; Rainnie et al. 1991; Woodruff et al. 2006; Woodruff and Sah 2007a, b). Although the BLA does not have a laminar organization, principal neurons are not a single population either, but can be separated into distinct populations by their firing properties (Faber et al. 2001; Washburn and Moises 1992).

The CeA is the dominant output-structure of the amygdala. In contrast to the BLA, the CeA is a striatal-like structure that exclusively contains GABAergic neurons (de Olmos et al. 1985), and is divided into lateral (CeL) and medial (CeM) divisions (Cassell et al. 1986). Such a segregation of these two subregions was made possible by the identification of distinct peptide expression (Cassell et al. 1986) and intrinsic firing properties (Dumont et al. 2002; Lopez de Armentia and Sah 2004; Martina et al. 1999). Moreover, the two subregions of the CeA not only differ in their chemical and electrophysiological properties, but also in their connectivity. In the CeL, neurons receive excitatory inputs from the BLA as well as thalamic and cortical regions (Sah et al. 2003). Stimulation of BLA inputs provide excitation to these neurons (Lopez de Armentia and Sah 2004), and is often accompanied by a disynaptic inhibitory response (Amano et al. 2010; Lopez de Armentia and Sah 2004; Royer et al. 1999). This inhibition comes from two distinct sources: first, a cluster of GABAergic neurons interposed between the BLA and CeL, the intercalated cells (ITC) (Millhouse 1986), that receive excitatory input from the BLA and project to the CeL (Delaney and Sah 2001; Royer et al. 1999; Strobel et al. 2015). Secondly, neurons within the CeL, which are all GABAergic, are extensively interconnected, thereby providing strong local inhibition (Haubensak et al. 2010; Lopez de Armentia and Sah 2004). Recordings from the CeL in vivo have shown that following fear conditioning, some neurons increase their response to the CS, while others are inhibited, suggesting that different cells within the CeA receive different types of input. In contrast to the CeL, much less in understood about neurons in the CeM. However, CeM neurons project to different downstream targets, mediate different physiological responses (Pare and Duvarci 2012), and can be separated on electrophysiological as well as pharmacological grounds (Viviani et al. 2011).

The prefrontal cortex (PFC) is neocortical frontal lobe structure that is involved in a variety of higher cognitive processes such as decision making and attention (Heidbreder and Groenewegen 2003). For emotional learning in humans and primates, two PFC structure are crucially involved, namely the dorsolateral PFC (dlPFC), and the medial PFC (mPFC). However, in rodents, the PFC is much less developed, and the mPFC is the crucial player in fear learning in rodents, where it is cytoarchitectonically divided into four distinct regions from dorsal to ventral: medial precentral cortex, anterior cingulate cortex (ACC), prelimbic (PL) and infralimbic prefrontal cortex (IL) (Heidbreder and Groenewegen 2003).

In primates, the PFC is a regular 6 layered structure, whereas the mPFC of rodents appears to lack the granular cell layer (layer IV). Hence, pyramidal cells are located in layers II/III and layers V/VI (Yang et al. 1996), and in acute brain slices, these neurons show a range of different intrinsic firing properties (Wang et al. 2006) similar to those described for other neocortical regions (Connors and Gutnick 1990). As with the BLA, the mPFC also contains a variety of types of interneuron (Van De Werd et al. 2010), with the expected distribution of interneuronal markers (Markram et al. 2004).

The hippocampus (HPC) has long been identified as the main source for the storage and retrieval of explicit memory. Located in the medial temporal lobe in humans, it can be divided into multiple sub-regions, of which the subiculum and the CA1 region play important roles in integrating fear-related information. The CA1 region can further be separated into the dorsal and ventral portions that play distinct roles. Anatomically, the hippocampal formation is a three layered structure with pyramidal neurons restricted to the cell body layer, and dendritic trees spreading into two layers separating the basal and apical dendrites. Similar to the amygdala and mPFC, a variety of interneurons can be found in the hippocampal formation (Acoli et al. 2008; Freund and Buzsaki 1996). Moreover, the hippocampus is extensively connected with both the mPFC and the amygdala. Until recently, hippocampal efferents to the mPFC, and the reciprocal connections with the amygdala were thought to be exclusively glutamatergic. However, recent studies are suggesting that some GABAergic neurons in the hippocampus may also form long distance connections (McDonald and Mott 2017), but the roles and connections of these projections are largely unknown.

2.5 Functional Roles

2.5.1 The Amygdala

Within the amygdala, anatomical studies indicate that the nuclei are extensively interconnected (Pitkänen et al. 1997), and both CS and US information enters the amygdala at the level of the BLA where it is first processed. These inputs form classical dual component glutamatergic synapses containing alpha-amino-3-hydroxy-5-methyl-4-isoxalepropionic acid (AMPA) and N-Methyl-D-Aspartate (NMDA) receptors (Mahanty and Sah 1996; Weisskopf and LeDoux 1999). Blocking glutamatergic transmission within the BLA by infusion of non-NMDA receptor antagonists blocks fear conditioning, and post-learning infusions block expression of learnt fear (Falls et al. 1992; Kim et al. 1993). These pharmacological manipulations also block fear extinction (Kim et al. 1993), confirming that the BLA is an essential component of the neural circuit that mediates fear conditioning and extinction. In contrast, infusion of selective NMDA receptor antagonists into the amygdala block fear conditioning and extinction learning but have no effect on previously learnt fear (Goosens and Maren 2004; Miserendino et al. 1990). Together, these results have led to the current model in which learning during fear conditioning and extinction requires NMDA-receptor-dependent plasticity of inputs to neurons within the BLA (Mayford et al. 2012). In classical fear conditioning, CS information that is typically either auditory (via auditory cortex and auditory thalamus) or visual (via pulvinar and inferior temporal area), as well as US information (via thalamic inputs from the posterior thalamus) arrives at the BLA (Farb and Ledoux 1999; Lanuza et al. 2008; Sah et al. 2003). This convergent input to neurons within the BLA (Windels et al. 2016), coupled with the associative presentation of CS and US, is thought to result in long term potentiation of inputs carrying CS information to BLA principal neurons (Izquierdo et al. 2016; Pape and Pare 2010). Extinction training also requires NMDA-receptor dependent plasticity of glutamatergic input to BLA principal neurons (Falls et al. 1992), however, how this plasticity is initiated during repetitive CS presentation is not known. Single unit recordings during fear conditioning and extinction suggest that, following fear learning, the CS activates a population of principal neurons that have been called "fear" neurons (Herry et al. 2008). These "fear neurons" in turn, project directly to the central amygdala (CeA), and downstream projections from the CeA initiate the physiological responses seen in conditioned fear (Ehrlich et al. 2009; Pape and Pare 2010; Pare and Duvarci 2012). In extinction, fear neurons loose their CS-evoked activity, and a new set of neurons, called "extinction" neurons are instead driven by the CS (Herry et al. 2008). The activity of these neurons effectively inhibits the fear response.

Projections from the BLA enter the central amygdala at the level of the CeL, and single unit studies have shown that following fear conditioning, the CS drives a population of neurons called 'ON neurons'. These cells locally inhibit a different population of tonically active 'OFF neurons'(Ciocchi et al. 2010; Haubensak et al. 2010). These 'OFF' cells are GABAergic, and project to the CeM and the overall impact is disinhibition of neurons in the CeM by the CS (Ciocchi et al. 2010; Haubensak et al. 2010). Thus, following fear conditioning, the CS evokes activity of CeM neurons to trigger a fear response (Ciocchi et al. 2010; Haubensak et al. 2010). Following extinction training, there is a reduction in the activity in 'fear neurons' in the BLA, and 'extinction neurons' become active (Herry et al. 2008). As described above, the ITC neurons form a set of GABAergic neurons that provide feed-forward inhibition to the CeA. Synaptic input from the BLA to ITC neurons also show NMDA receptor-dependent plasticity (Royer and Paré 2002), and it has been proposed that following extinction, plasticity of these inputs results in an increase in disynaptic inhibition to the CeA, effectively reducing the activity of ON neurons, thus inhibiting the fear response (Amano et al. 2010). An attractive possibility is that BLA extinction neurons are selectively engaged in driving this feed-forward inhibition of the CeL. Together, these finding show that within the amygdala, distinct circuits mediate fear expression and extinction, and the population of neurons engaged by fear learning and extinction form distinct sets that are driven by distinct inputs.

2.5.2 *Medial Prefrontal Cortex*

In addition to the direct projections that carry CS information to the amygdala, "top-down" information from the prefrontal cortex also modulates amygdala activity. The first data that suggested a role for the mPFC in fear learning came from experiments in which this structure was ablated, resulting in a deficit in extinction memory (Morgan et al. 1993). It was therefore suggested that the mPFC is required for consolidation of extinction. Subsequent stimulation and inactivation studies of the mPFC have established that this region is involved in both fear conditioning and extinction (Burgos-Robles et al. 2007; Corcoran and Quirk 2007; Laurent and Westbrook 2008; Sotres-Bayon and Quirk 2010), with the infralimbic (IL) and prelimbic mPFC (PL) having distinct roles (Burgos-Robles et al. 2007). While the amygdala is engaged during acquisition and expression of learnt fear, the PL plays a key role in consolidation and recall of fear memory (Maren and Quirk 2004; Quirk and Mueller 2008). Thus, inactivation of the PL after fear acquisition results in reduced fear responses (Corcoran and Quirk 2007). The PL in turn sends direct glutamatergic projections to the BLA, and injection of anterograde tracer into the PL labels terminals largely limited to the basal nucleus of the BLA (McDonald 1998; McDonald et al. 1996). As described above, activity of BLA neurons is required for fear expression, and this excitatory input from the PL to the BLA is thought to modulate this activity and fear expression. However, the relationship between direct CS sensory input to the BLA and that mediated via the PL, and how these interact, is not clear.

In contrast to the PL, the IL does not appear to have a significant role in either fear or extinction learning, but is required for consolidation, and perhaps expression of extinction memory. Recent experiments using optogenetics to either enhance or silence neural activity has provided direct proof for the role of the IL in fear extinction (Do-Monte et al. 2015). Moreover, in support of chemical inactivation studies, electrophysiological recordings demonstrate that following extinction training, neurons in the IL show enhanced responses to the CS (Milad and Quirk 2002). Interestingly, infusion of NMDA receptor antagonists into the IL, either before or immediately after extinction training, impair extinction learning (Burgos-Robles et al. 2007), again suggesting that in addition to the BLA, synaptic plasticity in the IL may also be also required in consolidation of extinction learning. Memory consolidation is well known to require gene transcription and protein synthesis (Lubin et al. 2011), and supporting the role of mPFC in the consolidation of memory for fear extinction, evidence exists for the necessity of protein synthesis and gene transcription within the mPFC during the establishment of long-lasting fear extinction memories (Mamiya et al. 1993; Santini et al. 2004).

While the role of the IL in extinction memory is well established, the neural circuits between the IL and the amygdala that mediate this action remain controversial. Injection of anterograde tracers into the IL show extensive labelling in the lateral amygdala as well as intermediate capsule, a region between the BLA and CeA (McDonald 1998; McDonald et al. 1996). Several lines of evidence using neuronal tracing, specific lesioning and neuronal activity markers have shown that ITC neurons are active during extinction, leading to a model in which IL activity in extinction drives ITC neurons thereby inhibiting the output of the CeA (Amano et al. 2010; Freedman et al. 2000; Likhtik et al. 2008; McDonald et al. 1996; Pinto and Sesack 2008). However, recent studies suggest that afferents from the IL do not, in fact, innervate ITC neurons directly but rather target BLA neurons, which in turn target ITC neurons (Pinard et al. 2012; Strobel et al. 2015). Thus, how IL activity in extinction inhibits amygdala outputs that mediate fear responses remains unclear.

As described above, the mPFC sends projections to the amygdala that are involved in fear and extinction. In return, the mPFC receives afferents from the amygdala as well as a number of cortical and subcortical regions (Conde et al. 1995). Thus, amygdala and mPFC activity during fear learning and extinction are likely mediated by reciprocal synaptic connections between them (Quirk and Mueller 2008; Sotres-Bayon and Quirk 2010). In support of this proposal, single unit recordings in vivo show that within the BLA, both 'fear neurons' and 'extinction neurons' have connections with the mPFC, with 'fear neurons' only sending projections to the mPFC while 'extinction neurons' appear to be reciprocally connected to the mPFC (Herry et al. 2008). Moreover, inactivation of the BLA can reduce the response of PL neurons to the CS (Sotres-Bayon et al. 2012). Together, these results suggest complex interactions between the amygdala and PL in processing a conditioned stimulus.

2.5.3 Hippocampus

The hippocampal formation forms a major part of the medial temporal lobe system, and has been linked with emotional regulation since the first studies of Papez (1937). In agreement with this, the hippocampus has extensive connections with both the mPFC and the amygdala (McDonald and Mott 2016). In fear conditioning, subjects are placed in a particular environment where the CS is contingently paired with the US. In this paradigm, subjects learn to associate both the cue (tone) and the context (the environment) with an aversive event. Following fear conditioning, subjects show defensive behaviours (e.g., freezing) to the context in which learning took place (contextual fear memory), but also to the cue (the tone—cued conditioning) in contexts different from those where learning took place. The hippocampus is well known to be involved in processing information regarding space, and lesions of the dorsal hippocampus impair contextual fear memory but have little effect on cued conditioning to an auditory stimulus (Maren and Holt 2000). Thus, it is likely that the conditioned response to the context and the cue use distinct neural circuits with the hippocampus playing a major role in defining contextual cues.

Unlike cued fear conditioning, fear extinction is highly context dependent as the fear memory trace can be retrieved outside the fear extinction context (renewal). This context dependency in extinction is dependent on the ventral hippocampus (Hobin et al. 2006; Ji and Maren 2007). As described above, the IL plays a key role in fear extinction, and in agreement with the role of the hippocampus in fear extinction, there are extensive projections from the hippocampus to the mPFC (Parent et al. 2010). These arise mainly in the CA1 region and subiculum (Hoover and Vertes 2007), and innervate both pyramidal neurons and interneurons in the PL and IL (Parent et al. 2010). In the PL, inactivation studies show that hippocampal input can inhibit activity of pyramidal neurons (Sotres-Bayon et al. 2012). However, the effect of hippocampal input on the IL is not known, and how hippocampal activity modulates extinction via the IL is currently not understood. These circuits that mediate fear learning and extinction are summarised in Fig. 2.2.

Fig. 2.2 (continued) frontal cortex (PLPFC) and the hippocampus (HPC) are also involved in fear expression, however, their exact roles in driving fear neurons are not well understood. Following extinction, CS drive of "fear neurons" in the BLA is weakened while activity of a different set of neurons, now called "extinction neurons" is enhanced most likely due to synaptic plasticity of inputs to these cells. It is likely that extinction neurons, in turn drive a set of inhibitory neurons, the intercalated cell masses (ITCs) that then inhibit response of neurons in the central amygdala thereby reducing the fear response. In extinction, the infralimbic prefrontal cortex (ILPFC) and the HPC are also engaged, both of which send projections to the amygdala, and are important for extinction expression

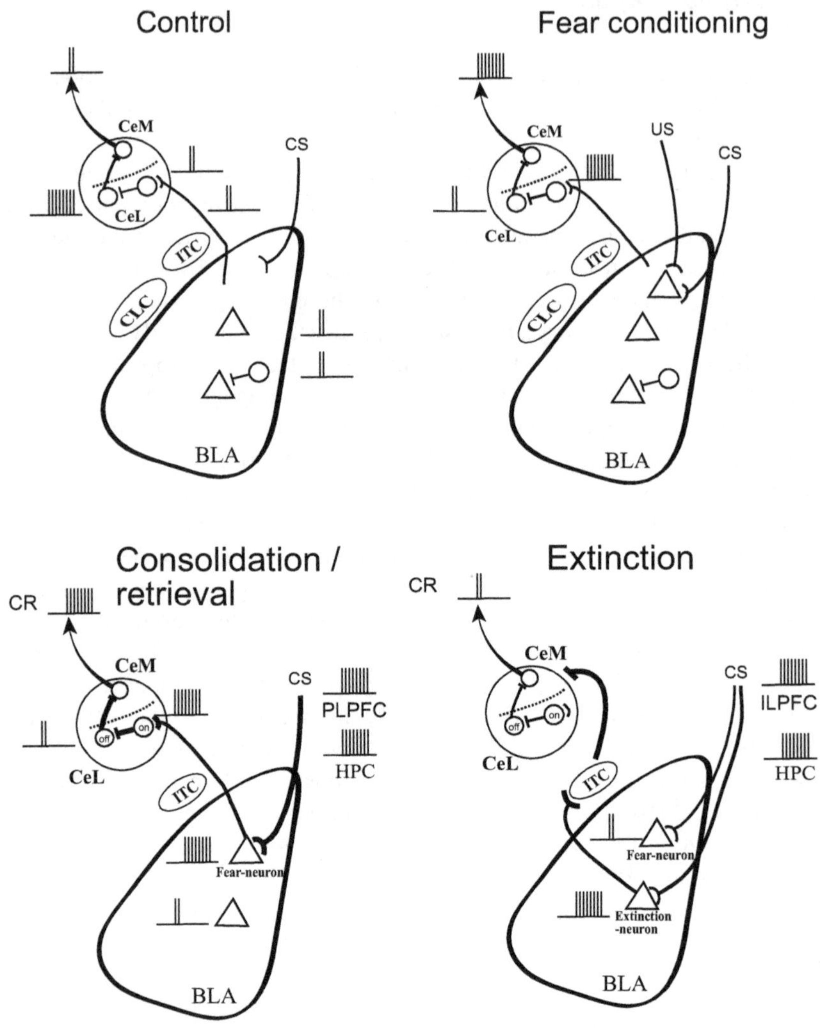

Fig. 2.2 Schematic of the neural circuitry involved in fear and its extinction. Under control conditions, sensory information which will be the conditioned stimulus (CS) reaches the amygdala at the level of the basolateral amygdala (BLA). This incoming input innervate excitatory principal neurons (triangles) and local inhibitory interneurons (circles). The BLA in turn projects to the central amygdala that contains inhibitory neurons with extensive local connections. Input arrives in the lateral division of the central amygdala (CeL), that has inhibitory connections with the medial division (CeM). Neurons in the central amygdala show different levels of tonic activity. During fear conditioning, CS inputs as well as the aversive unconditioned inputs (US) converge on principal neurons in the BLA, and as a result CS inputs are potentiated. Following consolidation, subsequent presentation of the CS enhances the activity of one set of principal neurons in the BLA that are now called "fear neurons". As result input to the central amygdala is larger and drives the activity of a set of neurons called "on neurons", these neurons in turn locally inhibit neurons in the CeL labelled "off neurons". These "off neurons" are thought to project to the CeM and disinhibition of the neurons in the CeM mediates the physiological response during fear expression. The prelimbic pre

2.6 Conclusions

Fear conditioning and extinction are two well-preserved learning paradigms seen in all mammalian species and involve the storage, consolidation and retrieval of a memory trace. It is widely believed that unravelling the mechanisms that underlie these learnt responses will provide a detailed understanding of learning and memory formation in the mammalian brain. The functional similarity between fear and anxiety disorders, and the fact that extinction recapitulates treatment strategies for these disorders, suggests that understanding the mechanisms that underpin these behaviours will lead to the development of treatments for human anxiety-related disorders. The neural circuits that mediate these two behaviours and the synaptic, biochemical and epigenetic changes that accompany them are beginning to be understood; however, there are clearly many gaps in our understanding. The development of new techniques to interrogate neural circuits in awake behaving animals holds much promise for the future.

Acknowledgments This work for funded by grants from the National Health and Medical Research Council of Australia and the Centre for Integrative Brain Function from the Australian Research Council (CE140100007).

References

Acoli, GA et. al., Petilla terminology: nomenclature of features of GABAergic interneurons of the cerebral cortex. Nat. Rev. Neuroscience. 2008; 9:557-568.

Amano T, Unal CT, Pare D. Synaptic correlates of fear extinction in the amygdala. Nat Neurosci. 2010;13:489–94.

Burgos-Robles A, Vidal-Gonzalez I, Santini E, Quirk GJ. Consolidation of fear extinction requires NMDA receptor-dependent bursting in the ventromedial prefrontal cortex. Neuron. 2007;53:871–80.

Cassell MD, Gray TS, Kiss JZ. Neuronal architecture in the rat central nucleus of the amygdala: a cytological, hodological, and immunocytochemical study. J Comp Neurol. 1986;246:478–99.

Ciocchi S, Herry C, Grenier F, Wolff SB, Letzkus JJ, Vlachos I, Ehrlich I, Sprengel R, Deisseroth K, Stadler MB, et al. Encoding of conditioned fear in central amygdala inhibitory circuits. Nature. 2010;468:277–82.

Conde F, Maire-Lepoivre E, Audinat E, Crepel F. Afferent connections of the medial frontal cortex of the rat. II. Cortical and subcortical afferents. J Comp Neurol. 1995;352:567–93.

Connors BW, Gutnick MJ. Intrinsic firing patterns of diverse neocortical neurons. Trends Neurosci. 1990;13:99–103.

Corcoran KA, Quirk GJ. Activity in prelimbic cortex is necessary for the expression of learned, but not innate, fears. J Neurosci. 2007;27:840–4.

de Olmos J, Hardy H, Heimer L. Amygdala. In: Paxinos G, editor. The rat nervous system. Sydney: Academic; 1985. p. 317–223.

Delaney AJ, Sah P. Pathway-specific targeting of GABA(A) receptor subtypes to somatic and dendritic synapses in the central amygdala. J Neurophysiol. 2001;86:717–23.

Do-Monte FH, Manzano-Nieves G, Quinones-Laracuente K, Ramos-Medina L, Quirk GJ. Revisiting the role of infralimbic cortex in fear extinction with optogenetics. J Neurosci. 2015;35:3607–15.

Dumont EC, Martina M, Samson RD, Drolet G, Paré D. Physiological properties of central amygdala neurons: species differences. Eur J Neurosci. 2002;15:544–52.
Ehrlich I, Humeau Y, Grenier F, Ciocchi S, Herry C, Luthi A. Amygdala inhibitory circuits and the control of fear memory. Neuron. 2009;62:757–71.
Faber ESL, Callister RJ, Sah P. Morphological and electrophysiological properties of principal neurons in the rat lateral amygdala in vitro. J Neurophysiol. 2001;85:714–23.
Falls WA, Miserendino MJ, Davis M. Extinction of fear-potentiated startle: blockade by infusion of an NMDA antagonist into the amygdala. J Neurosci. 1992;12:854–63.
Farb CR, Ledoux JE. Afferents from rat temporal cortex synapse on lateral amygdala neurons that express NMDA and AMPA receptors. Synapse. 1999;33:218–29.
Freedman LJ, Insel TR, Smith Y. Subcortical projections of area 25 (subgenual cortex) of the macaque monkey. J Comp Neurol. 2000;421:172–88.
Freund TF, Buzsaki G. Interneurons of the hippocampus. Hippocampus. 1996;6:347–470.
Goosens KA, Maren S. NMDA receptors are essential for the acquisition, but not expression, of conditional fear and associative spike firing in the lateral amygdala. Eur J Neurosci. 2004;20:537–48.
Haubensak W, Kunwar PS, Cai H, Ciocchi S, Wall NR, Ponnusamy R, Biag J, Dong HW, Deisseroth K, Callaway EM, et al. Genetic dissection of an amygdala microcircuit that gates conditioned fear. Nature. 2010;468:270–6.
Heidbreder CA, Groenewegen HJ. The medial prefrontal cortex in the rat: evidence for a dorso-ventral distinction based upon functional and anatomical characteristics. Neurosci Biobehav Rev. 2003;27:555–79.
Herry C, Ciocchi S, Senn V, Demmou L, Muller C, Luthi A. Switching on and off fear by distinct neuronal circuits. Nature. 2008;454:600–6.
Hobin JA, Ji J, Maren S. Ventral hippocampal muscimol disrupts context-specific fear memory retrieval after extinction in rats. Hippocampus. 2006;16:174–82.
Hoover WB, Vertes RP. Anatomical analysis of afferent projections to the medial prefrontal cortex in the rat. Brain Struct Funct. 2007;212:149–79.
Izquierdo I, Furini CR, Myskiw JC. Fear Memory. Physiol Rev. 2016;96:695–750.
Jasnow AM, Ehrlich DE, Choi DC, Dabrowska J, Bowers ME, McCullough KM, Rainnie DG, Ressler KJ. Thy1-expressing neurons in the basolateral amygdala may mediate fear inhibition. J Neurosci. 2013;33:10396–404.
Ji J, Maren S. Hippocampal involvement in contextual modulation of fear extinction. Hippocampus. 2007;17:749–58.
Kim M, Campeau S, Falls WA, Davis M. Infusion of the non-NMDA receptor antagonist CNQX into the amygdala blocks the expression of fear-potentiated startle. Behav Neural Biol. 1993;59:5–8.
Lanuza E, Moncho-Bogani J, Ledoux JE. Unconditioned stimulus pathways to the amygdala: effects of lesions of the posterior intralaminar thalamus on foot-shock-induced c-Fos expression in the subdivisions of the lateral amygdala. Neuroscience. 2008;155:959–68.
Laurent V, Westbrook RF. Distinct contributions of the basolateral amygdala and the medial prefrontal cortex to learning and relearning extinction of context conditioned fear. Learn Mem. 2008;15:657–66.
Likhtik E, Popa D, Apergis-Schoute J, Fidacaro GA, Pare D. Amygdala intercalated neurons are required for expression of fear extinction. Nature. 2008;454:642–5.
Lopez de Armentia M, Sah P. Firing properties and connectivity of neurons in the rat lateral central nucleus of the amygdala. J Neurophysiol. 2004;92:1285–94.
Lubin FD, Gupta S, Parrish RR, Grissom NM, Davis RL. Epigenetic mechanisms: critical contributors to long-term memory formation. Neuroscientist. 2011;17:616–32.
Mahanty NK, Sah P. The physiology of excitatory synapses in the lateral and basolateral amygdala. Soc Neurosci. Abstracts 22; 1996.
Mamiya N, Goldenring JR, Tsunoda Y, Modlin IM, Yasui K, Usuda N, Ishikawa T, Natsume A, Hidaka H. Inhibition of acid secretion in gastric parietal cells by the Ca2+/calmodulin-dependent protein kinase II inhibitor KN-93. Biochem Biophys Res Commun. 1993;195:608–15.

Marek, R. Strobel, C., Bredy, TW., Pankaj Sah, P. The amygdala and medial prefrontal cortex: partners in the fear circuit. The Journal of Physiology. 2013; 591(10):2381–2391
Maren S, Holt W. The hippocampus and contextual memory retrieval in Pavlovian conditioning. Behav Brain Res. 2000;110:97–108.
Maren S, Quirk GJ. Neuronal signalling of fear memory. Nat Rev Neurosci. 2004;5:844–52.
Markram H, Toledo-Rodriguez M, Wang Y, Gupta A, Silberberg G, Wu C. Interneurons of the neocortical inhibitory system. Nat Rev Neurosci. 2004;5:793–807.
Martina M, Royer S, Pare D. Physiological properties of central medial and central lateral amygdala neurons. J Neurophysiol. 1999;82:1843–54.
Mayford M, Siegelbaum SA, Kandel ER. Synapses and memory storage. Cold Spring Harb Perspect Biol. 2012;4(6):a005751.
McDonald AJ. Neurons of the lateral and basolateral amygdaloid nuclei: a golgi study in the rat. J Comp Neurol. 1982;212:293–312.
McDonald AJ. Projection neurons of the basolateral amygdala: a correlative Golgi and retrograde tract tracing study. Brain Res Bull. 1992;28:179–85.
McDonald AJ. Cortical pathways to the mammalian amygdala. Prog Brain Res. 1998;55:257–332.
McDonald AJ, Mascagni F. Colocalization of calcium-binding proteins and GABA in neurons of the rat basolateral amygdala. Neuroscience. 2001;105:681–93.
McDonald AJ, Mott DD. Functional neuroanatomy of amygdalohippocampal interconnections and their role in learning and memory. J Neurosci Res. 2017;95(3):797–820.
McDonald AJ, Mascagni F, Guo L. Projections of the medial and lateral prefrontal cortices to the amygdala: a Phaseolus vulgaris leucoagglutinin study in the rat. Neuroscience. 1996;71:55–75.
Milad MR, Quirk GJ. Neurons in medial prefrontal cortex signal memory for fear extinction. Nature. 2002;420:70–4.
Millhouse OE. The intercalated cells of the amygdala. J Comp Neurol. 1986;247:246–71.
Miserendino MJD, Sananes CB, Melia KR, Davis M. Blocking of acquisition but not expression of conditioned fear-potentiated startle by NMDA antagonists in the amygdala. Nature. 1990;345:716–8.
Morgan MA, Romanski LM, LeDoux JE. Extinction of emotional learning: contribution of medial prefrontal cortex. Neurosci Lett. 1993;163:109–13.
Pape HC, Pare D. Plastic synaptic networks of the amygdala for the acquisition, expression, and extinction of conditioned fear. Physiol Rev. 2010;90:419–63.
Papez JW. A proposed mechanism of emotion. Arch Neurol Psychiatry. 1937;38:725–43.
Pare D, Duvarci S. Amygdala microcircuits mediating fear expression and extinction. Curr Opin Neurobiol. 2012;22:717–23.
Parent MA, Wang L, Su J, Netoff T, Yuan LL. Identification of the hippocampal input to medial prefrontal cortex in vitro. Cereb Cortex. 2010;20:393–403.
Pavlov IP. Conditioned reflexes. New York: Dover; 1927.
Pinard CR, Mascagni F, McDonald AJ. Medial prefrontal cortical innervation of the intercalated nuclear region of the amygdala. Neuroscience. 2012;205:112–24.
Pinto A, Sesack SR. Ultrastructural analysis of prefrontal cortical inputs to the rat amygdala: spatial relationships to presumed dopamine axons and D1 and D2 receptors. Brain Struct Funct. 2008;213:159–75.
Pitkänen A, Savander V, LeDoux JE. Organization of intra-amygdaloid circuitries in the rat: an emerging framework for understanding functions of the amygdala. Trends Neurosci. 1997;20:517–23.
Price JL, Russchen FT, Amaral DG. The limbic region. II: the amygdaloid complex. In: Bjorklund A, Hökfelt T, Swanson LW, editors. Handbook of chemical neuroanatomy, vol. 5, Integrated systems of the CNS, part I. Amsterdam: Elsevier Science; 1987.
Quirk GJ, Mueller D. Neural mechanisms of extinction learning and retrieval. Neuropsychopharmacology. 2008;33:56–72.
Rainnie DG, Asprodini EK, Schinnick-Gallagher P. Inhibitory transmission in the basolateral amygdala. J Neurophysiol. 1991;66:999–1009.
Royer S, Paré D. Bidirectional synaptic plasticity in intercalated amygdala neurons and the extinction of conditioned fear responses. Neuroscience. 2002;115:455–62.

Royer S, Martina M, Paré D. An inhibitory interface gates impulse traffic between the input and output stations of the amygdala. J Neurosci. 1999;19:10575–83.
Sah P, Faber ES, Lopez De Armentia M, Power J. The amygdaloid complex: anatomy and physiology. Physiol Rev. 2003;83:803–34.
Santini E, Ge H, Ren K, Pena de Ortiz S, Quirk GJ. Consolidation of fear extinction requires protein synthesis in the medial prefrontal cortex. J Neurosci. 2004;24:5704–10.
Sotres-Bayon F, Quirk GJ. Prefrontal control of fear: more than just extinction. Curr Opin Neurobiol. 2010;20:231–5.
Sotres-Bayon F, Sierra-Mercado D, Pardilla-Delgado E, Quirk GJ. Gating of fear in prelimbic cortex by hippocampal and amygdala inputs. Neuron. 2012;76:804–12.
Spampanato J, Polepalli J, Sah P. Interneurons in the basolateral amygdala. Neuropharmacology. 2011;60:765–73.
Strobel C, Marek R, Gooch HM, Sullivan RK, Sah P. Prefrontal and auditory input to intercalated neurons of the amygdala. Cell Rep. 2015. https://doi.org/10.1016/j.celrep.2015.02.008. [Epub ahead of print].
Van De Werd HJ, Rajkowska G, Evers P, Uylings HB. Cytoarchitectonic and chemoarchitectonic characterization of the prefrontal cortical areas in the mouse. Brain Struct Funct. 2010;214:339–53.
Viviani D, Charlet A, van den Burg E, Robinet C, Hurni N, Abatis M, Magara F, Stoop R. Oxytocin selectively gates fear responses through distinct outputs from the central amygdala. Science. 2011;333:104–7.
Wang Y, Markram H, Goodman PH, Berger TK, Ma J, Goldman-Rakic PS. Heterogeneity in the pyramidal network of the medial prefrontal cortex. Nat Neurosci. 2006;9:534–42.
Washburn MS, Moises HC. Electrophysiological and morphological properties of rat basolateral amygdaloid neurons in vitro. J Neurosci. 1992;12:4066–79.
Weisskopf MG, LeDoux JE. Distinct populations of NMDA receptors at subcortical and cortical inputs to principal cells of the lateral amygdala. J Neurophysiol. 1999;81:930–4.
Windels F, Yan S, Stratton PG, Sullivan R, Crane JW, Sah P. Auditory Tones and Foot-Shock Recapitulate Spontaneous Sub-Threshold Activity in Basolateral Amygdala Principal Neurons and Interneurons. PLoS One. 2016;11:e0155192.
Woodruff AR, Sah P. Inhibition and synchronization of basal amygdala principal neuron spiking by parvalbumin-positive interneurons. J Neurophysiol. 2007a;98:2956–61.
Woodruff AR, Sah P. Networks of parvalbumin-positive interneurons in the basolateral amygdala. J Neurosci. 2007b;27:553–63.
Woodruff AR, Monyer H, Sah P. GABAergic excitation in the basolateral amygdala. J Neurosci. 2006;26:11881–7.
Yang CR, Seamans JK, Gorelova N. Electrophysiological and morphological properties of layers V-VI principal pyramidal cells in rat prefrontal cortex in vitro. J Neurosci. 1996;16:1904–21.

Chapter 3
The Hippocampal Ensemble Code for Spatial Navigation and Episodic Memory

Susumu Takahashi

3.1 Introduction

An initial cure for epilepsy involved surgically resecting the bilateral medial temporal lobe, which includes the hippocampal formation. In one case, the patient's epileptic seizures disappeared following the operation; however, he developed severe anterograde amnesia, a loss of the ability to create memories experienced after the surgery (Scoville and Milner 1957). Specifically, the mnemonic ability for the patient to use non-declarative memory acquired through perceptual, motor and stimulus–response learning was intact; however, he could not consciously recollect facts or events experienced post-surgery. Because the patient could recall events experienced before the surgery, the impaired memory recollection in this case is considered to involve the damage of an ability to recall memory and/or consolidate memory as a recallable form to stabilize and store it permanently. Such compelling clinical evidence strongly supports the hypothesis that the medial temporal lobe's function is, in part, to consolidate verbal knowledge, termed as declarative memory.

In another clinical case, in which the damage was restricted to the hippocampus, the patient could recall memory on facts and verbal knowledge, termed semantic memory. However, the ability to recall what occurred during an episode, where the episode took place, and when the episode happened was impaired. This memory is categorized as episodic memory. Based on these lines of clinical evidence, it is widely accepted that the hippocampus is involved in creating and/or storing episodic memory.

A third piece of clinical evidence that investigated the function of the hippocampus involved magnetic resonance imaging (MRI) of taxi drivers. In this study, the

S. Takahashi (✉)
Graduate School of Brain Science, Doshisha University, Kyoto, Japan
e-mail: stakahas@mail.doshisha.ac.jp

A. Cheung-Hoi Yu, L. Li (eds.), *Systems Neuroscience*, Advances in Neurobiology 21, https://doi.org/10.1007/978-3-319-94593-4_3

grey matter volume in the posterior hippocampus of taxi drivers, whose goal is to find an optimal driving route to a destination from various paths in everyday life, is larger than that of non-taxi drivers (Maguire et al. 2000). This suggests that prospective information on paths from self-location to the destination is encoded in the posterior hippocampus, and that the hippocampus is critically involved in the capacity for spatial navigation.

The next question to discuss is how hippocampal neurons process events that occur along the spatiotemporal continuum. Similar to the clinical cases, a surgical ablation of the hippocampus in rodents impaired their capacity for spatial navigation (Morris et al. 1982). In the dorsal hippocampus of the rodent brain, which corresponds to the human posterior hippocampus, neurons have been found that generate action potentials when animals pass through a specific location (O'Keefe and Dostrovsky 1971; O'Keefe and Nadel 1978). These neurons are called place cells. The presence alone of such cells, however, only indicates that the hippocampus represents self-location. Spatial information must be necessary to form memory of episodes that occur along space and time. Is the reason that damage to the hippocampus impairs episodic memories recall and spatial navigation just due to the loss of spatial information encoded therein? In this chapter, I address this question through discussing the role of place cells in spatial navigation and episodic memory retrieval. Specifically, I will review reports that extensively examine place functions.

3.2 Discovery of Place Cells

In 1971, O'Keefe and Dostrovsky published the first report on the presence of place cells in the hippocampus of freely behaving rats. They used an electrophysiological recording technique, now called multi-unit recording (O'Keefe and Dostrovsky 1971). They defined place cells as cells that maximally fired while an animal ran in a particular location. The space where place cells generate action potentials is spread to a certain extent, as shown in Fig. 3.1, thus it is referred to as the 'place field'. O'Keefe and Dostrovsky wrote their initial opinions on place cells in detail in a book entitled "The hippocampus as a cognitive map"(O'Keefe and Nadel 1978). In this book, they argued that the firings of place cells are a neuronal substrate of a cognitive map, which is a theory introduced by psychologist Edward C. Tolman (Tolman 1948). Without using any geographic maps, we are capable of finding an optimal route from our current location to a destination in our mind. This ability, called spatial navigation, is considered to be realized by using a mental representation of physical locations in the brain. The reason why spatial navigation is critically dependent on the hippocampus may be that the mental representation is in fact created by firings of hippocampal place cells (Buzsáki and Moser 2013).

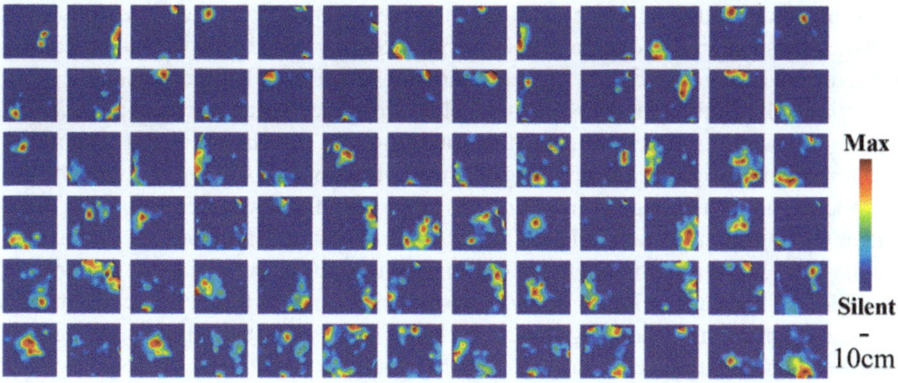

Fig. 3.1 A typical place field of 78 place cells in the hippocampal CA1 of freely behaving rats. The rats foraged food pellets randomly thrown down in a square-shaped open field. Each panel represents the spatial distribution of the firing for one place cell in the open field. The panels are arranged depending on their spatial information in a descending order (from left to right, top to bottom). The normalized firing rates were coded on a color scale from blue (silent) to red (maximum). Reproduced from Takahashi and Sakurai (2007)

3.3 Formation of Place Fields

How is the place field formed in the hippocampus? McNaughton and colleagues recorded place cell firings in the hippocampal CA1, a sub-region of the hippocampal formation, of rats during a task performance in which the rats traversed a linear track to receive rewards at one end (Gothard et al. 1996a). As expected, the place cells generated action potentials within a place field on the track. In the experiment, the start side was systematically moved toward the reward side to examine whether the length of the running path affected the firing location of the place field. Results revealed that the central location of the place field moved according to the actual length of the running path. This and subsequent reports thus supported the theory that place fields could emerge by path integration (McNaughton et al. 2009).

According to the path integration theory, place fields emerge when self-location is continuously formulated using an animal's heading direction and the number of steps. Such navigational processes, referred to as dead reckoning, have long been well understood and were used in early ship routing before the development of global positioning systems (GPS). During dead reckoning, the relative locations of landmarks can compensate for accumulated errors of estimated position. The requirements for the emergence of the place field under the path integration theory are that cells represent: (1) the heading direction, (2) the self-location, (3) the number of running steps of the animal, and (4) that self-location in response to the amount of movement can be updated (Redish and Touretzky 1997). Since place cell firings represent self-location and are modulated by an animal's running speed, the hippocampus only satisfies conditions (2) and (3). Thus, the hippocampus is not a core of path integration.

What, then, fulfills all the conditions in the requirements for path integration? In 1984, James Ranck Jr. found a cell showing a receptive field in a particular heading direction in the presubiculum of freely moving rats. He called these cells the head direction cell (Taube et al. 1990). Place fields of place cells can be remapped according to the heading direction under the condition that the source of the remapping predominantly originates from firings of head direction cells in the upstream of the hippocampus. In fact, head direction cells were found across several brain regions: the anterior thalamus, retrosplenial cortex, lateral mammillary nucleus, dorsal tegmental nucleus, striatum, and the entorhinal cortex (EC). Some of these regions are included in the Papez circuit. Some subpopulations of head direction cells strictly represent an animal's heading direction, while others additionally encode an animal's current location. This indicates that head direction cell firings can satisfy conditions (1) and (2).

May-Britt Moser and colleagues found another type of cell that specifically fires at physical locations in the medial EC, one synapse upstream structure of the hippocampus. Unlike place cells, these cells, termed grid cells, periodically fire at some locations that form lattice point patterns on a hexagonal grid spread in the environment (Hafting et al. 2005). Amazingly, the description given by grid cell place fields in the spatial environment is a hexagonal grid pattern. The preferred direction of and spatial gap between the place fields of each grid cell are different from each other, even between neighboring cells. Consequently, one synapse downstream neuron that is innervated from the ensemble of grid cells can sum up the displaced hexagonal grid patterns. Ideally, the largest overlapping location produced by the displacement indicates an animal's current location. Therefore, the ensemble of grid cells can be considered to contain information on self-location.

Similar to place cells, grid cell place fields are remapped according to an animal's heading direction and running speed. Taken together, the firings of grid cells satisfy all conditions in the path integration theory requirements, suggesting that the EC is a core region of path integration. Anatomically, the EC receives inputs from several sensory cortices, such as the somatosensory cortex and the vestibular system. Such internally generated and externally received multimodal information that gathers in the medial EC might contribute to the emergence of a hexagonal receptive field of grid cells.

An enormous number of reports investigating place cells have examined firings recorded from the dorsal hippocampal CA1. This large number is mainly due to the relative ease of inserting recording electrodes into it compared with corresponding ventral parts of the hippocampus. Place cells have been found in the hippocampal CA3 and the dentate gyrus (DG), which are direct input sources of the hippocampal CA1 (Jung and McNaughton 1993). The place field of place cells in the CA3 and DG is smaller than that in the CA1, suggesting that they represent accurate spatial information of self-location.

The main pathways from the EC to the CA1 can be divided into two types: indirect tri-synaptic pathway (EC–DG–CA3–CA1–EC) and direct pathway (EC–CA1–EC). Which pathway is the main source of the place field formation in the CA1?To exclusively inactivate the tri-synaptic pathway, Edvard Moser and colleagues dissected axonal projections from the CA3 to the CA1 and then extracellularly recorded place cell activity. In their results, they observed no obvious change in the place field between before and after the dissection (Brun et al. 2002). Interestingly, in a Morris water maze where animals had to find a platform hidden in a circular pool filled with white water with clue markings on the inner wall, rats could recollect the location of the platform. In a different Morris water maze with no clues, where animals had to recall experienced paths from their current location to a hidden platform, the rats often failed to reach the platform. The results suggest that the signal from the EC to the CA1 via a tri-synaptic pathway only helps animals navigate to a destination within their spatial environment.

These lines of compelling evidence strongly indicate that the activity of grid cells in the medial EC encodes sufficient information for the emergence of a place field of hippocampal place cells. In addition to such place-specific information, the CA1 receives additional inputs transformed via the tri-synaptic pathway. Therefore, the ability of spatial navigation, which enables the recollection of paths to a goal from experienced spatial memories, can be realized in the hippocampus. As aforementioned, the hippocampus has an ability to receive cues in the spatial environment as well as non-spatial events that occur in space. This supports the relational memory hypothesis that the hippocampus forms several types of memories in association with an animal's current location.

The examination of place cells exclusively has focused on the dorsal hippocampus. Are there place cells in the ventral hippocampus? Using information theoretic measures, McNaughton and colleagues compared spatial information encoded in the dorsal part of the hippocampal CA1 with that in the ventral part (Jung et al. 1994). The results revealed that the spatial information of place cells in the ventral hippocampal CA1 is significantly smaller than that in the dorsal part. On the other hand, Sakurai reported that the activity of hippocampal pyramidal cells represents a match/non-match status and go/no-go response during a delayed non-matching to sample performance of rats (Sakurai 1994, 1990). In that study, he could not identify place specificity of the recorded cells because the experiment was conducted in a skinner box too small to examine the place field.

To address this fundamental question, Eichenbaum and colleagues designed a clever experiment, which clearly demonstrated that place cells represent non-spatial events including odor, match/non-match status, and go/no-go response in association with an animal's current place (Wood et al. 1999). In these studies, they recorded neuronal activity from the dorsal hippocampus where most place cells showed higher place specificity. The results, therefore, were biased to the spatial information. In fact, cells in the lateral EC tend to show low place specificity (Frank et al. 2000). Furthermore, anatomical projection patterns from the EC to the CA1 are divided into two pathways: medial EC–dorsal CA1 and lateral EC–ventral CA1. Taken together, the medial EC–dorsal CA1 pathway transfers non-spatial information organized within a spatial structure. In contrast, spatial information might be streamed in a framework with non-spatial events on the pathway from the lateral EC to the ventral CA1.

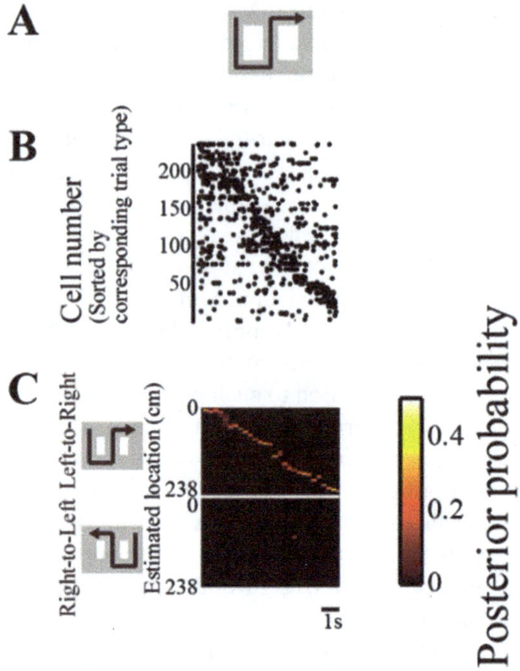

Fig. 3.2 Position reconstruction from the ensemble activity of place cells. (**a**) The path that a rat runs in the maze. (**b**) A representative raster plot shows spiking activity recorded from the rat in a single trial. Each dot represents an action potential. (**c**) The posterior probability of paths decoded by the Bayesian decoder involving left-to-right (upper) and right-to-left (lower) journeys (values indicated by color bar at right). Note that the decoded path matches the actual running path. Reproduced from Takahashi (2015)

3.4 Spatial Information of Place Fields

Is spatial information encoded in place cell activity used to understand self-location in a spatial environment? Wilson and McNaughton simultaneously monitored 148 pyramidal cells in the hippocampal CA1 of freely behaving rats as they foraged food pellets that were randomly scattered in a square open field. They used extracellular electrodes, which are a bundle of four twisted micro-wires called a tetrode (Gray et al. 1995). They decoded the location of rats from the ensemble of place cells using a population vector method that defined the place field of a cell as a vector weighted by the firing rate, and that identified the animals' current head position as the preferred location of vectors populated from the ensemble. The estimation error of the rat's head position ranged within 10 cm, provided that the ensemble consisted of over 80 place cells (Wilson and McNaughton 1993). Recently, using state-of-the-art technologies including the Bayesian decoder and large-scale multi-unit recording, the median error of the rat's head position estimated from 200 simultaneously monitored place cells dropped to under 6 cm (Takahashi 2015) (Fig. 3.2). This accuracy is far beyond the car navigation system based on the GPS.

Some reports suggested that the place-specific information encoded in place cell activity can be changed depending on the task demand. McNaughton and colleagues recorded place cell activity from the hippocampal CA1 of rats while they traversed a linear track (Gothard et al. 1996a). In this experiment, the spatial environment was invariant but the rat's heading direction was reversed between forward and backward journeys. The results demonstrated that each place cell fired within completely different place fields on forward and backward journeys. For instance, if the position was estimated from place cell firings on the backward journey based on the place field from the forward journey using the population vector method, the rat's estimated position would be very far from its actual physical location. Returning to the open field experiment conducted by Wilson and McNaughton, where the reward was randomly scattered by throwing food pellets down to the open field. The place cells fired whenever the rat entered the place field regardless of its heading direction, suggesting that place-specific firings are not simply influenced by an animal's heading direction. Was it the narrow space in the track that was responsible for the differing place fields between the forward and backward journeys? Interestingly, in an open field test where rewards were supplied at a fixed position, the place field could be dependent on an animal's heading direction (Gothard et al. 1996b). These lines of evidence suggest that under certain internal states, the heading direction information can be deposited into place cells regardless of external cues. In addition, several experiments reported that the firing rate of place cells within a place field was modulated in association with an animal's running speed (McNaughton et al. 1983).

To summarize, hippocampal place cells do not represent geographical information like that provided by a GPS, but rather stand for information based on egocentrically coordinated and accurate self-location. Since the discovery of place cells, a growing body of evidences strongly supports the seminal hypothesis postulated by O'Keefe that the ensemble of hippocampal cells appears to form a cognitive map.

3.5 Changes in the Location and Firing Rates of a Place Field

Can the change of a place field be manipulated? Muller and Kubie examined whether a place field was stable across three different conditions in a circular box: (A) landmarks in the box were rotated, (B) the size of the box was enlarged, and (C) the box was split by a wall set across the middle (Muller and Kubie 1987). Under each condition, the place field location was abruptly reassigned to a respective location in a dynamic fashion. This phenomenon, called 'remapping', was classified into two types: partial remapping and complete remapping.

During partial remapping, a subset of place cells in the recorded ensemble changed their place fields in response to the variation of the experimental box, whereas others were stable. In contrast, during complete remapping, the place field of all recorded place cells was completely reassigned. Under condition A, a subset of place cells rotated their place field in accordance with the landmark shifts.

Similarly, under condition B, the place field size was enlarged in association with the size of the experimental box. Partial remapping, therefore, can be induced by environmental manipulation. Under condition C, complete remapping was observed. We often feel something is wrong in a familiar room when a piece of furniture is misplaced. Partial remapping might correspond to such feelings. During complete remapping, an animal may perceive that the current environment is unknown. These remapping mechanisms of the place field thus support the cognitive map theory.

For instance, let's imagine a scenario where there are two rooms: A and B. The wall, furniture, and floor plan pattern are same in both rooms, but the size and arrangement are slightly different. When we visit room B after visiting room A, we may think we are in room A when actually in room B. Such associations might be due to the partial remapping of place fields in our hippocampus. By contrast, when we can easily differentiate the two rooms, complete remapping might be induced.

The change of place field can be measured in location and firing rates. Building from the formative remapping experiments, Moser and colleagues monitored the activity of place cells of freely moving rats in an experimental box (Leutgeb et al. 2005). They considered patterns of the wall of the room, and of the inner wall of the box as distal and proximal stimuli, respectively. In their results, altering distal stimuli induced global remapping, as manifested by the displacement of the place field and change of place field firing rates in spite of maintaining the same proximal stimuli. When the proximal stimuli were different but the distal stimuli were the same, the location of the place field remained unchanged, but firing rates were modulated. This phenomenon is referred to as 'rate remapping'. These results suggest that firing location and rate can induce remapping as independent variables. In accordance with this conclusion, the extent of remapping of the place field in CA1, CA3, and DG of the hippocampus differed significantly from each other (Leutgeb et al. 2007).

The extent of remapping is an effective tool to examine how we perceive the external world and determine whether it is novel. Marr classified pattern recognition ability into pattern separation and pattern completion (Marr 1969). Direct observations of this ability, however, have not been achieved because the internal cerebral representation of external patterns in the environment cannot be measured.

In the process of manipulating place field remapping and measuring the changes, it is possible to quantitatively measure the process of pattern recognition in the brain. Indeed, a few research teams have investigated pattern separation and completion through place field remapping. Since the remapping occurred in response to a minor environmental change, the DG showed the highest performance on pattern separation capacity. In contrast, since the extent to which the place field in the CA1 and CA3 were remapped linearly increased, the CA1 and CA3 showed high pattern completion performance. We can easily vary similar environments by virtue of the abilities of pattern separation and completion. The hippocampal CA1, CA3, and DG appear to play different roles on perceiving the external environment along the trisynaptic pathway. The induction of complete or global remapping suggests that the cognitive map in the hippocampus is not unique even if an animal confronts a common environment. If so, why is rate remapping induced? Overall, the location and firing rates of the place field might represent independent information.

3.6 Place Field Remapping in Terms of Past and Future Locations

As mentioned above, an animal's heading direction can shift the central location of a place field, and visual cues can trigger place field remapping in terms of the firing location and rates. This evidence supports the view that the representation encoded in place cell firings is associated with external events, including self-actions. Are such external events the only factor affecting the place field formation? To address this question, Eichenbaum and colleagues analyzed the place field during continuous spatial alternation task performance (Wood et al. 2000). They used a modified T maze constructed with a T—shaped junction with a central stem, right, and left arms, and return rails to a start position. After training rats to alternatively choose either the left or right arms (right—left—right etc.) at a junction of the central stem in order to receive rewards in the maze, they monitored the activity of hippocampal CA1 pyramidal cells using electrodes that consisted of two bundled micro-wires, called a stereotrode. In this experiment, as the rats were well trained to continuously exhibit spatially alternative responses, the heading direction and running speed were stable in the central stem. In contrast, while the firing location of some place cells were stable at a fixed position, firing rates were significantly modulated with respect to which arm the rats had previously chosen. Surprisingly, the firing rates were also substantially modulated as to which arm the rats will choose.

These results suggest that firings in the place field are affected by either retrospective memory, referring to events experienced in the past, or prospective memory, referring to remembering to perform planned actions in the future. The ability of spatial navigation enables us to flexibly find an optimal route from ones we have already experienced in our mind. Given that place cells play an important role in realizing such spatial navigation, the hippocampal ensemble activity necessarily forms a cognitive map in our mind. Place cell firings were substantially modulated in association with introspection, strongly supporting the view that the hippocampus plays a critical role in spatial navigation.

Changes in firing rates and location of the place field depend on past, present, and future memory. These temporal orders can also be observed in the firing timing rhythm typically seen in local field potentials. O'Keefe and Recce reported on the strong association between the firing timing of place cells and the theta band (4–12 Hz) of local field potentials (O'Keefe and Recce 1993). The phase is a variable that describes a neuron's firing timing as a degree, where the peak to peak, or valley to valley of a local field potential is defined as a 360 degree cycle. O'Keefe and Recce discovered the phenomenon called theta phase precession in which the phase of place cell's firing progressively shifts forward as an animal approaches the center of the place field. In other words, the phase of the place cell firings tells us whether the animal is approaching or departing its place field. Furthermore, we can predict where the animal is about to go next from the theta phase of the place cell firing, provided that the preferred place field location is known in advance.

Why does the firing rate in the place field increase as an animal runs faster? Even when the running speed varies, the theta wave is invariant. If an animal runs faster, the cycles of the theta wave increases. This may be why the firing rate is dependent on the running speed. The theta phase precession is a phenomenon operating at the level of a single neuron. So, how does the ensemble of downstream neurons receive meaningful information from those firing phases? Lisman proposed a theory where information flows on a gamma wave (approximately 40 Hz) (Lisman 1999). It postulates that past, present, and future places can be predicted from the theta phase of place cells that fire within a gamma cycle (10–20 ms).

In accordance with the theta phase precession theory, several reports using a cross-correlation analysis between spike trains suggested that information flows through the neuronal ensemble within a timeframe of 10–20 ms. According to the cell assembly hypothesis postulated by Hebb (1949), there is a proposed functional connectivity between two neurons that simultaneously generate action potentials, specifically when an animal performs a particular behavior. Sakurai argued that firing synchrony between hippocampal pyramidal cells is not coincidently observed, but rather is varied in association with the performance of memory tasks: reference memory and working memory (Sakurai 1994). In this experiment, the peak width of firing synchrony between two neurons ranged within approximately 15 ms.

Buzsaki and colleagues classified the hippocampal ensemble of place cells into two groups based on the extent of synchrony between them, which is called 'peer prediction weight' (Harris et al. 2003). They predicted the firing patterns of one of the place cells from those of others using their peer prediction weights and well-known behavioral factors that influence the emergence of the place field, such as location, running speed, and the heading direction of the animal. As a result, they ascertained that a 25 ms timescale of synchrony is optimal for the prediction.

Finally, Markram and colleagues demonstrated for the first time that the firing timing of neurons in a cell culture is critically important in forming the Hebbian plasticity between them. When the presynaptic neuron fired 10 ms before the postsynaptic neurons, long term depression could be observed between them. In contrast, long term facilitation was observed when the presynaptic neuron fired 10 ms after the postsynaptic neuron (Tsodyks and Markram 1997). These phenomena are now called 'spike timing dependent plasticity (STDP)' (Bi and Poo 1998). These reports unveiled that firings within a temporal duration of 10 ms were effective in dynamically reconfiguring the synapses among neurons in the brain.

3.7 Variation of Location and Firing Rate of the Place Field in Relation to Episodes

In the previous sections, the location and firing rates of the place field could be varied depending on where an animal came from and where it was about to go. Namely, the emergence of the place field is influenced by a combination of the where and when elements of episodic-like memory. The place field of the hippocampal pyramidal cells, however, is merely the tip of the iceberg. Previously, Vinogradova reported that the match/non-match status of the presented stimuli could change the hippocampal EEG (Vinogradova and Dudaeva 1972). Sakurai monitored hippocampal pyramidal cell activity during an auditory guided delayed non-matching to sample performance of rats (Sakurai 1990). He found that the firing rates of some neurons were significantly modulated with a match/non-match status of the auditory cues. Similarly, Otto and Eichenbaum reported that pyramidal cells in the hippocampal CA1 changed their firing rates in response to a match/non-match status of odor cues during an odor guided delayed non-matching to sample performance of rats (Otto and Eichenbaum 1992).

Building on these studies, we have investigated whether such modulation of firing timing and rates is preserved among closely neighboring neurons. Unfortunately, when two or more closely neighboring neurons simultaneously fire, the extracellularly recorded action potential waveforms overlap in a manner that makes sorting spikes of single neurons difficult. This is referred to as a spike overlapping problem. Thus, conventional methods could not correctly isolate spikes generated from closely neighboring neurons within a 1 ms precision from individual ones.

To overcome this limitation, I developed a novel spike sorting method, ICSort, in combination with independent component analysis (Hyvarinen 1999) and clustering (Takahashi et al. 2003a, b). Using the ICSort, we analyzed the synchronized activity of closely neighboring neurons within a 1 ms precision. We found that the occurrence rates of the sub-millisecond firing synchrony between closely neighboring pyramidal neurons in the hippocampus were substantially modulated with a match/non-match status of auditory cues and go/no-go motor response. Surprisingly, the information represented in the synchrony was comparable to that in the firing rates (Takahashi and Sakurai 2009a). Furthermore, using a support vector machine, which is a machine learning, we found that the ensemble activity of hippocampal pyramidal cells encodes more information on stimulus comparison than others (Takahashi and Sakurai 2009b). These lines of evidence suggest that hippocampal pyramidal cells not only encode spatial information, but also non-spatial information, such as what and how an animal is doing or has done. However, it was unknown whether the pyramidal cells examined in the aforementioned experiment were place cells.

Eichenbaum and colleagues elaborated on a task design to simultaneously examine the place-specificity and encoding of non-spatial events in the activity of hippocampal CA1 place cells (Wood et al. 1999). They found that some pyramidal cells maximally fired at a particular location regardless of non-spatial events, but that the firing rates in the place field of other cells were modulated in conjunction with a specific odor cue or match/non-match status of the odor cues. This suggests that place cells not only encode self-location, but also what happens at the self-location.

Astonishingly, Eichenbaum and colleagues found that hippocampal CA1 place cells tend to generate action potentials within a particular elapsed time at a specific location during delayed go/no-go performance in rats (MacDonald et al. 2011). Certain cells that only encoded a particular elapsed time regardless of the self-location were called time cells. Unlike retrospective and prospective memory coding in the hippocampus, this result suggests that the hippocampus encodes time. Furthermore, it indicates that the hippocampus represents conjunctive information on 'when and where'.

To investigate the firing patterns of hippocampal CA1 place cells under a condition in which the location, running speed, and heading direction remained unchanged, Buzsaki and colleagues trained rats to run on a running wheel for a 10 s delay period before choosing either left or right arm at the end of the central stem in a figure-eight shaped maze (Pastalkova et al. 2008). They found that some place cells maximally fired within a specific elapsed time during the delay period. Like the place field, the firings occupied a particular elapsed time so that the delay period could be filled with the firing fields of tens of cells. Interestingly, the future direction that the rat would choose at a decision point could be decoded from the ensemble activity pattern during the delay period. The firing duration of elapsed time was not associated with place-specific firing patterns in the maze even if the cell had a place field. Buzsaki and colleagues therefore supposed that the ensemble activity of place cells was involved in action planning. This is considered to be the first report that the hippocampal pyramidal cells encode information on time. However, Buzsaki stresses that the brain does not produce time per se. Rather, his understanding is that the time cells' presence is just the disguise of time tracking: cells coincidentally emerge by virtue of the sequential activation of place cells during episodic memory recollection (Buzsáki 2013). That is, space and time represented in the brain may be indistinguishable.

Finally, Mizumori and colleagues trained rats to navigate a plus-shaped maze (Smith and Mizumori 2006a). In this design, the rat's heading direction and running speed, as well as external landmarks, were common in the path from center to east during journeys from south and from north to east. The location and firing rates of the place field were changed. Although their results can be considered as retrospective memory encoding, they argued that the emergence of the place field can be affected by which task demand the animal confronts.

3.8 Context-Dependent Encoding Within the Place Field

On the basis that place-specific firings can be observed in hippocampal pyramidal cells and a hippocampal lesion impairs recalling episodic memory, Eichenbaum postulated a theory called conjunctive encoding, whereby the hippocampal place cells integrate multimodal sensory inputs and encountered memory (Eichenbaum et al. 1999). Following this theory, the emergence of place cells' firing fields is potentially triggered every moment by the mixture of varying external sensory inputs, such as visual and auditory cues, and internal events, including self-motion and experienced memory, from the past. On the other hand, considering that the ensemble activity of place cells utilizes a sequence of events as context, Mizumori argued that place cell firings are context-dependent (Smith and Mizumori 2006b).

According to this line of argument, the place field is direction-dependent only under certain circumstances because switching between spatial contexts is a prerequisite for receiving rewards. These theories continuously elucidate that place cell remapping is not only simply tied to the change of external inputs, but also is deeply involved in an animal's internal states.

3.9 Hippocampal Place Cells Hierarchically Organize Contexts in the Episodic-Like Memory Trace

Against the initial interpretations of place cells, the patterns of place cell firings can be modified in terms of the difference between past and future journeys, and between confronted task demands. Namely, the hippocampal place code is not only closely tied to the 'where' element of episodic-like memory, but also to the 'when and how'. However, since each report was conducted under different experimental conditions, the modifications have been interpreted as the remapping of the hippocampal cognitive map, whereby the relationship between place cell firings and episodic-like memory have been neglected. To confirm whether the changes and dynamics of the hippocampal place code are a signature of the neuronal underpinning of spatial and episodic-like memory, it is necessary to conduct a multifaceted experiment in which many hippocampal neurons are simultaneously monitored during repeated exposures to either spatiotemporal or non-spatial contexts in a constant spatial environment.

To carry-out this experimental design, I trained rats to navigate their way through a figure-eight maze in a continuous task that incorporated both visual discrimination and two types of memory-guided responses (Fig. 3.3a) (Takahashi 2013). By virtue of arrays of ten extracellular dodecatrodes (Takahashi and Sakurai 2005), a high-density bundled electrode consisting of twelve 8 micron micro-wires, 1119 pyramidal cells were monitored in the CA1 of the dorsal hippocampus of rats running in the maze. Although the rats encountered the same prominent environmental features in the same locations, the place codes were substantially different in both firing location and rate among different journeys, irrespective of either visually or mnemonically guided demands, in a global remapping manner (Fig. 3.3d). This result is consistent with the previous findings that hippocampal pyramidal cells show journey-dependent coding during internally or externally guided goal-directed behavior (Ferbinteanu et al. 2011). However, when the rats experienced subtask differences, the journey-dependent activity was only modulated by primarily changing the firing rates' intensity (Fig. 3.3d). The latter finding indicates that the extent to which place fields remap is demand-specific (Smith and Mizumori 2006a) in a rate remapping fashion.

In summary, the results suggested that: (1) the place field location was shifted according to where an animal had just been and where it was about to go next, and that (2) the firing rates within the place field were modified depending what demand

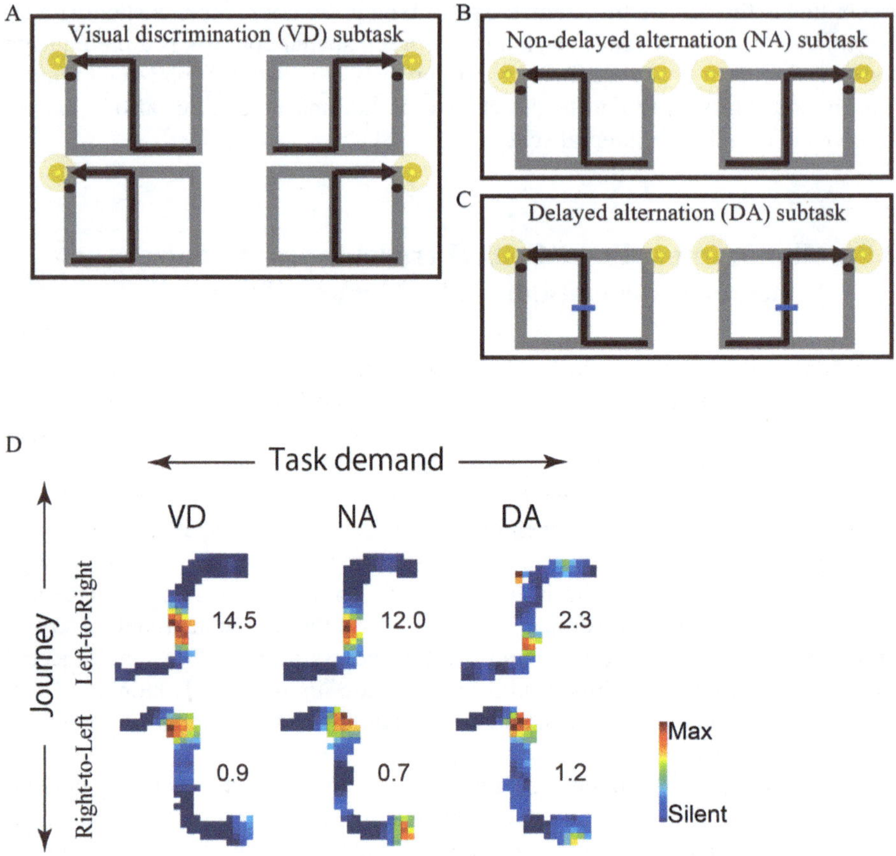

Fig. 3.3 Remapping of the place field. (**a**–**c**) Task design and configuration. Each subtask can only be identified at the decision point (upper junction) because the rats were not able to see the visual cues until they reached that point, even in the central maze stem. If correct, the rats received electrical stimulation of the medial forebrain bundle (MFB) as a reward once they reached the visual cue (yellow circle). After receiving a reward, the rats always briefly stopped within the start zones (black dots). In the VD (**a**) subtask, one of the visual cues was illuminated to guide the rat towards its goal. However, in the NA (**b**) and DA (**c**) subtasks, spatial memory retrieval was required for the rat to choose the opposite goal to the previous one, because both of the visual cues were illuminated. In the DA subtask, the rats had to wait for 5 s in the maze stem until the barrier wall (blue line) disappeared. (**d**) Representative color-coded rate maps of a single place cell for six trial types. Note that different journeys in the same place resulted in place fields in similar locations. However, different task-demands caused different firing intensities in those place fields. Reproduced from Takahashi (2013)

it was currently confronting. However, the association in the hippocampal neuronal ensemble codes remained unclear in the previous experiments because each piece of information was independently examined. I reported that all of this information could be simultaneously preserved in the hippocampal output to form episodes. In other words, my findings strongly support the seminal theories of conjunctive

(Eichenbaum et al. 1999) and context-dependent encoding (Anderson and Jeffery 2003; Smith and Mizumori 2006a).

In this experiment, the rats encountered eight trial types that varied in journey and task-demands. The trial type contained three elements of episodic-like memory that were interpreted as a single episode: where the animal had been, where it was about to go next and how it navigated its way. To depict the dynamics of neuronal ensemble activity in terms of episodes, I performed a neuronal trajectory analysis, (Harvey et al. 2012) in which the activity of n simultaneously monitored neurons was represented as a point in an n-dimensional space (Fig. 3.4a). To quantify the trajectory specificity, I calculated the distance from an individual's lap trajectory to the mean right-to-left/left-to-right trajectories. The neuronal trajectory was sufficient to separate the left-to-right journey from the right-to-left journey with high accuracy, and vice versa (Fig. 3.4b). Surprisingly, even in visually guided situations in which memory-guided decisions were not used, the prediction accuracy was sufficiently better than chance levels before the rat rotated its heading to the next direction (Fig. 3.4c).

Following these results, I concluded that the activity in the hippocampus could be considered to follow divergent, episode-specific trajectories. Indeed, using the Bayesian decoder in conjunction with a prediction method based on the firing rates, trial types were accurately estimated (>74%; Fig. 3.5). This phenomenon implies that the hippocampus begins to internally generate episodic-like memory before recollecting an episode in the mind.

In addition to the evidence from place field measurements, even when given task-demands differed, the classifier could sufficiently predict the journey from the ensemble activity patterns. This suggests that journey representation is generalized and that non-spatial demand-specific representation is hierarchically ranked at a lower level (Fig. 3.6). In computer science, it is widely accepted that the computational demand associated with searching the position of a target value can be remarkably reduced by exploiting predefined hierarchical structures. Building from this, I argue that the hippocampus plays a crucial role on the transformation of an encountered episode into the hierarchically organized contexts, in order to flexibly and quickly recall episodic-like memory.

3.10 Reactivation of Place Cell Activity While an Animal Briefly Pauses

Place cell activity can be sequentially reactivated in a temporally compressed fashion within sharp wave/ripples (SWRs) in local field potentials (LFPs) during slow-wave sleep or periods of awake immobility (Diba and Buzsáki 2007; Foster and Wilson 2006; Gupta et al. 2010; Karlsson and Frank 2009; O'Neill et al. 2006; Wilson and McNaughton 1994). The replay of place cell activity sequences appears to be a reflection of experienced paths. Indeed, during brief periods of immobility,

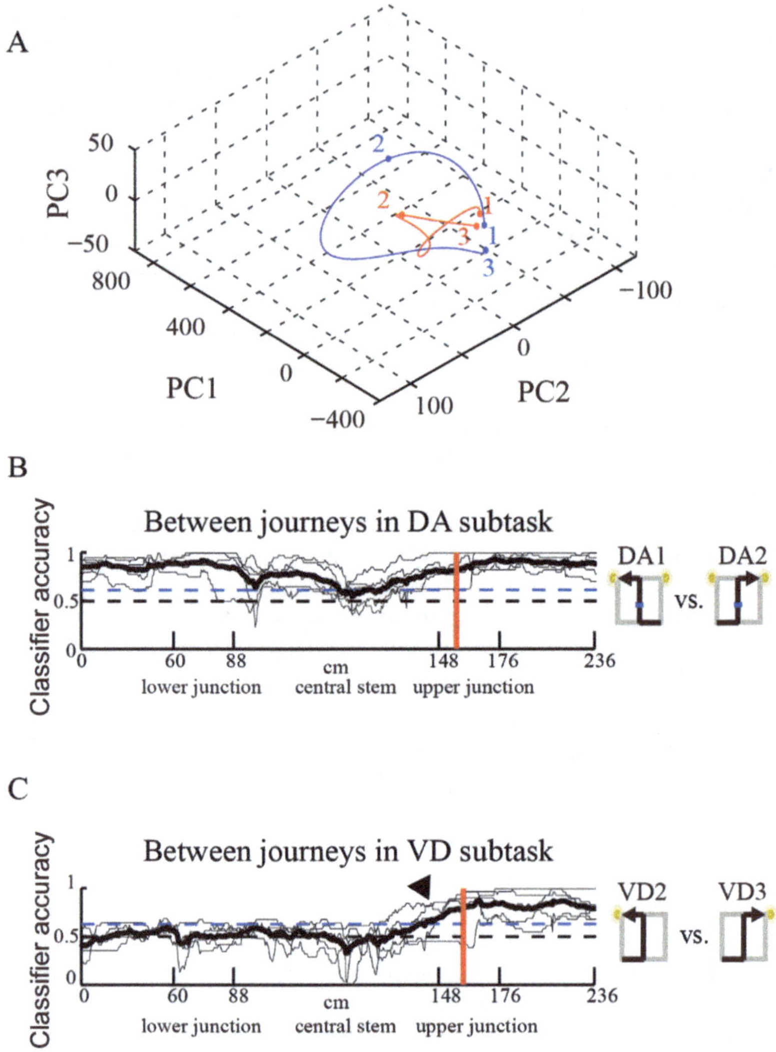

Fig. 3.4 Neuronal trajectory of the ensemble activity of place cells. (**a**) An example time course of mean, journey-specific trajectories for right-to-left (red) and left-to-right (blue) laps plotted on the 1st to 3rd principal component space. The points marked 1, 2, and 3 correspond to the mean locations where the rat entered the lower arm, turned, and entered the reward zone, respectively. (**b**) The classification accuracy of determining the journeys (shown in right) at different locations in the DA subtasks (solid black line, mean; blue dashed line, $p = 0.001$, binomial test; black dashed line, chance level; red vertical line, mean turn onset). The classifier was based on a distance-dependent classification scheme. (**c**) The same as for (**b**), except that the subtask condition was that the future direction was not predictable until arriving at the decision point. A high level of accuracy can be observed even before the onset of the turn (arrow heads). Reproduced from Takahashi (2013)

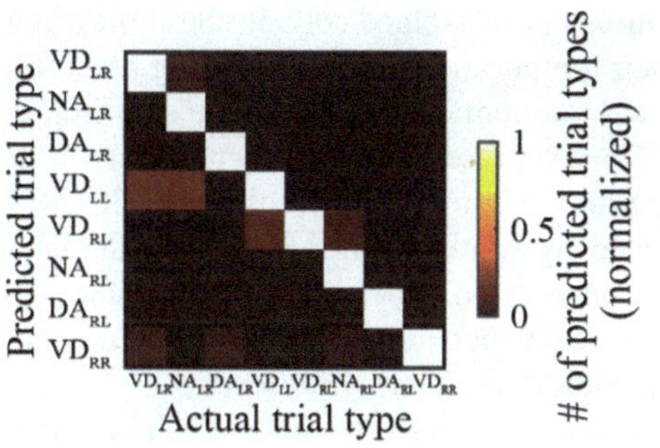

Fig. 3.5 Confusion matrices between predicted and actual trial types (values indicated by color bar at right). Note that the prominent diagonal line in the matrix shows that the predicted trial type matched the actual one across all trial types. Reproduced from Takahashi (2015)

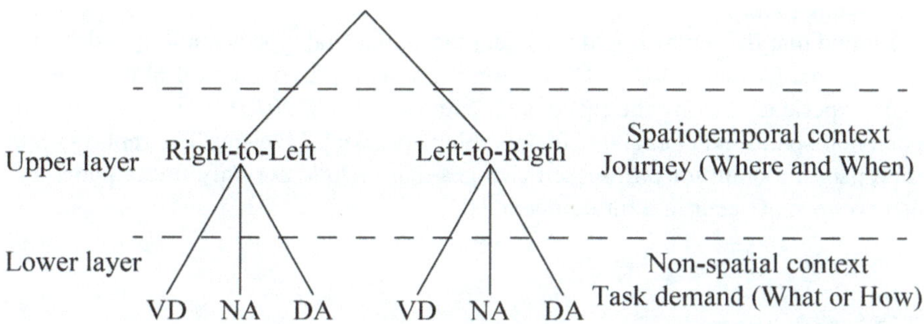

Fig. 3.6 Schematic of hierarchical organization of contexts. The journey representation was generalized irrespective of the demand-specific representation, suggesting that the information for journey and task-demand is not horizontally but hierarchically organized. In particular, the non-spatial demand-specific representation is ranked at a lower hierarchical level. Journey and task-demands can be considered to be spatiotemporal (where and when) and non-spatial (what or how) contexts, respectively. The spatiotemporal and non-spatial contexts are thus hierarchically organized into the hippocampal episodic code

pivotal paths to a remembered goal can be predicted, regardless of the location where the SWRs previously occurred (Pfeiffer and Foster 2013). The replay is thus considered to be a neuronal substrate for recalling memories of previously encountered path (Carr et al. 2011). However, since previous studies that investigated the content of replays primarily focused on the geography and timeline of an animals' running path in the environment, it was unclear whether the replay conveyed information on events (i.e. 'what' information) that occurred along the reactivated path.

In the task used in my previous study (Takahashi 2013), rats ran along similar spatial paths while performing different subtasks. The differences between subtasks can be interpreted as non-spatial 'what' information. The subtasks along the path in the maze can be examined in the temporally compressed replay. As expected, I found that the replay represents non-spatial information on subtasks, as well as spatial information concerning the path. While the path is encoded in temporally compressed firing timings across place cells in the replay, the accompanying subtask is encoded in their temporally compressed firing rates (Fig. 3.7). These results revealed that global and rate remapping mechanisms during running (Leutgeb et al. 2005; Takahashi 2013) are preserved in the temporally compressed replays that occur during brief periods of immobility (Takahashi 2015). I also found that the replay is only enhanced by spatial working memory demand. Since the rat must retrieve episodic-like memory on which path it chose and which task it confronted during the working memory performance, the results suggest that awake replay plays a key role in episodic-like memory retrieval.

While an animal pauses, it often looks ahead. Such behavior is termed 'vicarious trial-and-error' (VTE) because it suggests internal exploration of future possibilities (Hu and Amsel 1995). Since place cell activity sequences represent potential future paths even in completely unfamiliar places (Gupta et al. 2010), they may embody the VTE event. Unfortunately, reports that discuss this connection only examined the running path.

I found that the replay could represent the actual trial type accurately only when the trial type was identifiable. Otherwise, it evenly represented all trial types previously experienced along the reactivated path (Takahashi 2015). This result implies that a non-spatial event encoded in the replay is also a VTE event. The replays seem to prepare the brain for unforeseen changes that include not only future paths, but also events that occur in a future place.

3.11 Summary

In this chapter, I explored the neuronal underpinning of spatial navigation and episodic-like memory in the ensemble activity of hippocampal place cells. Whereas place cells generate action potentials at a particular location, the firing location can be changed depending on external cues and internal states, including action planning. These lines of evidence strongly support the seminal view that the hippocampus forms cognitive maps that enable spatial navigation.

Other lines of evidence, however, revealed that place cell firings can show episodic-like memory traces of what happened at a specific place and time. They suggest that the hippocampus does not assign a single map to a physical space, but rather forms a cognitive map for each episode. During spatial navigation, we have to find an optimal path by recalling what happened at a specific place and time from past memories in our mind—so-called 'mental time travel'. Spatial navigation is, therefore, considered to be the ability to use episodic memory retrieval. I found that the reactivation of sequential activity in hippocampal place cells encodes episodic-

Fig. 3.7 (continued) decoded paths (**b**) showed similar patterns (arrows), the firing rates (**a**, right) for DA_{RL} were greater than those for VD_{RL}, suggesting that the firing rate encodes the subtask-difference. Note that the method based on firing rates accurately predicted the trial type from the replay. Reproduced from Takahashi (2015)

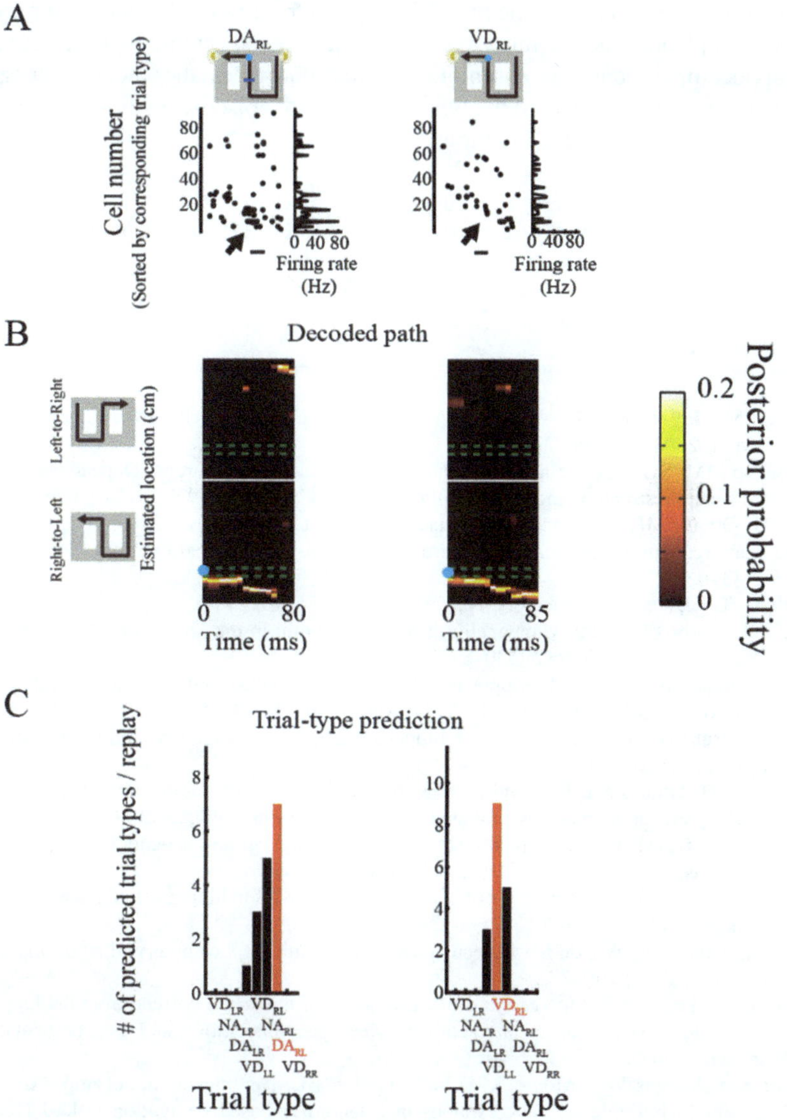

Fig. 3.7 Decoded path and predicted trial type during periods of awake immobility. Graphs are arranged into two columns for each trial type. Each sub-column consists of a raster plot of the spiking activity of place cells (**a**, left), the corresponding firing rates (**a**, right), the posterior probability of decoded paths (**b**), and the predicted trial types (**c**) for representative replays during periods of immobility. The scale bar indicates 10 ms. (**b**) Values are indicated by color bars (middle, right). In the decoded paths (middle), the maze's upper junctions are enclosed by two green dotted lines. The rat's physical location when the replay occurred is indicated by a solid blue circle. (**c**) In the trial-type prediction, red bars indicate the most frequently predicted trial type. Red labels indicate when the most frequently predicted and the actual trial types matched. The replays depicted an upcoming path to a memory-guided goal. While the spike raster plots (**a**, left) and the

like memory experienced in the past. The ensemble activity of place cells may correspond to episodic-like memory retrieval. I therefore speculate that the activity of the hippocampal neuronal ensemble not only provides the faculty for spatial navigation but also is linked to the typical abilities of episodic memory: mental time travel and foreseeing future situations.

Acknowledgements This work was supported by MIC SCOPE (152107008) and JPSP KAKENHI (JP25560435, JP16H02840, JP16H01623, JP16K13115, JP16H06543).

References

Anderson MI, Jeffery KJ. Heterogeneous modulation of place cell firing by changes in context. J Neurosci. 2003;23(26):8827–35.

Bi GQ, Poo MM. Synaptic modifications in cultured hippocampal neurons: Dependence on spike timing, synaptic strength, and postsynaptic cell type. J Neurosci. 1998;18(24):10464–72.

Brun VH, Otnass MK, Molden S, Steffenach H-A, Witter MP, Moser M-B, et al. Place cells and place recognition maintained by direct entorhinal-hippocampal circuitry. Science. 2002; 296:2243–6.

Buzsáki G. Time, space, and memory. Nature. 2013;497:568–9.

Buzsáki G, Moser EI. Memory, navigation and theta rhythm in the hippocampal-entorhinal system. Nat Neurosci. 2013;16(2):130–8.

Carr MF, Jadhav SP, Frank LM. Hippocampal replay in the awake state: a potential substrate for memory consolidation and retrieval. Nat Neurosci. 2011;14:147–53.

Diba K, Buzsáki G. Forward and reverse hippocampal place-cell sequences during ripples. Nat Neurosci. 2007;10:1241–2.

Eichenbaum H, Dudchenko P, Wood E, Shapiro M, Tanila H. The hippocampus, memory, and place cells: is it spatial memory or a memory space? Neuron. 1999;23(2):209–26.

Ferbinteanu J, Shirvalkar P, Shapiro ML. Memory modulates journey-dependent coding in the rat hippocampus. J Neurosci. 2011;31(25):9135–46.

Foster DJ, Wilson MA. Reverse replay of behavioural sequences in hippocampal place cells during the awake state. Nature. 2006;440:680–3.

Frank LM, Brown EN, Wilson M. Trajectory encoding in the hippocampus and entorhinal cortex. Neuron. 2000;27(1):169–78.

Gothard KM, Skaggs WE, McNaughton BL. Dynamics of mismatch correction in the hippocampal ensemble code for space: interaction between path integration and environmental cues. J Neurosci. 1996a;16(24):8027–40.

Gothard KM, Skaggs WE, Moore KM, McNaughton BL. Binding of hippocampal CA1 neural activity to multiple reference frames in a landmark-based navigation task. J Neurosci. 1996b;16(2):823–35.

Gray CM, Maldonado PE, Wilson M, McNaughton B. Tetrodes markedly improve the reliability and yield of multiple single-unit isolation from multi-unit recordings in cat striate cortex. J Neurosci Methods. 1995;63(1–2):43–54.

Gupta AS, van der Meer MAA, Touretzky DS, Redish AD. Hippocampal replay is not a simple function of experience. Neuron. 2010;65:695–705.

Hafting T, Fyhn M, Molden S, Moser M, Moser EI. Microstructure of a spatial map in the entorhinal cortex. Nature. 2005;436(7052):801–6.

Harris KD, Csicsvari J, Hirase H, Dragoi G, Buzsaki G. Organization of cell assemblies in the hippocampus. Nature. 2003;424(6948):552–6.

Harvey CD, Coen P, Tank DW. Choice-specific sequences in parietal cortex during a virtual-navigation decision task. Nature. 2012;484(7392):62–8.

Hebb D. The organization of behavior. New York: Wiley; 1949.

Hu D, Amsel A. A simple test of the vicarious trial-and-error hypothesis of hippocampal function. Proc Natl Acad Sci U S A. 1995;92:5506–9.

Hyvarinen A. Fast and robust fixed-point algorithms for independent component analysis. IEEE Trans Neural Netw. 1999;10(3):626–34.

Jung MW, McNaughton BL. Spatial selectivity of unit-activity in the hippocampal granular layer. Hippocampus. 1993;3(2):165–82.

Jung MW, Wiener SI, Mcnaughton BL. Comparison of spatial firing characteristics of units in dorsal and ventral hippocampus of the rat. J Neurosci. 1994;14(12):7347–56.

Karlsson MP, Frank LM. Awake replay of remote experiences in the hippocampus. Nat Neurosci. 2009;12:913–8.

Leutgeb JK, Leutgeb S, Moser MB, Moser EI. Pattern separation in the dentate gyrus and CA3 of the hippocampus. Science. 2007;315(5814):961–6.

Leutgeb S, Leutgeb JK, Barnes CA, Moser EI, McNaughton BL, Moser MB. Independent codes for spatial and episodic memory in hippocampal neuronal ensembles. Science. 2005; 309(5734):619–23.

Lisman JE. Relating hippocampal circuitry to function: recall of memory sequences by reciprocal dentate-CA3 interactions. Neuron. 1999;22(2):233–42.

MacDonald CJ, Lepage KQ, Eden UT, Eichenbaum H. Hippocampal "time cells" bridge the gap in memory for discontiguous events. Neuron. 2011;71(4):737–49.

Maguire EA, Gadian DG, Johnsrude IS, Good CD, Ashburner J, Frackowiak RS, et al. Navigation-related structural change in the hippocampi of taxi drivers. Proc Natl Acad Sci U S A. 2000;97(8):4398–403.

Marr D. A theory of cerebellar cortex. J Physiol. 1969;202(2):437–70.

McNaughton BL, Barnes CA, O'keefe J. The contributions of position, direction, and velocity to single unit-activity in the hippocampus of freely-moving rats. Exp Brain Res. 1983;52(1):41–9.

McNaughton BL, Battaglia FP, Jensen O, Moser EI, Moser M-B. Path integration and the neural basis of the "cognitive map". Nat Rev Neurosci. 2009;7(8):663–78.

Morris RG, Garrud P, Rawlins JN, O'Keefe J. Place navigation impaired in rats with hippocampal lesions. Nature. 1982;297(5868):681–3.

Muller RU, Kubie JL. The effects of changes in the environment on the spatial firing of hippocampal complex-spike cells. J Neurosci. 1987;7(7):1951–68.

O'Keefe J, Dostrovsky J. The hippocampus as a spatial map. Preliminary evidence from unit activity in the freely-moving rat. Brain Res. 1971;34(1):171–5.

O'Keefe J, Recce ML. Phase relationship between hippocampal place units and the EEG theta rhythm. Hippocampus. 1993;3(3):317–30.

O'keefe L, Nadel L. The hippocampus as a cognitive map. Oxford: Clarendon; 1978.

O'Neill J, Senior T, Csicsvari J. Place-selective firing of CA1 pyramidal cells during sharp wave/ripple network patterns in exploratory behavior. Neuron. 2006;49(1):143–55.

Otto T, Eichenbaum H. Neuronal activity in the hippocampus during delayed non-match to sample performance in rats: evidence for hippocampal processing in recognition memory. Hippocampus. 1992;2(3):323–34.

Pastalkova E, Itskov V, Amarasingham A, Buzsaki G. Internally generated cell assembly sequences in the rat hippocampus. Science. 2008;321(5894):1322–7.

Pfeiffer BE, Foster DJ. Hippocampal place-cell sequences depict future paths to remembered goals. Nature. 2013;497:74–9.

Redish AD, Touretzky DS. Cognitive maps beyond the hippocampus. Hippocampus. 1997; 7(1):15–35.

Sakurai Y. Hippocampal cells have behavioral correlates during the performance of an auditory working memory task in the rat. Behav Neurosci. 1990;104(2):253–63.

Sakurai Y. Involvement of auditory cortical and hippocampal-neurons in auditory working-memory and reference memory in the rat. J Neurosci. 1994;14(5):2606–23.

Scoville WB, Milner B. Loss of recent memory after bilateral hippocampal lesions. J Neurol Neurosurg Psychiatry. 1957;20(1):11–21.

Smith DM, Mizumori SJ. Learning-related development of context-specific neuronal responses to places and events: the hippocampal role in context processing. J Neurosci. 2006a;26(12):3154–63.

Smith DM, Mizumori SJY. Hippocampal place cells, context, and episodic memory. Hippocampus. 2006b;16(9):716–29.

Takahashi S. Hierarchical organization of context in the hippocampal episodic code. elife. 2013;2:e00321.

Takahashi S. Episodic-like memory trace in awake replay of hippocampal place cell activity sequences. elife. 2015;4:e08105.

Takahashi S, Anzai Y, Sakurai Y. A new approach to spike sorting for multi-neuronal activities recorded with a tetrode—how ICA can be practical. Neurosci Res. 2003b;46(3):265–72.

Takahashi S, Anzai Y, Sakurai Y. Automatic sorting for multi-neuronal activity recorded with tetrodes in the presence of overlapping spikes. J Neurophysiol. 2003a;89(4):2245–58.

Takahashi S, Sakurai Y. Real-time and automatic sorting of multi-neuronal activity for sub-millisecond interactions in vivo. Neuroscience. 2005;134(1):301–15.

Takahashi S, Sakurai Y. Coding of spatial information by soma and dendrite of pyramidal cells in the hippocampal CA1 of behaving rats. Eur J Neurosci. 2007;26(7):2033–45.

Takahashi S, Sakurai Y. Sub-millisecond firing synchrony of closely neighboring pyramidal neurons in hippocampal CA1 of rats during delayed non-matching to sample task. Front Neural Circuits. 2009a;3:9.

Takahashi S, Sakurai Y. Information in small neuronal ensemble activity in the hippocampal CA1 during delayed non-matching to sample performance in rats. BMC Neurosci. 2009b;10:115.

Taube JS, Muller RU, Ranck JB. Head-direction cells recorded from the postsubiculum in freely moving rats. I. Description and quantitative analysis. J Neurosci. 1990;10:420–35.

Tolman EC. Cognitive maps in rats and men. Psychol Rev. 1948;55(4):189–208.

Tsodyks MV, Markram H. The neural code between neocortical pyramidal neurons depends on neurotransmitter release probability. Proc Natl Acad Sci U S A. 1997;94(2):719–23.

Vinogradova OS, Dudaeva KI. Comparator function of the hippocampus. Dokl Akad Nauk SSSR. 1972;202(2):486–9.

Wilson MA, McNaughton BL. Dynamics of the hippocampal ensemble code for space. Science. 1993;261(5124):1055–8.

Wilson MA, McNaughton BL. Reactivation of hippocampal ensemble memories during sleep. Science. 1994;265(5172):676–9.

Wood ER, Dudchenko PA, Eichenbaum H. The global record of memory in hippocampal neuronal activity. Nature. 1999;397(6720):613–6.

Wood ER, Dudchenko PA, Robitsek RJ, Eichenbaum H. Hippocampal neurons encode information about different types of memory episodes occurring in the same location. Neuron. 2000;27(3):623–33.

Chapter 4
Context-Dependent Adjustments in Executive Control of Goal-Directed Behaviour: Contribution of Frontal Brain Areas to Conflict-Induced Behavioural Adjustments in Primates

Farshad A. Mansouri and Mark J. Buckley

4.1 Introduction

Humans and animals frequently face the dilemma of selecting one out of several potential options to achieve their behavioural goal. Sometimes the choice is straightforward because all the available information about the benefit and cost of each option clearly indicate the suitability and priority of one of them for attaining the goal. However, in many occasions, the most appropriate option might not be immediately clear and therefore making a choice would require resolving the competition between the potential options. The concept of 'conflict' emerges when a decision should be made between such competing options. The conflict might emerge at sensory level between two or more sources of information or between competing actions (responses) or even between two or more behavioural strategies. In a changing/volatile environment the relative value (in terms of outcomes) of behavioural rules that guide actions in achieving goals might change and therefore selection of the most appropriate rule/behaviour would depend on updated estimates of the integrated cost and benefit of each option. In a changing environment, assigning value to each option would require consideration of alterations in factors such as the internal state of the subject (e.g. hunger, taste and the urge to achieve the goal), contextual information such as the associated cost and benefit of actions associated with each option and the recent outcome histories of decisions for the various available options. Imagine a person used to driving a car on the left side of streets in Japan

F. A. Mansouri (✉)
Cognitive Neuroscience Laboratory, Department of Physiology, Monash Biomedicine Discovery Institute, Monash University, Clayton, VIC, Australia
e-mail: farshad.mansouri@monash.edu

M. J. Buckley
Department of Experimental Psychology, Oxford University, Oxford, UK
e-mail: buckley@psy.ox.ac.uk

A. Cheung-Hoi Yu, L. Li (eds.), *Systems Neuroscience*, Advances in Neurobiology 21, https://doi.org/10.1007/978-3-319-94593-4_4

moves to Korea and now faces the challenge of driving on the right side. In Japan, driving on the left side was the most appropriate option in fulfilling the contextual requirements and achieving the goal of safe driving. Such routinely performed and beneficial behaviour would become more habitual and act as a potentially valuable option in guiding behaviour. However, in a new environment the previous rule would no longer be beneficial and should be abandoned in favour of a new rule (right-side driving). Therefore competition/conflict would arise between the previous rule and the currently appropriate rule. Psychophysical studies have shown that such conflict between behavioural options adversely affects humans' behaviour in terms of accuracy and response time and appears as 'conflict cost' in various cognitive tasks. The behavioural effects of conflict are not limited to the current trial wherein the subjects experience the conflict between the behavioural options, but also extend to the following trials. Indeed, it is robust observation that after experiencing conflict, accuracy and response time are enhanced in the following trial when the subjects face the conflict again. Such an extended effect of conflict has been referred to as 'conflict adaptation' and is seen in many cognitive tasks. Theoretical models have emerged to explain the behavioural effects of conflict and the possible impact of conflict-induced behavioural modulation on adaptability of human behaviour in a changing condition. Influential models (Botvinick et al. 2004; Kerns et al. 2004; Carter and van Veen 2007) suggest that conflict in information processing is detected by brain areas such anterior cingulate cortex (ACC) and then conveyed to areas such as dorsolateral prefrontal cortex (DLPFC) to adjust the allocation of executive control to enhance resolving the conflict in the upcoming occasions. The conflict monitoring hypothesis explains the findings in various imaging studies (Kerns et al. 2004; Carter and van Veen 2007) that activation in ACC correlates with the magnitude of conflict experienced in the current trial and with the magnitude of behavioural adjustments and activation in DLPFC in the following trial; in addition, the activation in DLPFC in the following trials correlated with the magnitude of behavioural adaptation. Another alternative hypothesis proposed that ACC itself regulates the allocation of executive control based on the level of experienced conflict (Paus et al. 1998; Paus 2001; Posner and Rothbart 1998). An important implication of the conflict monitoring hypothesis is that it could effectively explain when and how allocation of cognitive resources was adjusted to enhance resolving the conflict between behavioural options and consequently support adaptive behaviour. The model was also expanded to explain error related changes in behaviour and the event-related potentials during conflict tasks. Imaging studies such as functional magnetic resonance imaging (fMRI) and positron emission tomography (PET) are correlational and do not necessarily show whether a brain area is indispensable for a particular cognitive process or ability. However, additional studies also supported the conflict monitoring hypothesis by showing that activity of single cells in ACC, in patients undergoing surgical treatment, represent conflict in information processing (Davis et al. 2005) and that some patients with lesions involving ACC showed impaired conflict-induced behavioural adjustments (di Pellegrino et al. 2007).

However, findings from other studies in humans have not supported the proposed neural substrate of conflict detection and resolution processes. In separate studies, neuropsychological examination of patients with ACC lesions showed that the conflict-induced behavioural modulations were within the normal range (Vendrell et al. 1995; Stuss et al. 2001; Fellows and Farah 2005). These studies questioned the crucial role of ACC in mediating the conflict-induced behavioural adjustment. In addition, the results of some imaging studies indicated that while the participants performed more trials of conflict tasks, the behavioural effect of conflict was sustained while ACC activation gradually disappeared suggesting that ACC activation was not necessarily associated with the behavioural effects of conflict (Milham et al. 2003).

Studies in animal models provide the opportunity to implement various detailed neurobiological techniques such as single cell recording to examine the neuronal correlate of behaviour. In addition, lesion-behavioural studies in suitable animal models can reveal whether a particular brain area have an essential role in supporting particular cognitive ability. Whilst the conflict monitoring hypothesis gained support from numerous imaging studies in humans, some single cell recording studies in macaque monkeys could not find neuronal correlate of conflict in ACC (Ito et al. 2003; Nakamura et al. 2005). This discrepancy between findings in humans and monkeys led to intensive debate about the role of ACC in conflict monitoring. Currently, the neural substrate and the underlying mechanisms of conflict-induced behavioural and executive control adjustments still remain unclear. However, recent studies in humans and monkeys are starting to shed more light on the involvement of various brain areas and the underlying neural mechanisms in the conflict processing.

4.2 Conflict-Induced Behavioural Adjustment

4.2.1 Conflict Tasks Used in Psychophysical Studies in Humans

A well-studied paradigm used in humans to examine the behavioural effects of conflict is the Stroop test (MacLeod 1991; Botvinick et al. 2004; Kerns et al. 2004; Carter and van Veen 2007), in which participants are presented with the name of a colour printed in coloured ink and they must identify the colour of the ink as fast and as accurately as possible. In incongruent (high conflict) conditions, the colour's name differs from the ink colour, however in congruent (low-conflict) conditions the colour name matches the ink colour and in neutral condition, the word is not colour-related. A consistent observation is that the subjects are less accurate and slower in incongruent conditions (conflict cost). It is assumed that information regarding the ink colour and information regarding the word are processed separately, leading to distinct competing motor responses. The conflict cost has been reported in other tasks such as flanker test, Simon test and Go-No-go tasks (Carter and van Veen 2007; Mansouri et al. 2009). Conflict can be also evoked between the emotional content and other attributes of stimuli and influence the behaviour (Braem et al. 2013; Etkin et al. 2011). The effects of conflict are not limited to the current trial and can also influence performance in the upcoming trial when the participants are required to resolve the conflict between competing choices again. To estimate the conflict-adaptation effect the difference in mean accuracy or response time is compared between high-conflict trials that are preceded by low-conflict trials (LH condition) and high-conflict trials that are preceded by high-conflict trials (HH condition). Conflict-adaptation effects appear as improved performance in resolving the conflict in HH trials (Kerns et al. 2004; Carter and van Veen 2007; Egner 2007; Mansouri et al. 2009). The conflict-adaptation effect has been observed in various conflict tasks such Stroop test. Flanker and Simon tests (Erickson et al. 2004; Mansouri et al. 2009).

4.2.2 Neural Substrate and Underlying Mechanisms of Conflict-Induced Behavioural Modulations

4.2.2.1 Imaging Studies in Humans

The conflict monitoring hypothesis (Botvinick et al. 2004; Carter and van Veen 2007) emerged from a series of imaging and event-related potential studies showing that in Stroop test and other conflict tasks, ACC was more active during high conflict trials; this led to the conclusion that the ACC was involved in the conflict-detection process (Botvinick et al. 2004; Carter and van Veen 2007). The hypothesis also proposed that other regions such as DLPFC which were more active in HH than in LH conditions were involved in mediating the executive-control adjustment required to deal effectively with sustained conflict (Botvinick et al. 2004; Kerns et al. 2004; Carter and van Veen 2007). Further studies (Fan et al. 2003; Kerns et al. 2004; Egner and Hirsch 2005a, b; Liston et al. 2006) provided support for this hypothesis by showing that in high-conflict trials the magnitude of ACC activity predicted the degree of behavioural adjustment and the activation level in the DLPFC on the subsequent trial. In addition, they showed that ACC activity in the second trial of the HH conditions was lower than that in the LH conditions and that the increase in DLPFC activity observed in HH trials tended to correlate with greater degrees of behavioural adjustment. These findings fit with the idea that conflict is detected by the ACC and signals adjustments in control, mediated by DLPF, that serve to effectively decrease the conflict in the 2nd trial of HH conditions. Although the conflict monitoring hypothesis has been focused mainly on the role of ACC and DLPFC in conflict detection and resolution, activation of other brain areas have also been shown in the conflict tasks.

Parietal Cortex Several imaging studies have shown activation changes in parietal cortex in conflict tasks (Casey et al. 2000; Barch et al. 2001; Adleman et al. 2002; Milham et al. 2003; Durston et al. 2003; Fan et al. 2003; Egner and Hirsch 2005b; Liston et al. 2006; Roelofs et al. 2006; Krebs et al. 2015). It has been suggested that parietal cortex might be involved in detection of conflict at the level of sensory processing (Liston et al. 2006). A recent imaging study (Krebs et al. 2015) showed that irrelevant incongruent information (words), which elicited conflict cost, led to improved subsequent memory for the relevant target stimuli (faces) and that the conflict induced memory benefit was selectively associated with activity modulations in the DLPFC and the parietal cortex suggesting that DLPFC and parietal cortex were involved in conflict-induced behavioural enhancement.

Insula Neuroanatomical and functional imaging studies suggest that the ACC and the insular cortex comprise a closely related functional network both during active task performance and at rest (Augustine 1996; Dosenbach et al. 2007; Nelson et al. 2010). Egner and Hirsch (2005a) reported that in the context of a face-word Stroop like test the behavioural performance was enhanced in HH trials and that fMRI signal in the right DLPFC and the left anterior insula were stronger in the HH condition than in the LH condition, suggesting that in addition to the DLPFC, insula may also be involved in conflict adaptation. Neuroimaging activation changes in insular cortex have also been reported in other versions of the Stroop test (Banich et al. 2001).

Cortical Areas around Inferior Frontal Sulcus Studies also suggest that areas around the junction of the inferior frontal sulcus and the inferior precentral sulcus also change activation in conflict tasks suggesting their involvement in cognitive control processes (Derrfuss et al. 2005; Sundermann and Pfleiderer 2012).

Cerebellum Alteration in cerebellar activation has also been observed in conflict tasks (Casey et al. 2000; Egner and Hirsch 2005b) and patients with cerebellar lesions exhibit higher conflict costs in the absence of task-switching costs (Schweizer et al. 2007). These findings suggest that the cerebellum may also play a crucial role in conflict processing, however the exact functional role of cerebellum remains unclear.

Orbitofrontal Cortex (OFC) The activation of orbitofrontal cortex during performance of the Stroop test has been seen in human imaging studies. Bench et al. (1993) conducted a PET study in humans performing Stroop test and observed activation in right OFC and cingulate cortex. Mitchell (2005) also reported activation of OFC during performance of Stroop test in a fMRI study. In addition, Goldstein et al. (2011) conducted a PET study in control and drug-addicted human subjects and showed that a higher activation in orbitofrontal gyrus was associated with higher conflict level. They suggested that OFC might be involved in the evaluation of conflict level to increase the mental efforts to improve the behaviour.

These studies in humans suggest that a distributed network of brain regions is involved in processing the conflict information and possibly mediating the behavioural effects of conflict. These imaging studies have greatly contributed to our understanding of the activation patterns in different brain regions but the findings are correlational and do not necessarily indicate whether these brain regions have any indispensable role in conflict detection or resolution. Further detailed neurobiological assessment are necessary to examine the essential function of these areas in conflict processing.

4.2.2.2 Studies in Non-Human Primates

Animal models provide the opportunity to conduct various detailed neurobiological investigations, however it was first crucial to show that conflict exerts similar behavioural modulations in such animal models. Conflict-related behavioural modulations have been reported in monkeys in the context of various tasks (Stoet and Snyder 2009). Lauwereyns et al. (2000) examined macaque monkeys' behaviour in an analog of Stroop test and showed conflict-related behavioural modulations. Ito et al. (2003) trained macaque monkeys to perform a saccade countermanding task in which the monkeys initiated a saccadic eye movement upon receiving a go signal. However, in a smaller proportion of trials a stop signal was presented and the monkeys had to stop their initiated or planned saccade. By varying the time between the go and stop signals the difficulty of saccade inhibition could be controlled. The animals' behaviour indicated a conflict cost that was presumably associated with competition between gaze-shifting and gaze-holding processes. The neuronal activity was recorded in ACC, however no modulation directly linked to the conflict level was found in the ACC cell activities. Nakamura et al. (2005) recorded ACC activity in the context of a conflict task requiring saccadic responses but did not find encoding of conflict in ACC cell activities. Mansouri et al. trained monkeys to perform a conflict task in which the conflict emerged between two behavioural rules (Fig. 4.1). Monkeys' behaviour showed a significant conflict cost as well as a robust conflict adaptation effect, however bilateral lesions within ACC neither affected conflict cost nor conflict adaptation. In contrast, conflict adaptation was significantly attenuated in DLPFC- and OFC-lesioned monkeys (Mansouri et al. 2007, 2014).

The absence of conflict encoding in ACC cell activity and intact conflict-induced behavioural modulations in ACC-lesioned monkeys were inconsistent with the predictions of conflict monitoring hypothesis. These apparently contradictory findings between studies in humans and monkeys led some investigators to conclude that conflict monitoring by ACC is a unique property of human brain function and non-human primates basically do not perceive and process conflict as humans do (Cole et al. 2009, 2010). They proposed that the reported conflict-related behavioural modulations in monkeys and other animals performing conflict tasks might be related to the other aspects of the task (Cole et al. 2009, 2010). A crucial question also emerged as to whether conflict is a separate entity that could be encoded in neuronal activity independent from other aspects of the task. Mansouri et al. (2007,

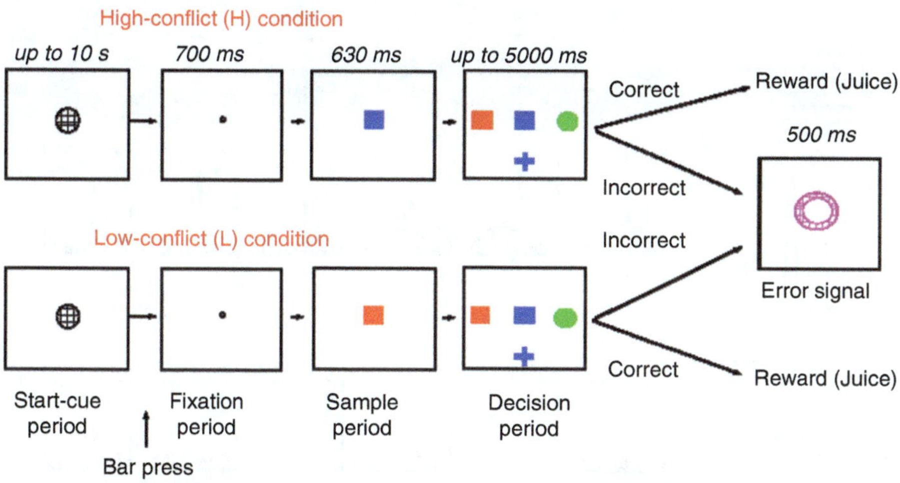

Fig. 4.1 In each trial, a start cue appeared when an inter-trial interval was over. The bar pressing changed the start cue to a fixation point. If the monkey kept pushing the bar and maintained its gaze on the fixation point for 700 ms, a sample stimulus replaced the fixation point. If the monkey maintained eye fixation and bar press for another 630 ms, three test items appeared (to the left, right and below the sample). The relevant rule for matching (matching by shape or matching by color) was consistent within a block of trials, and it changed without any notice to the monkey when a criterion of 85% correct performance was achieved. The relevant rule was not cued and the monkeys were only able to identify it by applying a rule and then interpreting the reward or error feedback in the context of the applied rule. Twelve and twenty four samples were shown in the low-conflict and high-conflict conditions, respectively

2009, 2014) recorded neuronal activity in DLPFC and OFC of monkeys performing a variant of Wisconsin Card Sorting Test (WCST) in which the competition between rules led to conflict cost and also conflict adaptation in monkeys' behaviour. These studies showed that activity of single neurons in DLPFC and also in OFC were significantly different between low-conflict and high-conflict conditions and that the conflict was encoded independently of the other aspects of the task such as the features of the visual stimuli or the behavioural rule or the upcoming actions. These findings suggested that both DLPFC and OFC cell activity encoded conflict as a separate variable (Fig. 4.2). The encoded conflict information was also maintained in the neuronal activity across the trials in DLPFC, but not OFC, cell activities (Fig. 4.3). This suggested that information of experienced conflict was retained by mnemonic processes in the neurocircuitry of DLPFC even after the conflicting situation was already over; hence such maintained information could be potentially used to evoke behavioural adjustment in the upcoming trials (Mansouri et al. 2007, 2009, 2014, 2015).

A few studies in humans have been able to record single ACC cell activity in patients undergoing surgeries. Davis et al. (2005) recorded ACC cell activity while the subjects performed versions of Stroop test (Counting and emotional interference) and found that neuronal activity was significantly different between low- and

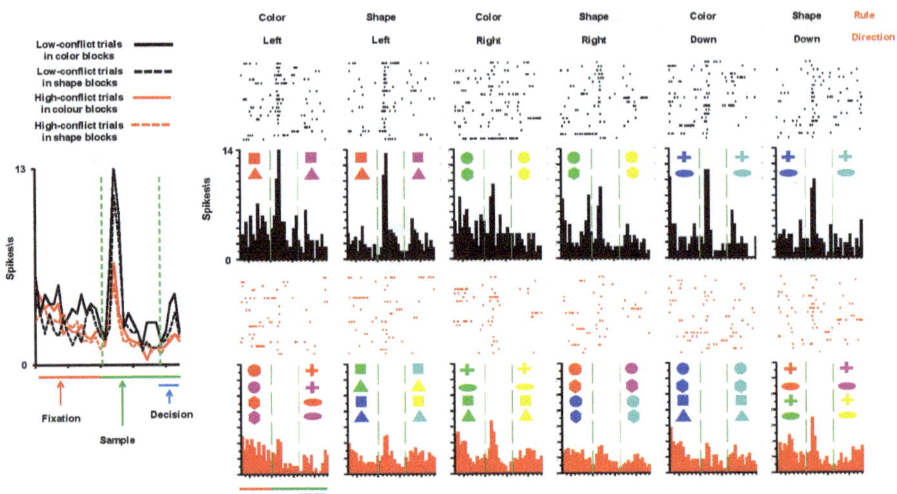

Fig. 4.2 Representation of conflict level in orbitofrontal cortex cell activity. The leftmost peri-stimulus time histograms (PSTH) show activities in low-conflict and high-conflict trials when colour- or shape-matching rules were relevant. Each column of the right histograms shows activities in low-conflict (upper) and high-conflict (lower) trials that required the application of the same rule and responses in the same direction. The raster-grams show spikes in individual trials. Samples presented in each condition are shown above individual histograms. Bin width is 50 ms for the leftmost PSTH, and 20 ms for other PSTH. Left and right vertical broken lines indicate the sample and test items onset, respectively. Only correct trials were included. The difference in activity between the low- and high-conflict conditions was seen independent of the rule, stimulus identity or the upcoming response direction

high-conflict conditions suggesting that conflict was encoded in ACC cell activities. In a recent comprehensive study, Sheth et al. (2012) conducted imaging, single-cell recording and lesion-behavioural studies in humans performing a conflict task. All the patients expressed behavioural effects of conflict that appeared as conflict cost and conflict adaptation and the fMRI showed a higher activation in ACC and in DLPFC in the high-conflict condition. The activated foci in ACC were then targeted for single-cell recording and subsequently for lesion study. Importantly, after selective lesions were made in the same regions within ACC, the behavioural effect of conflict in the current trial (conflict cost) remained intact but, the conflict adaptation effect was significantly impaired. This study provided solid evidence for encoding of conflict information in human ACC neurocircuitry. These finding appeared in contradiction to the findings in monkeys, however two recent studies have shown that in monkeys, ACC cells encode conflict independent of the other aspects of the cognitive task.

Ebitz and Platt (2015) trained monkeys to perform a conflict task in which conflict emerged between two different oculomotor responses or between task relevant and task irrelevant information. They found that both types of conflict influenced the monkeys' behaviour and the activity of ACC cells encoded the conflict level and errors. In addition, the ACC cell activity conveyed information about the current

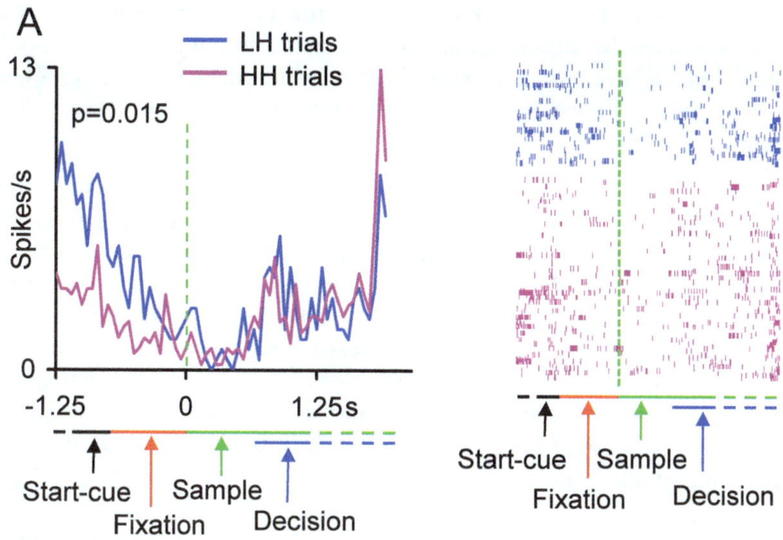

Fig. 4.3 Representation of history of conflict level in dorsolateral prefrontal cortex cell activity. Activities in high-conflict trials after low-conflict trials (LH, blue) and those in high-conflict trials after high-conflict trials (HH, pink) are shown for a single cell. The mean activities are aligned at sample onset. Only activities in correct trials that were preceded by correct trials were included. The p values show the significance level of activity difference in the fixation period between HH and LH trials. Bin size is 55 ms

pupil size or upcoming adjustments in the pupil size. This suggested that the ACC cell activity conveyed information about conflict-related adaptive changes in pupil size and presumably the arousal level. In another study, Michelet et al. (2016) trained monkeys to perform a conflict task in which the conflict arose between associated colour of a particular object with its presented colour. Behavioural effects of conflict was detected in the current (conflict cost) and in the following trial (conflict adaptation). Activities of a small but significant proportion of ACC cells conveyed information about the conflict level, however these conflict-related activity modulations were seen only in correct trials. Although conflict adaptation was seen in the monkeys' behaviour, information regarding conflict level in the previous trial was not represented in the ACC cell activity. These two studies, in the context of different conflict tasks, have clearly shown that ACC cells in monkeys convey information about the conflict independent of the other task-relevant events.

Conclusions made through neuropsychological examination of patients with ACC damage have necessarily been limited by the heterogeneity and inconsistency of lesions across patients. Furthermore, patients receive recordings/stimulation/lesions to ACC for clinical reasons indicative of significantly disturbed (from normal) brain function so inferring strong conclusions from such patients about normal brain function needs to be done with great caution. In this respect it is crucial and highly informative that we now have a few lesion-behavioural studies in animal models wherein lesions that are more circumscribed and reproducible across animals, and which can be introduced into animals with normal pre-lesion brain function; these studies have to-date examined the role of a number of different cortical regions in conflict-cost and conflict-induced behavioural adaptation and their findings are.

Conflict Cost Mansouri et al. (2007, 2009, 2014, 2015) reported that bilateral lesions in DLPFC or ACC or OFC or superior part of dorsal lateral prefrontal cortex (sdlPFC) or posterior cingulate cortex (PCC) or frontal pole cortex do not impair the behavioural effects of conflict in the current trial (conflict cost). This indicates that other brain areas might mediate the conflict cost or it might result from mutual inhibitory effects between the neural processing related to competing options (responses). Different sensory-perceptual processes or the related actions might compete for controlling the behaviour and mutual inhibition between such parallel processing pathways might lead to slowing in reaching a final decision about one of the competing options.

Conflict Adaptation Mansouri et al. (2007, 2009, 2014, 2015) showed that bilateral lesions in ACC or PCC or sdlPFC or frontal pole cortex did not impair behavioural effects of conflict in the upcoming trials (conflict adaptation). However, bilateral lesions in DLPFC or OFC significantly attenuated the conflict adaptation indicating that DLPFC and OFC play an indispensable role in mediating the conflict-induced behavioural modulation and presumably in conflict-induced executive control adjustments.

4.3 Conclusion

Limitations in cognitive resources necessitate adaptive adjustment in allocation of these resources to optimize behaviour for achieving goals in changing environments. Conflict-induced behavioural adjustment has been extensively studied in the last two decades leading to influential hypothesis regarding context-dependent executive control adjustment. Studies in humans and non-human primates indicate great similarities in conflict-induced behavioural modulations. Recent studies in humans and monkeys have further advanced our knowledge regarding the neural substrate and underlying mechanisms of conflict processing and related behavioural alterations. Taken together these studies now indicate that a wider than previously appreciated distributed neural network might be involved in representation of conflict and mediating its effects, however within this network both DLPFC and OFC play indispensable roles. While ACC has long been considered a key component, the activations of ACC in conflict tasks might not be necessarily related to its role in conflict monitoring but may instead reflect ACC function in other cognitive domains such as adjusting the concomitant autonomic and affective aspects of the task and/or a role in action valuation and selection processes (Rushworth et al. 2004;

Kennerley et al. 2006, 2011; Euston et al. 2012; Shenhav et al. 2013; Heilbronner and Hayden 2016; Kolling et al. 2016).

Conflict is an abstract entity that might emerge in different contexts and between different elements of cognitive processes. DLPFC and OFC might be crucially involved in extracting and encoding conflict information in different contexts. In addition, mnemonic processes, mainly mediated through DLPFC, might support conflict-induced behavioural modulations by maintaining conflict information within and across trials. Studying conflict-induced behavioural modulations has opened a window to better understand the executive functions in primate brain and to gain insight to deficits in executive functions that is a hallmark of major neuropsychological disorders.

References

Augustine JR. Circuitry and functional aspects of the insular lobe in primates including humans. Brain Res Brain Res Rev. 1996;22(3):229–44.

Adleman NE, Menon V, Blasey CM, White CD, Warsofsky IS, Glover GH, Reiss AL. A developmental fMRI study of the stroop color-word task. NeuroImage 2001;16(1):61–75.

Banich MT, Milham MP, Jacobson BL, Webb A, Wszalek T, Cohen NJ, et al. Attentional selection and the processing of task-irrelevant information: insights from fMRI examinations of the Stroop task. Prog Brain Res. 2001;134:459–70.

Barch DM, Braver TS, Akbudak E, Conturo T, Ollinger J, Snyder A. Anterior cingulate cortex and response conflict: effects of response modality and processing domain. Cereb Cortex. 2001;11(9):837–48.

Bench CJ, Frith CD, Grasby PM, Friston KJ, Paulesu E, Frackowiak RSJ, Dolan RJ. Investigations of the functional anatomy of attention using the stroop test. Neuropsychologia 1993;31(9):907–922.

Botvinick MM, Cohen JD, Carter CS. Conflict monitoring and anterior cingulate cortex: an update. Trends Cogn Sci. 2004;8(12):539–46.

Braem S, King JA, Korb FM, Krebs RM, Notebaert W, Egner T. Affective modulation of cognitive control is determined by performance-contingency and mediated by ventromedial prefrontal and cingulate cortex. J Neurosci. 2013;33(43):16961–70.

Carter CS, van Veen V. Anterior cingulate cortex and conflict detection: an update of theory and data. Cogn Affect Behav Neurosci. 2007;7(4):367–79.

Casey BJ, Thomas KM, Welsh TF, Badgaiyan RD, Eccard CH, Jennings JR, et al. Dissociation of response conflict, attentional selection, and expectancy with functional magnetic resonance imaging. Proc Natl Acad Sci U S A. 2000;97(15):8728–33.

Cole MW, Yeung N, Freiwald WA, Botvinick M. Cingulate cortex: Diverging data from humans and monkeys. Trends Neurosci. 2009;32(11):566–574.

Cole MW, Yeung N, Freiwald WA, Botvinick M. Conflict over Cingulate cortex: Between-Species differences in cingulate may support enhanced cognitive flexibility in humans. Brain Behav Evol. 2010;75(4):239–240.

di Pellegrino G, Ciaramelli E, Ladavas E. The regulation of cognitive control following rostral anterior cingulate cortex lesion in humans. J Cogn Neurosci. 2007;19(2):275–86.

Davis KD, Taylor KS, Hutchison WD, Dostrovsky JO, Mcandrews MP, Richter EO, et al. Human anterior cingulate cortex neurons encode cognitive and emotional demands. J Neurosci. 2005;25(37):8402–6.

Derrfuss J, Brass M, Neumann J, von Cramon DY. Involvement of the inferior frontal junction in cognitive control: meta-analyses of switching and Stroop studies. Hum Brain Mapp. 2005;25(1):22–34.

Dosenbach NU, Fair DA, Miezin FM, Cohen AL, Wenger KK, Dosenbach RA, et al. Distinct brain networks for adaptive and stable task control in humans. Proc Natl Acad Sci U S A. 2007;104(26):11073–8.

Durston S, Davidson MC, Thomas KM, Worden MS, Tottenham N, Martinez A, et al. Parametric manipulation of conflict and response competition using rapid mixed-trial event-related fMRI. NeuroImage. 2003;20(4):2135–41.

Ebitz RB, Platt ML. Neuronal activity in primate dorsal anterior cingulate cortex signals task conflict and predicts adjustments in pupil-linked arousal. Neuron. 2015;85(3):628–40.

Egner T. Congruency sequence effects and cognitive control. Cogn Affect Behav Neurosci. 2007;7(4):380–90.

Egner T, Hirsch J. Cognitive control mechanisms resolve conflict through cortical amplification of task-relevant information. Nat Neurosci. 2005a;8(12):1784–90.

Egner T, Hirsch J. The neural correlates and functional integration of cognitive control in a Stroop task. NeuroImage. 2005b;24(2):539–47.

Erickson KI, Milham MP, Colcombe SJ, Kramer AF, Banich MT, Webb A, et al. Behavioral conflict, anterior cingulate cortex, and experiment duration: implications of diverging data. Hum Brain Mapp. 2004;21(2):98–107.

Etkin A, Egner T, Kalisch R. Emotional processing in anterior cingulate and medial prefrontal cortex. Trends Cogn Sci. 2011;15(2):85–93.

Euston DR, Gruber AJ, Mcnaughton BL. The role of medial prefrontal cortex in memory and decision making. Neuron. 2012;76(6):1057–70.

Fan J, Flombaum JI, Mccandliss BD, Thomas KM, Posner MI. Cognitive and brain consequences of conflict. NeuroImage. 2003;18(1):42–57.

Fellows LK, Farah MJ. Is anterior cingulate cortex necessary for cognitive control? Brain. 2005;128:788–96.

Goldstein RZ, Volkow ND. Dysfunction of the prefrontal cortex in addiction: Neuroimaging findings and clinical implications. Nat Rev Neurosci. 2011;12(11):652–669.

Heilbronner SR, Hayden BY. Dorsal anterior cingulate cortex: a bottom-up view. Annu Rev Neurosci. 2016;39:149–70.

Ito S, Stuphorn V, Brown JW, Schall JD. Performance monitoring by the anterior cingulate cortex during saccade countermanding. Science. 2003;302(5642):120–2.

Kennerley SW, Walton ME, Behrens TE, Buckley MJ, Rushworth MF. Optimal decision making and the anterior cingulate cortex. Nat Neurosci. 2006;9(7):940–7.

Kennerley SW, Behrens TEJ, Wallis JD. Double dissociation of value computations in orbitofrontal and anterior cingulate neurons. Nat Neurosci. 2011;14(12):1581–U119.

Kerns JG, Cohen JD, Macdonald AW 3rd, Cho RY, Stenger VA, Carter CS. Anterior cingulate conflict monitoring and adjustments in control. Science. 2004;303(5660):1023–6.

Kolling N, Behrens T, Wittmann MK, Rushworth M. Multiple signals in anterior cingulate cortex. Curr Opin Neurobiol. 2016;37:36–43.

Krebs RM, Boehler CN, De Belder M, Egner T. Neural conflict-control mechanisms improve memory for target stimuli. Cereb Cortex. 2015;25(3):833–43.

Lauwereyns J, Koizumi M, Sakagami M, Hikosaka O, Kobayashi S, Tsutsui K. Interference from irrelevant features on visual discrimination by macaques (Macaca fuscata): a behavioral analogue of the human Stroop effect. J Exp Psychol Anim Behav Process. 2000;26(3):352–7.

Liston C, Matalon S, Hare TA, Davidson MC, Casey BJ. Anterior cingulate and posterior parietal cortices are sensitive to dissociable forms of conflict in a task-switching paradigm. Neuron. 2006;50(4):643–53.

Macleod CM. Half a century of research on the Stroop effect: an integrative review. Psychol Bull. 1991;109(2):163–203.

Mansouri FA, Buckley MJ, Tanaka K. Mnemonic function of the dorsolateral prefrontal cortex in conflict-induced behavioral adjustment. Science. 2007;318(5852):987–90.
Mansouri FA, Tanaka K, Buckley MJ. Conflict-induced behavioural adjustment: a clue to the executive functions of the prefrontal cortex. Nat Rev Neurosci. 2009;10(2):141–52.
Mansouri FA, Buckley MJ, Tanaka K. The essential role of primate orbitofrontal cortex in conflict-induced executive control adjustment. J Neurosci. 2014;34(33):11016–31.
Mansouri FA, Rosa MG, Atapour N. Working memory in the service of executive control functions. Front Syst Neurosci. 2015;9:166.
Michelet T, Bioulac B, Langbour N, Goillandeau M, Guehl D, Burbaud P. Electrophysiological correlates of a versatile executive control system in the monkey anterior cingulate cortex. Cereb Cortex. 2016;26(4):1684–97.
Milham MP, Banich MT, Claus ED, Cohen NJ. Practice-related effects demonstrate complementary roles of anterior cingulate and prefrontal cortices in attentional control. NeuroImage. 2003;18(2):483–93.
Mitchell RLC. The BOLD response during Stroop task-like inhibition paradigms: Effects of task difficulty and task-relevant modality. Brain Cogn. 2005;59(1):23–37.
Nakamura K, Roesch MR, Olson CR. Neuronal activity in macaque SEF and ACC during performance of tasks involving conflict. J Neurophysiol. 2005;93(2):884–908.
Nelson SM, Dosenbach NU, Cohen AL, Wheeler ME, Schlaggar BL, Petersen SE. Role of the anterior insula in task-level control and focal attention. Brain Struct Funct. 2010;214(5–6):669–80.
Paus T. Primate anterior cingulate cortex: where motor control, drive and cognition interface. Nat Rev Neurosci. 2001;2(6):417–24.
Paus T, Koski L, Caramanos Z, Westbury C. Regional differences in the effects of task difficulty and motor output on blood flow response in the human anterior cingulate cortex: a review of 107 PET activation studies. Neuroreport. 1998;9(9):R37–47.
Posner MI, Rothbart MK. Attention, self-regulation and consciousness. Philos Trans R Soc Lond B Biol Sci. 1998;353(1377):1915–27.
Roelofs A, van Turennout M, Coles MGH. Anterior cingulate cortex activity can be independent of response conflict in Stroop-like tasks. Proc Natl Acad Sci U S A. 2006;103(37):13884–9.
Rushworth MF, Walton ME, Kennerley SW, Bannerman DM. Action sets and decisions in the medial frontal cortex. Trends Cogn Sci. 2004;8(9):410–7.
Schweizer TA, Oriet C, Meiran N, Alexander MP, Cusimano M, Stuss DT. The cerebellum mediates conflict resolution. J Cogn Neurosci. 2007;19(12):1974–82.
Shenhav A, Botvinick MM, Cohen JD. The expected value of control: an integrative theory of anterior cingulate cortex function. Neuron. 2013;79(2):217–40.
Sheth SA, Mian MK, Patel SR, Asaad WF, Williams ZM, Dougherty DD, et al. Human dorsal anterior cingulate cortex neurons mediate ongoing behavioural adaptation. Nature. 2012;488(7410):218–21.
Stoet G, Snyder LH. Neural correlates of executive control functions in the monkey. Trends Cogn Sci. 2009;13(5):228–34.
Stuss DT, Floden D, Alexander MP, Levine B, Katz D. Stroop performance in focal lesion patients: dissociation of processes and frontal lobe lesion location. Neuropsychologia. 2001;39(8):771–86.
Sundermann B, Pfleiderer B. Functional connectivity profile of the human inferior frontal junction: involvement in a cognitive control network. BMC Neurosci. 2012;13:119.
Vendrell P, Junque C, Pujol J, Jurado MA, Molet J, Grafman J. The role of prefrontal regions in the Stroop task. Neuropsychologia. 1995;33(3):341–52.

Chapter 5
Synaptic Excitatory-Inhibitory Balance Underlying Efficient Neural Coding

Shanglin Zhou and Yuguo Yu

5.1 Introduction

Neural information coding is one of the central topics in neuroscience. The brain utilizes some features of action potential sequences (spike trains) to encode sensory and cognitive information. The algorithm operating within those features is the called the neural code. Half a century ago, Perkel and Bullock (1968) noted that a potential neural code must serve at least four functions: stimulus representation, interpretation, transformation and transmission. Stimulus representation indicates that the neural activity should be altered by the stimulus properties needed to be coded, and therefore, the neural code can represent this stimulus (Perkel and Bullock 1968; Kumar et al. 2010). Due to its basic role, neural representation has been extensively studied using experimental and theoretical approaches. Barlow in 1961 proposed a theoretical framework which hypothesized that the action potentials in the sensory neurons formed a neural code for efficiently representing sensory information. By efficient Barlow meant that the code minimized the number of neurons and spikes needed to represent an input signal. This is the origin of sparse coding or efficient neural coding (Barlow 1961). Barlow's model treats the sensory pathway as a communication channel where neuronal spiking is an efficient code for representing sensory signals. The spiking code aims to maximize available channel capacity by minimizing the redundancy between representational units (Simoncelli and Olshausen 2001). In addition, one of the major components of a typical neural code also include reliable information transmission. The brain is highly modular, and a successful neural code should be able to be transmitted (propagated) from one module to another with high fidelity (Perkel and Bullock 1968; Kumar et al. 2010).

S. Zhou · Y. Yu (✉)
State Key Laboratory of Medical Neurobiology, School of Life Science and Human Phenome Institute, Institutes of Brain Science, Institute of Science and Technology for Brain-Inspired Intelligence, Fudan University, Shanghai, China
e-mail: yuyuguo@fudan.edu.cn

A. Cheung-Hoi Yu, L. Li (eds.), *Systems Neuroscience*, Advances in Neurobiology 21, https://doi.org/10.1007/978-3-319-94593-4_5

The transmission property of neural coding has also drawn significant attention recently (Diesmann et al. 1999; Kistler and Gerstner 2002; van Rossum et al. 2002; Litvak et al. 2003; Vogels and Abbott 2005; Kumar et al. 2008, 2010).

The balance of excitatory and inhibitory synaptic membrane currents (E/I balance) received by a neuron underlying its spontaneous firing and/or responding to sensory inputs has been widely observed (van Vreeswijk and Sompolinsky 1996; Brunel 2000; Shu et al. 2003; Wehr and Zador 2003; Froemke et al. 2007; Murphy and Miller 2009). Here, E/I balance generally refers to excitatory-inhibitory balance in terms of either overall global balance or temporal balance on a fine time scale. Global E/I balance refers to the bulk measurement of relative contributions of excitatory and inhibitory synaptic currents received by a specific neuron. It is called global E/I balance if across a range of spatio-temporal conditions of interest, the ratio between the synaptic excitation and inhibitioin is kept approximately constant. In some situations, even the measurement of firing rates of excitatory and inhibitory neurons or excitatory and inhibition synaptic conductances received by a neuron can represent E/I balance for individual neurons within the cortical network circuit. Temporal balance indicates that the relative magnitudes of excitatory and inhibitory synaptic currents are matched in a point-to-point manner on a fast time scale. Global E/I balance is often used to examine pathological or dysfunctional brain states, whereas temporal E/I balance can be used to examine the effect of synaptic correlation on spiking timing to sensory input and stimulus feature seletivity. Bothe global and temporal E/I balances enable cortical operation in a precise manner to represent sensory inputs. The disruption of the cortical E/I balance has been demonstrated to cause cognitive dysfunction, such as schizophrenia (Yizhar et al. 2011; Murray et al. 2014). Because the E/I balance may be the key structure underlying the neural code and cognition, multiple questions arise: (1) How is the E/I balance achieved? (2) Why does the neural system choose such a scenario to function? (3) How does the E/I balance evolve during neural plasticity and coding? Specifically, how does the E/I balance influence information representation and propagation across regions?

Recently, more and more studies are conducted to answer these questions. Here, we briefly summarize the evidence for the existence of an E/I balance in the cortex and the mechanisms by which the E/I balance is achieved. We then review the experimental and computational development on the impact of the E/I balance on neural coding, especially the processes of stimulus representation and information propagation.

5.2 E/I Balance Is Ubiquitous in Cortical Circuits

Over the last decades, E/I balance has been found to exist in many situations including ongoing spontaneous activity, sensory-evoked activity and storage of memories. Synaptic plasticity at both excitatory and inhibitory synapses is suggested to play a central role in balancing the excitatory and inhibitory inputs to a targeted cell during the training or learning process (Vogels et al. 2011; Yu et al. 2014). The level of the developed balance depends on the time scale of the correlation between the excitatory and inhibitory inputs to the cell, ranging from a global balance, either without a correlation or with a correlation at a slower time scale, to a fine-scale balance for strong correlations with a fast time scale.

Global balance is quantified by using global measures of excitatory and inhibitory synaptic currents, including measuring spontaneous or ongoing excitatory and inhibitory postsynaptic currents (mEPSC and mIPSC) and the field potential, which is considered a rough signiture of the relative timing and magnitude of excitation and inhibition. In fact, it is difficult to simultaneously measure excitatory and inhibitory current inputs on the same neuron. However, researchers can overcome this by measuring the excitatory and inhibitory currents separately at different holding potentials and then calculating the average conductance of both currents (g_E and g_I) (Shu et al. 2003; Haider et al. 2006; Monier et al. 2008). By using this method, Shu et al. (2003) found that the received synaptic conductance values of g_E and g_I were always balanced with a certain ratio during the up state generated by recurrent connection patterns in the *in vitro* brain slice (Fig. 5.1a). Other experimental results also support the idea that the ratio of g_E and g_I of a given neuron remains constant across different conditions and in many systems (Wehr and Zador 2003; Haider et al. 2006; Xue et al. 2014). Additionally, many studies have demonstrated that the E/I balance still exists even when the system is driven by external inputs (Anderson et al. 2000; Martinez et al. 2002; Tan et al. 2004, 2011; Wilent and Contreras 2005; Cardin et al. 2007; Wu et al. 2008; Tan and Wehr 2009; Runyan et al. 2010; Liu et al. 2011). In fact, some studies have shown that the tuning curves of the excitatory and inhibitory conductance are similar to one another (Anderson et al. 2000; Wehr and Zador 2003; Cardin et al. 2007; Runyan et al. 2010).

To understand the E/I balance on the fine time scale (the fine-scale balance), researchers have tried to simultaneously record the time series of both the excitatory and inhibitory currents and then obtain the correlation between them. Since adjacent neurons in the cortex generally receive strongly correlated synaptic inputs, researchers can record both excitatory and inhibitory currents separately and simultaneously, each in a single neuron in a pair of neighboring cells, and the correlation between the excitatory and inhibitory currents onto a single cell can be inferred from the correlation between the time series from the two cells (Okun and Lampl 2008). Based on this method, researchers have found that the excitatory and inhibitory inputs from ongoing spontaneous activity or sensory-evoked activity are strongly correlated with one another, with inhibitory currents tracking excitatory currents closely with a few milliseconds of a delay (Fig. 5.1b) (Okun and Lampl 2008). More evidence has also shown that a fine-scale E/I balance exists during oscillations in the gamma and beta frequencies (Atallah and Scanziani 2009; Poo and Isaacson 2009).

Recently, in an interesting study using in vivo recordings with dense multielectrodes in the neocortex of higher level mammals (including human and primate), Dehghani et al. (2016) found that excitatory and inhibitory ensembles are well-balanced and co-fluctuate instantaneously in all states of the wake-sleep cycle (wake, slow-wave sleep and rapid-eye movement sleep) at different temporal scales

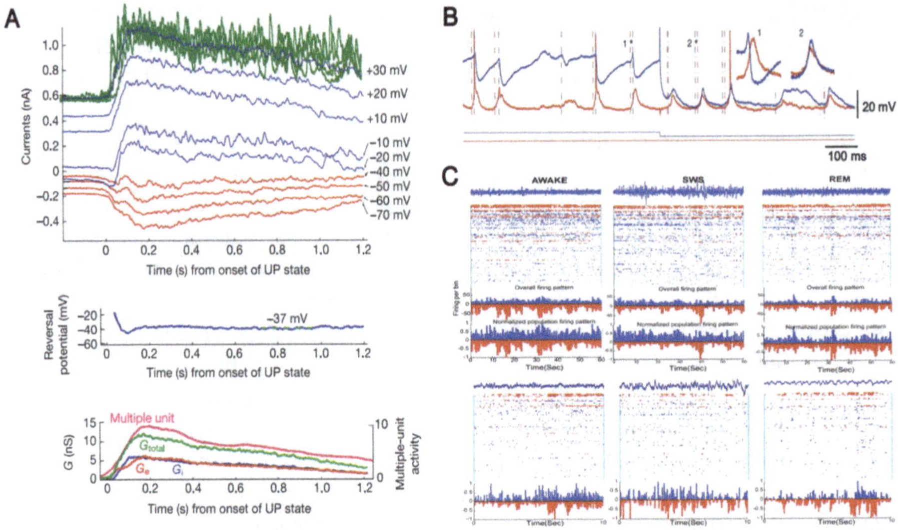

Fig. 5.1 Experimental evidence of the E/I balance. (**a**) Average currents during the up state in recordings clamped at different membrane potentials from in vitro brain slices (top, red and blue curves showing the average currents, the green curves showing the raw traces at +30 mV), the reversal potential of the average synaptic currents (middle), and additional conductances during the up state (bottom). Adapted from Shu et al. (2003). (**b**) Simultaneous in vivo recordings from two cortical cells. One cell (red) was continuously recorded in a hyperpolarized mode, and the other cell (blue) was switched between depolarized and hyperpolarized modes (current depicted below the traces). Dashed lines mark the onset of synaptic events. Insets show examples of two events (marked by asterisks). Adapted from Okun and Lampl (2008). (**c**) Recordings in humans during awake (left), slow-wave sleep (SWS) (middle) and rapid-eye movement (REM) (right) states. Top row shows 60-s windows; bottom row shows a 10-s window of the same state. Putative inhibitory neurons (FS cells) are shown in red. Putative excitatory neurons (RS) are depicted in blue. At the top of each panel, a sample LFP trace (in blue) accompanies the spiking activity. Histograms show the overall activity of the RS (blue) and FS (red) cells. Adapted from Dehghani et al. (2016)

(Fig. 5.1c). Beyond the temporal view of E/I balance, the spatial properties of E/I balance are also important in information processing. For instance, researchers observed that local E/I imbalance coexisting with overall balance facilitaates neural network creating novel features selectivity (Wu et al., 2008). However, we will mainly focus on temporal E/I balance in the following discussion.

5.3 Mechanisms to Achieve E/I Balance

To achieve global E/I balance in a dynamic neural network, several theoretical studies have shown that a neural network needs to be equipped with the following properties: (1) neurons in the network must be connected sparsely (the number of connections per neuron should be much smaller than the total number of neurons in the network) and randomly; (2) the strength of the inhibitory connections must be higher than the excitatory connections (van Vreeswijk and Sompolinsky 1996, 1998; Brunel 2000). Under such conditions, the average of the excitatory and inhibitory synaptic currents could be well-balanced, and the network dynamics could be stable. In such a balanced network, the membrane potentials and spike trains of the individual neurons may be highly uncorrelated (van Vreeswijk and Sompolinsky 1996, 1998; Brunel 2000). Beyond the two aspects mentioned above, synaptic plasticity may also play a vital role in the formation of the E/I balance (Froemke 2015).

In contrast to requirements for global balance of a network, Renart et al. (2010) proposed a neural network with random and dense connections (with the number of connections per neuron comparable to the total number of neurons) to achieve a more fine-scale E/I balance by setting the synaptic conductances and connection structure in the optimal range. In such a network, the excitatory and inhibitory currents received by each neuron are strongly correlated on a fast time scale. If excitatory and inhibitory currents cancel each other, then the net input current will be highly random, resulting in highly variable neural responses. Boerlin et al. (2013) have demonstrated that both variable neural responses and balanced excitation/inhibition are necessary consequences of neural networks that represent information efficiently in their spike trains. However, the Boerlin model assumes instantaneous synapses (transmission without delays) and only achieves balance because of this assumption. Further work allowed to relax this assumption and introduce realistic synapses (Koren and Deneve 2017). With realistic synapses it is however required that parameters that weight the cost on spiking are fine-tuned. Those parameters can be interpreted in biological terms as determining the excitability of the network.

Furthermore, a theoretical study proved that a network with synaptic plasticity of inhibitory synapses could evolve into a fine-scale E/I-balanced state with sparse connections (Vogels et al. 2011), and a later experimental study demonstrated the existence of this form of synaptic plasticity (D'amour and Froemke 2015). Further information about inhibitory synanptic plasticity could be found in other works (Vogels et al., 2011).

Beyond the theoretical work, experimental studies have provided some additional insights into the development of E/I balance (Froemke 2015). For example, Liu (2004) found that the ratio of the number of the excitatory and inhibitory synapses on the dendrites of cultured hippocampal neurons remained constant along different developmental stages, which suggests that the E/I balance may be related to an anatomical basis. In addition, sensory experiences at different developmental stages may play important roles in shaping the final E/I balance level (Froemke 2015).

5.4 E/I Balance and Information Representation

One of the fundamental functions of the neural systems is to represent the sensory information and make use of it for guiding action, a process termed neural coding. Representing neural signals is the process of interpreting prominent features of external sensory inputs with individual or population neuronal activity. Experimental and theoretical studies have demonstrated that action potential generation is an energy-expensive process (Attwell and Laughlin 2001; Alle et al. 2009; Yu et al. 2012). Therefore, efficient coding can be defined as coding with as few neurons and action potentials as possible, while not losing fidelity in representation of certain stimulus features (equivalent to sparse coding as defined by Barlow, 1961) or minimized coding error during stimulus representation (Deneve and Machens, 2016). Although there is no strict theoretical proof, the minimal coding error and information maximization or redundancy reduction are related in some aspects. Intuitively, coding effor reduction means a decrease in noise information, which would increase the mutual information between the neural response and the input signal, thus increasing the information coding efficiency. Because the E/I balance is ubiquitous in neural systems, there must be some strategic benefits of the E/I balance for efficient representation. Here, we summarize the evidence for this as follows.

5.4.1 *Irregular Spike Trains and Global E/I Balance*

A typical well-known property of the firing pattern of an individual neuron recorded in vivo is its irregularity or stochasticity, which is similar to Poisson-like time sequences. Revealing how individual neurons establish such irregular firing patterns is important for understanding the network states with spontaneous firing and how such states could be used to represent stimulus inputs. In fact, it has been widely shown that irregular firing patterns could be achieved by a neuron with balanced excitatory and inhibitory synaptic inputs on multiple time scales (Shadlen and Newsome 1994, 1998; van Vreeswijk and Sompolinsky 1996; Amit and Brunel 1997; Brunel 2000). There is an intuitive explanation to why such a globally balanced network would lead to the irregular firing of a single neuron. Imagine that there is a neural network where each neuron is bombarded with noisy, Poisson-distributed synaptic inputs from both excitatory and inhibitory sources. When the excitatory input values exceed the inhibitory inputs, then the net mean positive input would depolarize the neuron to fire quasi-regularly. However, if the excitatory inputs and inhibitory inputs cancel each other out in a slow time scale without correlation on a fast time scale, then the membrane potential of each neuron would randomly cross the threshold dependent on the fast noise, resulting in a spiking pattern with a high level of irregularity (Denève and Machens 2016).

Although such a network architecture would capture the irregularity of the neural firing pattern, the network behavior would become very sensitive to even a small perturbation due to its chaotic dynamics (Shadlen and Newsome 1998; Brunel 2000). This hypothesis suggests a low reliability of the neural network in response to sensory input, which makes such an E/I-balanced network represent stimulus features in a poor fidelity.

5.4.2 Sparse Coding, E/I Balance and Energy Efficiency

Sparse coding implies that only a small fraction of cells in the network displays a transient response to an input signal (Vinje and Gallant 2000) and such an energy-efficient paradigm greatly extends the coding capacity of a large family of sensory inputs (Olshausen and Field 1996; Dhawale et al. 2010; Wolfe et al. 2010; Koulakov and Rinberg 2011). Yu et al. (2013, 2014) implemented a large-scale olfactory bulb model with mitral cell and granule cell connected by dendro-dendritic synapses with regular LTP/LTD synaptic plasticity, and they found that balanced excitation/inhibition in strongly activated mitral cells leads to a sparse representation of odorant inputs (Fig. 5.2a). They further found that such a network with synaptic plasticity could always evolve into a sparsely oscillatory state to represent the input signal efficiently. During the evolving process, global synaptic excitation and inhibition gradually reach an optimal balance with which the network produces firing patterns with the highest level of sparseness (Fig. 5.2b) (Yu et al. 2014). The optimal level of synaptic excitation and inhibition could produce the highest level of sparseness and decorrelation in the network response and reduce energy cost (Nawroth et al. 2007).

Interestingly, the formation of response sparseness in such an olfactory bulb network does not depend on a specific type of synaptic plasticity, meaning either Hebbian or non-Hebbian rules can both develop the network dynamics into sparseness during the training process (Migliore et al. 2010; Yu et al. 2013, 2014). In a recent work by Vogels et al. (2011), the sparsely connected network endowed with plasticity of inhibitory synapses could evolve to a sparse response to natural stimuli. In addition, this type of network can accommodate synaptic memories with activity similar to the background activity; and same activity can be reactivated by external stimuli (Vogels et al. 2011).

Note that the above approaches assumed sparse network connection, i.e., neurons receive few connections K compared to the size of the network N, so that $K \ll N$. Such a sparse connectivity usually leads to uncorrelated excitation and inhibition, resulting in random fluctuations as input to neurons within network. Indeed, achieving efficient representation of input signal does not have to set the network to be sparsely connected. Recent theoretical works investigated network with dense connections. In such a scenario excitation and inhibition received in a neuron are strongly correlated while neuronal output spike trains are highly uncorrelated. Such a balanced network could represent information efficiently in their spikes (Boerlin et al. 2013). To ensure the network performance, two assumptions are required: (1) information on dynamic variables can be read out linearly from spike trains, and (2) neurons fire a spike only if it improves the representation of

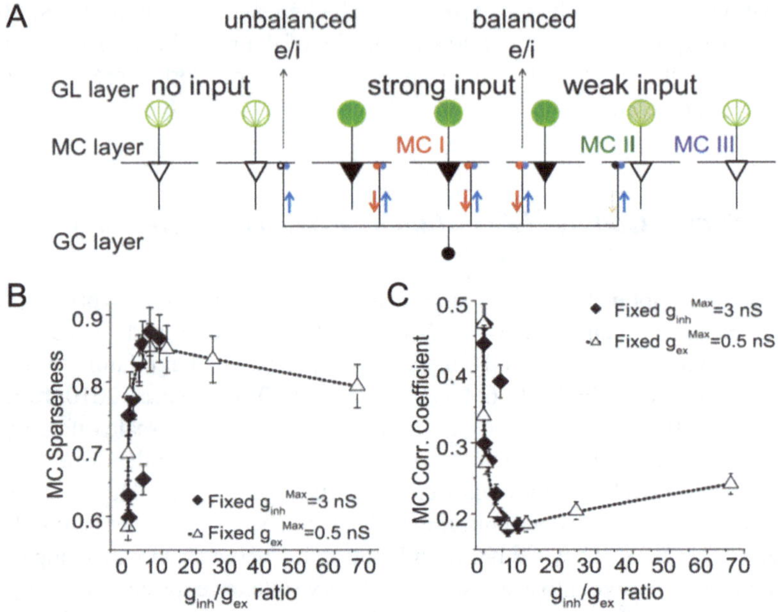

Fig. 5.2 The correlation of firing sparseness and mitral cell spiking in a large-scale olfactory bulb model. (**a**) Schematic representation of balanced and unbalanced excitation and inhibition in the MC–GC circuit. Three activated middle MCs (solid black triangles) receive strong input from glomeruli (solid deep green color); through back-propagation of APs in their lateral dendrites, they distribute the excitation (red) through reciprocal synapses, activating lateral inhibition in the surrounding MCs through the reciprocal inhibitory synapses. This mode of excitation and inhibition is balanced, and these MCs are called MC type I. The activated GCs (small blue spheres) deliver lateral inhibition to other surrounding MCs with weak or no excitatory inputs, making their reciprocal synapses unbalanced. These MCs are called MC type II. MCs that do not receive lateral inhibition are MC type III. (**b**) The MC network sparseness level as a function of reciprocal inhibitory weight to excitation weight ratio g_{inh}/g_{ex} for the cases of different g_{inh}^{Max} with a fixed g_{ex}^{Max} = 0.5 nS and different g_{ex}^{Max} with a fixed g_{inh}^{Max} = 0.3 nS. (**c**) Same in (**b**) but shows the mitral cell spiking correlation. Adapted from Yu et al. (2014)

dynamical variables. When a network satisfies these conditions, it evolves naturally to achieve the objective representation of a time-varying input with a minimum number of spikes with maximal efficiency (Boerlin et al. 2013; Denève and Machens 2016). In the following work, they further revealed that the maximally efficient network is right at the transition from synchronous to asynchronous network states (Koren and Deneve 2017). Moreover, there is a tight relationship between coding efficiency and energy efficiency, it was observed that there is a "sweet spot" where maximal coding efficiency coincides with rather low number of spikes (Boerlin et al. 2013; Koren and Deneve 2017).

Using single-compartment computational models with stochastic voltage-gated ion channels, Sengupta et al. (2013) calculated information content under either E/I-balanced or unbalanced conditions. They found that balanced synaptic currents evoke fewer spikes per second than the unbalanced conditions but with more information content in a single spike (bits/spike) in the balanced conditions. The total informative rate is similar in the two conditions (Sengupta et al. 2013). These results strongly support the hypothesis that E/I balance can promote both coding efficiency and energy efficiency.

Indeed, maximizing the ratio of the coding capacity to energy cost has been suggested to be one of the key principles chosen by the nervous system to evolve under selective pressure, and the metabolic energy efficiency demands of the nervous system could be sufficiently large to influence the design, function and evolution of the brain (Niven and Laughlin 2008). A recent theoretical work revealed a general rule for population coding in which the neuronal number should be sufficiently large to ensure reliable information transmission that is robust to the noisy environment but small enough to minimize energy cost (Yu et al. 2016). Experiments in cortex cultures, anesthetized rats, and awake monkeys, as well as computer models, have shown that balanced excitation/inhibition (E/I) could lead to a critical dynamic of avalanches in the cortical neural network (Shew et al. 2009; Poil et al. 2012; Yang et al. 2012). The number of metastable states (Haldeman and Beggs 2005) and the dynamic range to the input stimuli (Shew et al. 2009), as well as the information capacity and transmission (Beggs 2008; Shew et al. 2011) of the cortical neural networks, could be maximized at the critical point. The developed E/I balance within brain circuits during rest, learning and memory states may be beneficial for the brain to maintain an optimal state based on the theory of criticality. A large E/I ratio leads to a super-critical state whereby the neurons are highly activated and spikes among neurons are highly correlated. However, a small E/I ratio leads to a sub-critical state whereby the overall neural activity level drops and the spikes among neurons are random and not correlated (Yang et al. 2012). For information processing, highly correlated spikes reduce entropy in the former case, and in the latter case, the reduced correlation increases entropy, but this increase is counteracted by the concurrent drop in total information, resulting in maximal information transmission at a moderate E/I ratio (Shew et al. 2011). However, energy expenditure increases monotonically as the E/I ratio increases due to the increasing overall neural activity level. Therefore, a relatively large information transmission while relatively low energy cost, is expected to be maximized around an optimal E/I ratio (Poo and Isaacson 2009; Yu et al. 2014; Denève and Machens 2016).

5.4.3 Decorrelation and E/I Balance

The correlations among the spiking trains of all individual neurons in a network in response to sensory input can either help or harm the information transfer (Averbeck et al. 2006). More specifically, if positive signal correlations (i.e., neurons with similar selectivities of stimulus features) are linked to positive noise correlation, this would harm the information transfer. On the contrary, neurons have opposite stimulus selectivities, positive noise correlation help the information transfer. In many cases, correlations will not influence the information transfer (see for example, a theoretical study by Moreno-bote et al. 2014). To overcome the spiking correlation problem induced by correlated presynaptic input, Renart et al. (2010) built a densely connected neural network with excitatory and inhibitory currents canceling each other on a fast time scale (fine-scale balance). By using such mechanism, they showed that, theoretically, a fine-scale balanced network could generate an asynchronous state of population activity with a low mean spiking correlation despite correlated inputs (Renart et al. 2010). In the same study described above, Yu et al. (2014) found the E/I balance-induced sparse representation of odorant inputs was accompanied by a decorrelated state of mitral cell firing patterns, and the maximal decorrelation value existed at the optimal level for synaptic excitation and inhibition for the sparseness (Fig. 5.2c). In another interesting experimental work, researchers manipulated the excitation/inhibition ratio (E/I ratio) to obtain an optimal E/I ratio that maximized the information capacity by trading off between a lower correlation state (induced by low E/I ratio) and moderate activity (induced by a relatively high E/I ratio) (Shew et al. 2011).

5.5 E/I Balance and Information Propagation

Because the brain is highly modular, and spiking activity may carry a lot of neural information, it is important that the spiking activity can be transmitted from one module to another with high fidelity. Indeed, Perkel and Bullock (1968) noted that one of the major components of a typical neural code should be the inclusion of reliable information transmission or information propagation. The identification of the conditions under which spiking activity can propagate with high fidelity has attracted the attention of many theoretical researchers in the recent decade (Diesmann et al. 1999; Kistler and Gerstner 2002; van Rossum et al. 2002; Litvak et al. 2003; Kumar et al. 2008, 2010). Researchers usually address the propagation topic using a model of a cascade of neural assemblies in which a single neuron can participate at multiple levels (termed a feedforward network). To construct a more biologically oriented neural network, theoretical works tend to embed a feedforward sub-network into a larger recurrent neural network. However, neurons in the feedforward sub-network receive stronger correlated excitation than the rest of the recurrent network. This may destabilize the activity of the recurrent network. To solve this defect, Aviel et al. (2003) added inhibitory neurons into the subset network to balance the extra excitation.

Researchers have identified two modes of spiking activity propagation: the asynchronous mode (rate code, with information about the stimulus is carried and propagated in the firing rate of a neuron or the average of population activity) (van Rossum et al. 2002; Litvak et al. 2003; Vogels and Abbott 2009) and synchronous mode (temporal code, with information about the stimulus is carried and propagated by the precise timing of action potentials) (Aertsen et al. 1996; Diesmann et al. 1999; Gewaltig et al. 2001; Litvak et al. 2003; Kumar et al. 2008). For a network with a feedforward configuration, the firing rates at each layer could be stabilized to a constant level after an initial increase (Fig. 5.3a) (Litvak et al. 2003). The network inhibition precisely balanced with excitation plays a key role in modulating the mean firing rate level. Small deviations from the precise balance would result in a large fluctuation in the firing rate at each layer (Fig. 5.3b) (Litvak et al. 2003). Litvak et al. (2003) showed that the population synchrony could be formed after a few layers and then propagate stably through many layers in such a feedforward network with the excitation firing rate balanced with the inhibitory firing rate (Fig. 5.3c).

Beyond the fidelity of information propagation, the regulation of the spiking activity is also important for neural coding. A given module of the neural system has the potential to respond to several different signal pathways. To accomplish a single task, some mechanisms must exist to selectively block or boost some signal pathways. Recently, Vogels and Abbott (2009) showed that a detailed balance of excitation and inhibition in the target feedforward network group could be a potential

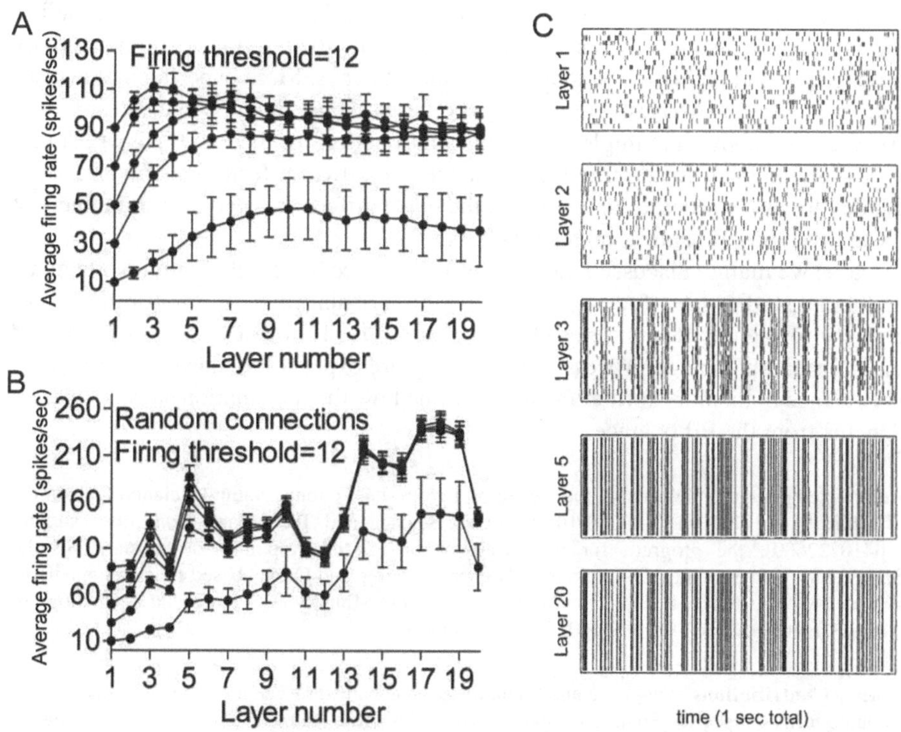

Fig. 5.3 Propagation of firing rate in a multilayer feedforward network. (**a**) Average firing rate of different layers in the precisely balanced network. (**b**) Average firing rate of different layers in the feedforward network with small deviations from the precise balance. (**c**) Raster plot showing the firing pattern of excitatory neurons in different layers in a feedforward network with balanced firing rates between excitation and inhibition. Adapted from Litvak et al. (2003)

gating mechanism; there information transmission can be gated ‘on’ by adjusting the excitatory and inhibitory gains to upset this detailed balance.

5.6 Conclusion

Stimulus representation and information propagation are two basic functions of neural coding. E/I balance, which is acknowledged as a fundamental paradigm for many brain functions, has been demonstrated to play a fundamental role in shaping the neural coding process. On one hand, the E/I balance can significantly increase the coding efficiency and energy efficiency to extend the coding capacity by promoting a sparse representation and signal decorrelation. Intuitively, an E/I ratio that is too high leads to excitatory dominance, resulting in high correlation (low level of coding efficiency) and activity (high level of energy consumed); however, an E/I ratio that is too low leads to suppressed activity with low information content. Therefore, the tradeoff between these two aspects requires the balance of the excitatory and inhibitory currents. On the other hand, based on recent theoretical studies, the E/I balance also plays a vital role in determining the fidelity of spiking activity propagation and gating of the multiple signal pathways. More experimental investigations are expected to test theoretical hypotheses and predictions in the near future. As discussed above, the implementations and functions of the two types of E/I balance—global balance and fine-scale balance, are different from one another. More studies are also needed to demonstrate the exact differences between the effects of these two types of balance on neural coding.

Here, we mainly discussed the roles of E/I balance in neural coding, while many additional studies have focused on the role of E/I balance in other brain functions, e.g., the formation of memories (Vogels et al. 2011; Lim and Goldman 2013) and the information storage process. More studies are expected to clarify the effects of E/I balance in memory formation and examine how the information storage process benefits from the E/I balance.

Acknowledgements YY thanks for the support from the National Natural Science Foundation of China (81761128011, 31571070), Shanghai Science and Technology Committee support (16410722600), the program for the Professor of Special Appointment (Eastern Scholar SHH1140004) at Shanghai Institutions of Higher Learning, and Omics-based precision medicine of epilepsy entrusted by the Key Research Project of the Ministry of Science and Technology of China (Grant No. 2016YFC0904400) for their support.

Author Contributions Yuguo Yu and Shanglin Zhou designed research; Yuguo Yu and Shanglin Zhou performed research; Shanglin Zhou and Yuguo Yu wrote the paper. All authors reviewed the manuscript.

Conflict of Interest The authors declare no competing financial interests.

References

Aertsen A, Diesmann M, Gewaltig MO. Propagation of synchronous spiking activity in feedforward neural networks. J Physiol Paris. 1996;90(3–4):243–7.

Alle H, Roth A, Geiger JRP. Energy-efficient action potentials in hippocampal mossy fibers. Science. 2009;325:1405–8. Available at: http://www.sciencemag.org/content/325/5946/1405

Amit DJ, Brunel N. A model of spontaneous activity and local delay activity during delay periods in the cerebral cortex. Cereb Cortex. 1997;7:237–52. Available at: http://cercor.oxfordjournals.org/content/7/3/237?related-urls=yes&legid=cercor;7/3/237

Anderson JS, Carandini M, Ferster D. Orientation tuning of input conductance, excitation, and inhibition in cat primary visual cortex. J Neurophysiol. 2000;84:909–26.

Atallah BV, Scanziani M. Instantaneous modulation of gamma oscillation frequency by balancing excitation with inhibition. Neuron. 2009;62:566–77. Available at: http://linkinghub.elsevier.com/retrieve/pii/S0896627309003511

Attwell D, Laughlin SB. An energy budget for signaling in the grey matter of the brain. J Cereb Blood Flow Metab. 2001;21:1133–45. Available at: http://www.ncbi.nlm.nih.gov/pubmed/11598490

Averbeck BB, Latham PE, Pouget A. Neural correlations, population coding and computation. Nat Rev Neurosci. 2006;7(5):358–66.

Aviel Y, Mehring C, Abeles M, Horn D. On embedding synfire chains in a balanced network. Neural Comput. 2003;15:1321–40. Available at: http://www.mitpressjournals.org/doi/abs/10.1162/089976603321780290

Barlow HBH. Possible principles underlying the transformation of sensory messages. In: Rosenblith W, editor. Sensory communication. Cambridge, MA: MIT Press; 1961. p. 217–34.

Beggs JM. The criticality hypothesis: how local cortical networks might optimize information processing. Philos Trans R Soc A Math Phys Eng Sci. 2008;366:329–43. Available at: http://rsta.royalsocietypublishing.org/content/366/1864/329.abstract

Boerlin M, Machens CK, Denève S. Predictive coding of dynamical variables in balanced spiking networks. PLoS Comput Biol. 2013;9:e1003258.

Brunel N. Dynamics of sparsely connected networls of excitatory and inhibitory neurons. J Comput Neurosci. 2000;8:183–208.

Cardin JA, Palmer LA, Contreras D. Stimulus feature selectivity in excitatory and inhibitory neurons in primary visual cortex. J Neurosci. 2007;27:10333–44.

D'Amour JA, Froemke RC. Inhibitory and excitatory spike-timing-dependent plasticity in the auditory cortex. Neuron. 2015;86:514–28.

Dehghani N, Peyrache A, Telenczuk B, Le Van Quyen M, Halgren E, Cash SS, Hatsopoulos NG, Destexhe A. Dynamic balance of excitation and inhibition in human and monkey neocortex. Sci Rep. 2016;6:1–12. Available at: http://arxiv.org/abs/1410.2610

Denève S, Machens CK. Efficient codes and balanced networks. Nat Neurosci. 2016;19:375–82. Available at: http://www.ncbi.nlm.nih.gov/pubmed/26906504

Dhawale AK, Hagiwara A, Bhalla US, Murthy VN, Albeanu DF. Non-redundant odor coding by sister mitral cells revealed by light addressable glomeruli in the mouse. Nat Neurosci. 2010;13:1404–12. Available at: http://www.pubmedcentral.nih.gov/articlerender.fcgi?artid=3208311&tool=pmcentrez&rendertype=abstract

Diesmann M, Gewaltig MO, Aertsen A. Stable propagation of synchronous spiking in cortical neural networks. Nature. 1999;402:529–33.

Froemke RC. Plasticity of cortical excitatory-inhibitory balance. Annu Rev Neurosci. 2015;38:195–219.

Froemke RC, Merzenich MM, Schreiner CE. A synaptic memory trace for cortical receptive field plasticity. Nature. 2007;450:425–9. https://doi.org/10.1038/nature06289.

Gewaltig MO, Diesmann M, Aertsen A. Propagation of cortical synfire activity: survival probability in single trials and stability in the mean. Neural Netw. 2001;14:657–73.

Haider B, Duque A, Hasenstaub AR, Mccormick DA. Behavioral/systems/cognitive neocortical network activity in vivo is generated through a dynamic balance of excitation and inhibition. J Neurosci. 2006;26:4535–45.

Haldeman C, Beggs JM. Critical branching captures activity in living neural networks and maximizes the number of metastable states. Phys Rev Lett. 2005;94(5):058101.

Kistler WM, Gerstner W. Stable propagation of activity pulses in populations of spiking neurons. Neural Comput. 2002;14:987–97. Available at: http://www.ncbi.nlm.nih.gov/pubmed/11972904

Koren V, Deneve S. Computational account of spontaneous activity as a signature of predictive coding. PLoS Comput Biol. 2017;13:e1005355.

Koulakov AA, Rinberg D. Sparse incomplete representations: a potential role of olfactory granule cells. Neuron. 2011;72:124–36.

Kumar A, Rotter S, Aertsen A. Conditions for propagating synchronous spiking and asynchronous firing rates in a cortical network model. J Neurosci. 2008;28:5268–80. Available at: http://www.ncbi.nlm.nih.gov/pubmed/18480283

Kumar A, Rotter S, Aertsen A. Spiking activity propagation in neuronal networks: reconciling different perspectives on neural coding. Nat Rev Neurosci. 2010;11:615–27. Available at: http://www.nature.com/doifinder/10.1038/nrn2886

Lim S, Goldman MS. Balanced cortical microcircuitry for maintaining information in working memory. Nat Neurosci. 2013;16:1306–14. Available at: http://www.ncbi.nlm.nih.gov/pubmed/23955560

Litvak V, Sompolinsky H, Segev I, Abeles M. On the transmission of rate code in long feedforward networks with excitatory-inhibitory balance. J Neurosci. 2003;23:3006–15.

Liu BH, Li YT, Ma WP, Pan CJ, Zhang LI, Tao HW. Broad inhibition sharpens orientation selectivity by expanding input dynamic range in mouse simple cells. Neuron. 2011;71:542–54.

Liu G. Local structural balance and functional interaction of excitatory and inhibitory synapses in hippocampal dendrites. Nat Neurosci. 2004;7:373–9.

Martinez LM, Alonso J-M, Reid RC, Hirsch JA. Laminar processing of stimulus orientation in cat visual cortex. J Physiol. 2002;540:321–33.

Migliore M, Hines ML, Mctavish TS, Shepherd GM. Functional roles of distributed synaptic clusters in the mitral-granule cell network of the olfactory bulb. Front Integr Neurosci. 2010;4:122.

Monier C, Fournier J, Frégnac Y. In vitro and in vivo measures of evoked excitatory and inhibitory conductance dynamics in sensory cortices. J Neurosci Methods. 2008;169:323–65.

Moreno-Bote R, Beck J, Kanitscheider I, Pitkow X, Latham P, Pouget A. Information-limiting correlations. Nat Neurosci. 2014;17:1410–7. https://doi.org/10.1038/nn.3807.

Murphy BK, Miller KD. Balanced amplification: a new mechanism of selective amplification of neural activity patterns. Neuron. 2009;61:635–48. https://doi.org/10.1016/j.neuron.2009.02.005.

Murray JD, Anticevic A, Gancsos M, Ichinose M, Corlett PR, Krystal JH, Wang XJ. Linking microcircuit dysfunction to cognitive impairment: effects of disinhibition associated with schizophrenia in a cortical working memory model. Cereb Cortex. 2014;24:859–72.

Nawroth JC, Greer CA, Chen WR, Laughlin SB, Shepherd GM. An energy budget for the olfactory glomerulus. J Neurosci. 2007;27:9790–800. Available at: http://www.jneurosci.org/cgi/doi/10.1523/JNEUROSCI.1415-07.2007

Niven JE, Laughlin SB. Energy limitation as a selective pressure on the evolution of sensory systems. J Exp Biol. 2008;211:1792–804. Available at: http://www.ncbi.nlm.nih.gov/pubmed/18490395

Okun M, Lampl I. Instantaneous correlation of excitation and inhibition during ongoing and sensory-evoked activities. Nat Neurosci. 2008;11:535–7. Available at: http://www.ncbi.nlm.nih.gov/pubmed/18376400

Olshausen BA, Field DJ. Emergence of simple-cell receptive field properties by learning a sparse code for natural images. Nature. 1996;381:607–9. https://doi.org/10.1038/381607a0.

Perkel DH, Bullock TH. Neural coding: a report based on an NRP work session. Neurosci Res Progr Bull. 1968;6:219–349.

Poil S-S, Hardstone R, Mansvelder HD, Linkenkaer-Hansen K. Critical-state dynamics of avalanches and oscillations jointly emerge from balanced excitation/inhibition in neuronal networks. J Neurosci. 2012;32:9817–23. Available at: http://www.jneurosci.org/content/32/29/9817.short

Poo C, Isaacson JS. Odor representations in olfactory cortex: "sparse" coding, global inhibition, and oscillations. Neuron. 2009;62:850–61. https://doi.org/10.1016/j.neuron.2009.05.022.

Renart A, de la Rocha J, Bartho P, Hollender L, Parga N, Reyes A, Harris KD. The asynchronous state in cortical circuits. Science. 2010;327:587–90.

Runyan CA, Schummers J, Van Wart A, Kuhlman SJ, Wilson NR, Huang ZJ, Sur M. Response features of parvalbumin-expressing interneurons suggest precise roles for subtypes of inhibition in visual cortex. Neuron. 2010;67:847–57.

Sengupta B, Laughlin SB, Niven JE. Balanced excitatory and inhibitory synaptic currents promote efficient coding and metabolic efficiency. PLoS Comput Biol. 2013;9(10):e1003263.

Shadlen MN, Newsome WT. Noise, neural codes and cortical organization. Curr Opin Neurobiol. 1994;4:569–79.

Shadlen MN, Newsome WT. The variable discharge of cortical neurons: implications for connectivity, computation, and information coding. J Neurosci. 1998;18:3870–96. Available at: http://www.ncbi.nlm.nih.gov/pubmed/9570816

Shew WL, Yang H, Petermann T, Roy R, Plenz D. Neuronal avalanches imply maximum dynamic range in cortical networks at criticality. J Neurosci. 2009;29:15595–600.

Shew WL, Yang H, Yu S, Roy R, Plenz D. Information capacity and transmission are maximized in balanced cortical networks with neuronal avalanches. J Neurosci. 2011;31:55–63. Available at: https://www.researchgate.net/publication/49731639_Information_Capacity_and_Transmission_Are_Maximized_in_Balanced_Cortical_Networks_with_Neuronal_Avalanches

Shu Y, Hasenstaub A, Mccormick DA. Turning on and off recurrent balanced cortical activity. Nature. 2003;423:288–93.

Simoncelli EP, Olshausen BA. Natural image statistics and neural representation. Annu Rev Neurosci. 2001;2001:1193–216.

Tan AY, Zhang LI, Merzenich MM, Schreiner CE. Tone-evoked excitatory and inhibitory synaptic conductances of primary auditory cortex neurons. J Neurophysiol. 2004;92:630–43. Available at: http://www.ncbi.nlm.nih.gov/entrez/query.fcgi?cmd=Retrieve&db=PubMed&dopt=Citation&list_uids=14999047

Tan AYY, Wehr M. Balanced tone-evoked synaptic excitation and inhibition in mouse auditory cortex. Neuroscience. 2009;163:1302–15.

Tan AYY, Brown BD, Scholl B, Mohanty D, Priebe NJ. Orientation selectivity of synaptic input to neurons in mouse and cat primary visual cortex. J Neurosci. 2011;31:12339–50.

van Rossum MCW, Turrigiano GG, Nelson SB. Fast propagation of firing rates through layered networks of noisy neurons. J Neurosci. 2002;22:1956–66.

van Vreeswijk C, Sompolinsky H. Chaos in neuronal networks with balanced excitatory and inhibitory activity. Science. 1996;274:1724–6. Available at: http://www.ncbi.nlm.nih.gov/pubmed/8939866

van Vreeswijk C, Sompolinsky H. Chaotic balanced state in a model of cortical circuits. Neural Comput. 1998;10:1321–71. Available at: http://www.ncbi.nlm.nih.gov/pubmed/9698348

Vinje WE, Gallant JL. Sparse coding and decorrelation in primary visual cortex during natural vision. Science. 2000;287:1273–6.

Vogels TP, Abbott LF. Signal propagation and logic gating in networks of integrate-and-fire neurons. J Neurosci. 2005;25:10786–95. Available at: http://www.jneurosci.org/content/25/46/10786

Vogels TP, Abbott LF. Gating multiple signals through detailed balance of excitation and inhibition in spiking networks. Nat Neurosci. 2009;12:483–91. Available at: http://www.ncbi.nlm.nih.gov/pubmed/19305402

Vogels TP, Sprekeler H, Zenke F, Clopath C, Gerstner W. Inhibitory plasticity balances excitation and inhibition in sensory pathways and memory networks. Science. 2011;334:1569–73. Available at: http://science.sciencemag.org/content/334/6062/1569.abstract

Wehr MS, Zador AM. Balanced inhibition underlies tuning and sharpens spike timing in auditory cortex. Nature. 2003;426:442–6. Available at: http://www.ncbi.nlm.nih.gov/pubmed/14647382

Wilent WB, Contreras D. Dynamics of excitation and inhibition underlying stimulus selectivity in rat somatosensory cortex. Nat Neurosci. 2005;8:1364–70.

Wolfe J, Houweling AR, Brecht M. Sparse and powerful cortical spikes. Curr Opin Neurobiol. 2010;20:306–12. https://doi.org/10.1016/j.conb.2010.03.006.

Wu GK, Arbuckle R, Liu BH, Tao HW, Zhang LI. Lateral sharpening of cortical frequency tuning by approximately balanced inhibition. Neuron. 2008;58:132–43.

Xue M, Atallah BV, Scanziani M. Equalizing excitation-inhibition ratios across visual cortical neurons. Nature. 2014;511:596–600. https://doi.org/10.1038/nature13321.

Yang H, Shew WL, Roy R, Plenz D. Maximal variability of phase synchrony in cortical networks with neuronal avalanches. J Neurosci. 2012;32:1061–72. Available at: http://www.jneurosci.org/content/32/3/1061.abstract

Yizhar O, Fenno LE, Prigge M, Schneider F, Davidson TJ, O'Shea DJ, Sohal VS, Goshen I, Finkelstein J, Paz JT, Stehfest K, Fudim R, Ramakrishnan C, Huguenard JR, Hegemann P, Deisseroth K. Neocortical excitation/inhibition balance in information processing and social dysfunction. Nature. 2011;477:171–8. https://doi.org/10.1038/nature10360.

Yu L, Zhang C, Liu L, Yu Y. Energy-efficient population coding constrains network size of a neuronal array system. Sci Rep. 2016;6:19369. Available at: http://www.nature.com/srep/2016/160119/srep19369/full/srep19369.html

Yu Y, Hill AP, Mccormick DA. Warm body temperature facilitates energy efficient cortical action potentials. PLoS Comput Biol. 2012;8:e1002456.

Yu Y, Mctavish TS, Hines ML, Shepherd GM, Valenti C, Migliore M. Sparse distributed representation of odors in a large-scale olfactory bulb circuit. PLoS Comput Biol. 2013;9:1–20.

Yu Y, Migliore M, Hines ML, Shepherd GM. Sparse coding and lateral inhibition arising from balanced and unbalanced dendrodendritic excitation and inhibition. J Neurosci. 2014;34:13701–13. Available at: http://www.jneurosci.org/cgi/doi/10.1523/JNEUROSCI.1834-14.2014

Chapter 6
Mapping Molecular Datasets Back to the Brain Regions They are Extracted from: Remembering the Native Countries of Hypothalamic Expatriates and Refugees

Arshad M. Khan, Alice H. Grant, Anais Martinez, Gully A. P. C. Burns, Brendan S. Thatcher, Vishwanath T. Anekonda, Benjamin W. Thompson, Zachary S. Roberts, Daniel H. Moralejo, and James E. Blevins

A. M. Khan (✉)
UTEP Systems Neuroscience Laboratory, University of Texas at El Paso, El Paso, TX, USA

Department of Biological Sciences, University of Texas at El Paso, El Paso, TX, USA

Border Biomedical Research Center, University of Texas at El Paso, El Paso, TX, USA
e-mail: amkhan2@utep.edu

A. H. Grant · A. Martinez
UTEP Systems Neuroscience Laboratory, University of Texas at El Paso, El Paso, TX, USA

Department of Biological Sciences, University of Texas at El Paso, El Paso, TX, USA

Graduate Program in Pathobiology, University of Texas at El Paso, El Paso, TX, USA

Gully A. P. C. Burns
Information Sciences Institute, Viterbi School of Engineering, University of Southern California, Marina del Rey, CA, USA

B. S. Thatcher · V. T. Anekonda · B. W. Thompson · Z. S. Roberts
VA Puget Sound Health Care System, Office of Research and Development Medical Research Service, Department of Veterans Affairs Medical Center, Seattle, WA, USA

D. H. Moralejo
Division of Neonatology, Department of Pediatrics, University of Washington School of Medicine, Seattle, WA, USA

J. E. Blevins
VA Puget Sound Health Care System, Office of Research and Development Medical Research Service, Department of Veterans Affairs Medical Center, Seattle, WA, USA

Division of Metabolism, Endocrinology, and Nutrition, Department of Medicine, University of Washington School of Medicine, Seattle, WA, USA

A. Cheung-Hoi Yu, L. Li (eds.), *Systems Neuroscience*, Advances in Neurobiology 21, https://doi.org/10.1007/978-3-319-94593-4_6

Abbreviations

ACB	nucleus accumbens
AchE	acetylcholinesterase
ADP	anterodorsal preoptic nucleus
AgRP	Agouti-Related Peptide
AHN	anterior hypothalamic nucleus
AHNa	anterior hypothalamic nucleus, anterior part
AHNc	anterior hypothalamic nucleus, central part
AHNd	anterior hypothalamic nucleus, dorsal part
AHNp	anterior hypothalamic nucleus, posterior part
AP	area postrema
ARH	arcuate hypothalamic nucleus
ATN	anterior nuclei, dorsal thalamus
AVP	anteroventral preoptic nucleus
AVPV	anteroventral periventricular nucleus hypothalamus
BST	bed nuclei of the stria terminalis
BSTal	bed nuclei of the stria terminalis, anterior division, anterolateral area
BSTam	bed nuclei of the stria terminalis, anterior division, anteromedial area
BSTdm	bed nuclei of the stria terminalis, anterior division, dorsomedial nucleus
BSTfu	bed nuclei of the stria terminalis, anterior division, fusiform nucleus
BSTif	bed nuclei of the stria terminalis, posterior division, interfascicular nucleus
BSTju	bed nuclei of the stria terminalis, anterior division, juxtacapsular nucleus
BSTmg	bed nuclei of the stria terminalis, anterior division, magnocellular nucleus
BSTov	bed nuclei of the stria terminalis, anterior division, oval nucleus
BSTpr	bed nuclei of the stria terminalis, posterior division, principal nucleus
BSTrh	bed nuclei of the stria terminalis, anterior division, rhomboid nucleus
BSTtr	bed nuclei of the stria terminalis, posterior division, transverse nucleus
BSTv	bed nuclei of the stria terminalis, anterior division, ventral nucleus
C.a.	anterior commissure
C.f.d.	fornix
CCK1R	cholecystokinin 1 receptor
Ch. Opt.	optic chiasm
CRH	corticotropin-releasing hormone
CTB	cholera toxin subunit b
DMH	dorsomedial hypothalamic nucleus
EGFP	enhanced green fluorescent protein
FG	fluorogold
fx	fornix
GFP	freen fluorescent protein
HNS	hypothalamo-neurohypophysial system
I	internuclear area, hypothalamic periventricular region

KO	knockout
LCM	laser-capture microdissection
LHA	lateral hypothalamic area
LHAai	lateral hypothalamic area, anterior region, intermediate zone
LHAav	lateral hypothalamic area, anterior region, ventral zone
LHAd	lateral hypothalamic area
LHAjd	lateral hypothalamic area, juxtadorsomedial region
LHAjp	lateral hypothalamic area, juxtaparaventricular region
LHAjvd	lateral hypothalamic area, juxtaventromedial region, dorsal zone
LHAjvv	lateral hypothalamic area, juxtaventromedial region, ventral zone
LHApc	lateral hypothalamic area, parvicellular region
LHAsfa	lateral hypothalamic area, subfornical region, anterior zone
LPO	lateral preoptic area
LS	lateral septal nucleus [Cajal]
LSc.d	lateral septal nucleus, caudal part, dorsal zone
LSc.v	lateral septal nucleus, caudal part, ventral zone
LSr.dl	lateral septal nucleus, rostral part, dorsolateral zone
LSr.m	lateral septal nucleus, caudal part, medial zone
LSr.vl	lateral septal nucleus, rostral part, ventrolateral zone
LSv	lateral septal nucleus, ventral part [Risold-Swanson]
MC4-R	melanocortin 4 receptor
ME	median eminence
MEex	median eminence, external lamina
MEin	median eminence, internal lamina
MEPO	median preoptic nucleus
MID	midline nuclei, dorsal thalamus
MM	medial mammillary nucleus, body
MNs	magnocellular neurons
MPN	medial preoptic nucleus
MPNc	medial preoptic nucleus, central part
MPNl	medial preoptic nucleus, lateral part
MPNm	medial preoptic nucleus, medial part
MPO	medial preoptic area
MS	medial septal nucleus [Cajal]
NDB	diagonal band nucleus [Broca]
NPY	neuropeptide Y
NTS	nucleus of the solitary tract
opt	optic tract
OT	oxytocin
PCR	polymerase chain reaction
PFA	paraformaldehyde
PMd	dorsal premammillary nucleus
PMv	ventral premammillary nucleus
POMC	pro-opiomelanocortin
PR	perireuniens nucleus
PSCH	suprachiasmatic preoptic nucleus

PT	paratenial nucleus
PVH	paraventricular hypothalamic nucleus
PVHd	paraventricular hypothalamic nucleus, descending division
PVHf	paraventricular hypothalamic nucleus, descending division, forniceal part
PVHm	paraventricular hypothalamic nucleus, magnocellular division
PVHmpd	paraventricular hypothalamic nucleus, medial parvicellular part, dorsal zone
PVHp	paraventricular hypothalamic nucleus, parvicellular division
PVHpv	paraventricular hypothalamic nucleus, periventricular part
PVi	periventricular hypothalamic nucleus, intermediate part
PVp	periventricular hypothalamic nucleus, posterior part
PVpo	preoptic periventricular nucleus
PVR	hypothalamic periventricular region
PVT	paraventricular thalamic nucleus
qPCR	quantitative polymerase chain reaction
RCH	retrochiasmatic area, lateral hypothalamic area
RE	nucleus reuniens [Malone]
REcd	nucleus reuniens, caudal division, dorsal part
REcm	nucleus reuniens, caudal division, medial part [Gurdjian]
REcp	nucleus reuniens, caudal division, posterior part
RIN	RNA integrity number
S.t.	infundibular stalk
SBPV	subparaventricular zone hypothalamus
SCH	suprachiasmatic nucleus [Spiegel-Zwieg]
SFO	subfornical organ
SMT	submedial nucleus thalamus
SO	supraoptic hypothalamic nucleus
SOr	supraoptic nucleus, retrochiasmatic part
sup	supraoptic commissures
T.M.	tractus Meynert (fasciculus retroflexus)
TH	tyrosine hydroxylase
TUi	tuberal nucleus, intermediate part
TUsv	tuberal nucleus, subventricular part
V.d'A.	tract of Vicq D'Azyr (mammillothalamic tract)
V3 h	third ventricle, hypothalamic part
vlt	ventrolateral hypothalamic tract
VMH	ventromedial hypothalamic nucleus
VMHa	ventromedial hypothalamic nucleus, anterior part
VMHc	ventromedial hypothalamic nucleus, central part
VMHdm	ventromedial hypothalamic nucleus, dorsomedial part
VMHvl	ventromedial hypothalamic nucleus, ventrolateral part
VP	vasopressin
VPL	ventral posterolateral nucleus thalamus, principal part
VPM	ventral posteromedial nucleus thalamus, principal part
μ-array	microarray

6.1 Introduction

6.1.1 Summary and Rationale

In this article, we envision ways in which molecular information extracted from the brain using methods such as transcriptomics, proteomics, and peptidomics can be anchored to locations in standardized atlas maps of the brain in order to preserve the provenance of the datasets and contextualize them with other datasets. We argue that whereas most researchers probe, dissect, mine, or interrogate the living brain and report back with valuable scientific information, such information would be worth more if it included *mapped locations* of where they traveled and what they found there. Mapping to a standardized reference allows current and future travelers to return to the same landscape with accuracy and precision, generate reproducible data from reproducible experiments, and allows them further to integrate and contextualize new data they gathered in that mapped location with other data gathered in the same space. By carefully documenting the locations, for example, of brain regions from which molecular information is extracted for large-scale analyses, scientists can contribute further to our collective history of the native landscape from which this expatriated molecular information originated.

6.1.2 Topic and Organization

We have chosen to use the hypothalamus as an exemplar structure to illustrate the possibilities of such an effort, a choice that is predicated in part on our own experiences in mapping and modeling multi-scale data for this brain region (e.g., Khan et al. 2006, 2017, 2018; Khan 2013; Zséli et al. 2016), and because a review of "-omics" work on the hypothalamus in the context of spatial mapping has not yet, to our knowledge, been attempted. So far, molecule extraction from hypothalamus has been focused primarily on mining either the whole hypothalamus or its well-defined sub-regions to the virtual exclusion of parts that are less well understood. If wider and more systematic sampling of areas within the hypothalamus were to be conducted, atlas mapping efforts will play an even greater role in helping us understand the organization of those areas that remain poorly defined. The additional benefit of mapping molecular data to a standardized atlas is that the data can be contextualized with multi-scale datasets mapped to the same reference map.

Below, following a brief exploration of the biological importance of location information in the brain (Sect. 6.2), we summarize the historical antecedents to current molecular extraction work done on the brain (Sect. 6.3) and the hypothalamus specifically (Sect. 6.4.1), focusing on those datasets that include spatial data about the regions extracted. We then survey studies that have examined the molecular landscape of the hypothalamus using transcriptomics, proteomics and peptidomics (Sect. 6.4.2). The rationale behind the separation of proteomics from its sub-domain, peptidomics, is based on the fact that the latter involves analytical procedures that are distinct from those in general proteomics, including more rigorous purification and more comprehensive identification procedures (Alzate 2010; Schrader et al. 2014; Romanova and Sweedler, 2015). The differences are great enough in methodology and concept that a separate consideration of peptidomic studies is warranted. The narrative then shifts to specific strategies that we envision will be required, especially the technique of laser-capture microdissection (LCM) (Sect. 6.5), to enable the accurate mapping of hypothalamic molecular datasets to a standardized atlas of the brain (Sect. 6.6), and the benefits of such mapping (Sect. 6.7). We conclude with a view to current and future directions for this research (Sect. 6.8).

6.2 Why Does Location Matter?

The brain is a very heterogeneous organ that contains diverse, non-repeating, and non-redundant sub-regions (e.g., see Balázs et al. 1972; Lehrer and Maker, 1972; Palay and Chan-Palay, 1972). Studies in many animal model systems have now revealed that brain region is a major determinant of gene expression patterns. Therefore, the location of areas sampled using "-omics" technologies will determine critically the complement of molecules expressed. Left- and right-handedness in cichlid fish, for example, is correlated strongly with hemispheric and regional asymmetry of gene expression (H. Lee et al. 2017). In songbirds, clustering analyses performed on retrieved sets of genes demonstrate a strong association of gene expression with brain region (Replogle et al. 2008; Drnevich et al. 2012; Balakrishnan et al. 2014). This also holds true for mammalian brain. Even between strains of mice (which can exhibit size differences for the whole brain and for individual brain regions: Badea et al. 2009), one report has estimated a 1% difference in baseline expression patterns in at least one brain region, and that gene expression differences in response to a physiological perturbation (in this case, seizure) produce marked differences in gene expression patterns in brain regions between strains (Sandberg et al. 2000). A re-analysis of the datasets of Sandberg et al. (2000) by Pavlidis and Noble (2001) reveals even greater differences in regional variation among the genes between the strains. These observations were extended by Nadler et al. (2006), who found, across ten inbred mouse strains, that there was a nearly 30% difference in gene expression in at least one brain region among those examined. Robust strain differences have also been documented for transcripts enriched in the rat hypothalamic neurohypophysial system (Hindmarch et al. 2007). Moreover, Dong et al. (2009) show that specific patterns of gene expression are associated with specific domains where distinct neural projection patterns emerge within the hippocampus, and Wolf et al. (2011) show that there is a strong predictive association of neural connections and gene expression within specific brain regions (also see Sun et al. 2012). Superimposed on this complexity are strain-dependent variations in the sexual dimorphism of certain brain nuclei (Robinson et al., 1985; Mathieson et al. 2000), and differences in how gene expression networks in the brain are modulated as a result of expression quantitative trait loci (eQTLs) that are sex-specific (Mozhui et al. 2012; also see Pandey and Williams, 2014; Hasin-Brumshtein et al. 2016). Thus, it is important to consider just what we as scientists lose if we endeavor to extract molecular information from the brain without attempting to preserve the provenance of where the extraction took place. Before addressing this issue more directly, it is useful to survey the history behind efforts to identify chemical and molecular information encoded in the brain.

6.3 Historical Antecedents

6.3.1 Heuristic Entry Points to Relevant History

Recent "-omics" work has been informed to various extents by seminal works conducted during the last 150 years which we have categorized heuristically along major research themes: *composition*, *communication*, *reaction* and *localization*. First, regarding *composition*, our current effort to understand dynamic changes in the expression of genes and proteins in the nervous system is predated by work that first identified its fundamental chemical (elemental) constituents (e.g., Thudicum 1884; Richter 1957; McIlwain 1959; Friede, 1966). Studies of the molecular constituents of neural machinery were motivated in part by contemporaneous questions concerning the ionic and chemical bases of muscle and nerve excitability (Helmholtz 1850; Nernst 1888; Overton 1902a, b; Hill 1932; Hodgkin and Huxley 1939, 1945, 1952a, b, c, d, e, f; Hodgkin et al. 1952; Fatt and Katz 1952; see various reviews by Boring 1942; Kleinzeller 1999; Häusser 2000; Bennett 2001; Huxley 2002; De Palma and Pareti 2011; Schwiening 2012; also see Khan 2009). Predating current work on proteomics and peptidomics, work on chemical composition was also marked by efforts in the 1980s by Tatemoto and colleagues to use chemical methods to isolate, identify and determine the sequence of neuropeptides such as galanin and neuropeptide Y (Tatemoto and Mutt 1980; Tatemoto et al. 1982, 1983; see Schrader et al. 2014).

Second, concerning *communication*, the mining of molecules coding for neurotransmitter and neuropeptide machinery in the nervous system finds its antecedents in both Bayliss and Starling's discovery of peptide hormone secretion from the pancreas (Bayliss and Starling 1902; also see Hirst 2004; Schrader et al. 2014), and Loewi's discovery (1921) of cholinergic neurotransmission in the peripheral nervous system. Ensuing efforts to gather evidence for a role for acetylcholine as a neurotransmitter in the central nervous system (e.g., Feldberg 1952) were facilitated by histochemical methods (Koelle and Friedenwald 1949; Anglade and Larabi-Godinot 2010; but see Levey et al. 1983), which helped contribute to the maturation of chemical neuroanatomy as a sub-discipline of neuroanatomy (also see: Jacobowitz and Palkovits, 1974; Palkovits and Jacobowitz, 1974). Importantly, histochemistry became useful to trace metabolic turnover in the brain, since it was performed on living or fresh frozen tissue and was based on enzymatic activities catalyzing the conversion of substrates to detectable products.

This work complemented contemporaneous studies—grouped thematically under *reaction*—that concerned the metabolism of living neural tissue, pioneered by

Warburg, McIlwain and others (e.g., see Warburg et al. 1924). Finally, a fourth long-standing body of work that informs "-omics" approaches concerns the historical quest to understand how various functions of the brain are derived from specific locations within its complex structure, the theme of *localization* (e.g., Flourens 1824; Ferrier 1873; Adrian 1939; also see Swanson 2007). This theme directly informs efforts to isolate portions of the nervous system for detailed study through careful extraction and sampling, a topic we delve into next.

6.3.2 Sampling at the Level of the Single Cell

Early interests in sampling very small portions of the central nervous system prefigure current interests in developing "spatially resolved" approaches (e.g., Crosetto et al. 2015) for transcriptomics and proteomics of neural tissues. Otto Deiters (1865) famously provided anatomical descriptions emphasizing the emergence of a single axon and multiple dendrites from motor neuron cell bodies in the spinal cord (also see Deiters and Guillery, 2013), which he isolated individually by hand from chromic acid- or potassium dichromate-hardened (i.e., fixed) tissue (Fig. 6.1A). Several investigators such as Hans Held and others followed suit using a variety of fixed preparations to study neurons in greater detail (see introductory comments in Chu (1954) for an overview).

Deiters' manual single-cell microdissection technique anticipates, by almost a century, single cell isolation from *fresh* neural tissue preparations pioneered by Giacobini (1956) for frog, rat and cat spinal cord and peripheral ganglia; and Hydén (1959) for the mammalian brain. Hydén, for example, used manual microdissection to isolate and chemically analyze (fittingly) the "giant neurons of Deiters" found in the lateral vestibular nucleus (Fig. 6.1B; Hydén 1959; also see Hydén 1967; Sotelo and Palay 1968; Rose 1999). Along with Giacobini's and Hydén's work using freshly microdissected neurons, related methods developed by Lowry (1953), Chu (1954), Roots and Johnston (1965), Johnston and Roots (1966) and others using fixed, freeze-dried, and reagent-impregnated tissues ushered in an era of "micro-chemical methods", in which a variety of *chemical assays* could be performed on single cells isolated from various regions of the central nervous system (cogently reviewed by Johnston and Roots 1970; also see Roberts and Baxter 1963; Osborne 1974). Eberwine et al. (1992) performed gene expression analysis on individual, freshly dissociated rat hippocampal neurons. More recently, single-cell isolation has now been conducted using laser-capture microdissection (LCM) methods (e.g., Williams et al. 2008; Blevins et al. 2009; Briski et al. 2010; Carreño et al. 2011; Landmann et al. 2012) or cell sorting methods (e.g., Draper et al. 2010; Henry et al. 2015; Macosko et al. 2015; R. Chen et al. 2017; S. Chung et al. 2017; Mickelsen et al. 2017; Romanov et al. 2017; also see Okaty et al. 2011; Poulin et al. 2016). Thus, single cell isolation methods first used for the purposes of morphological and structural investigation evolved for use in biochemical, molecular and functional analyses.

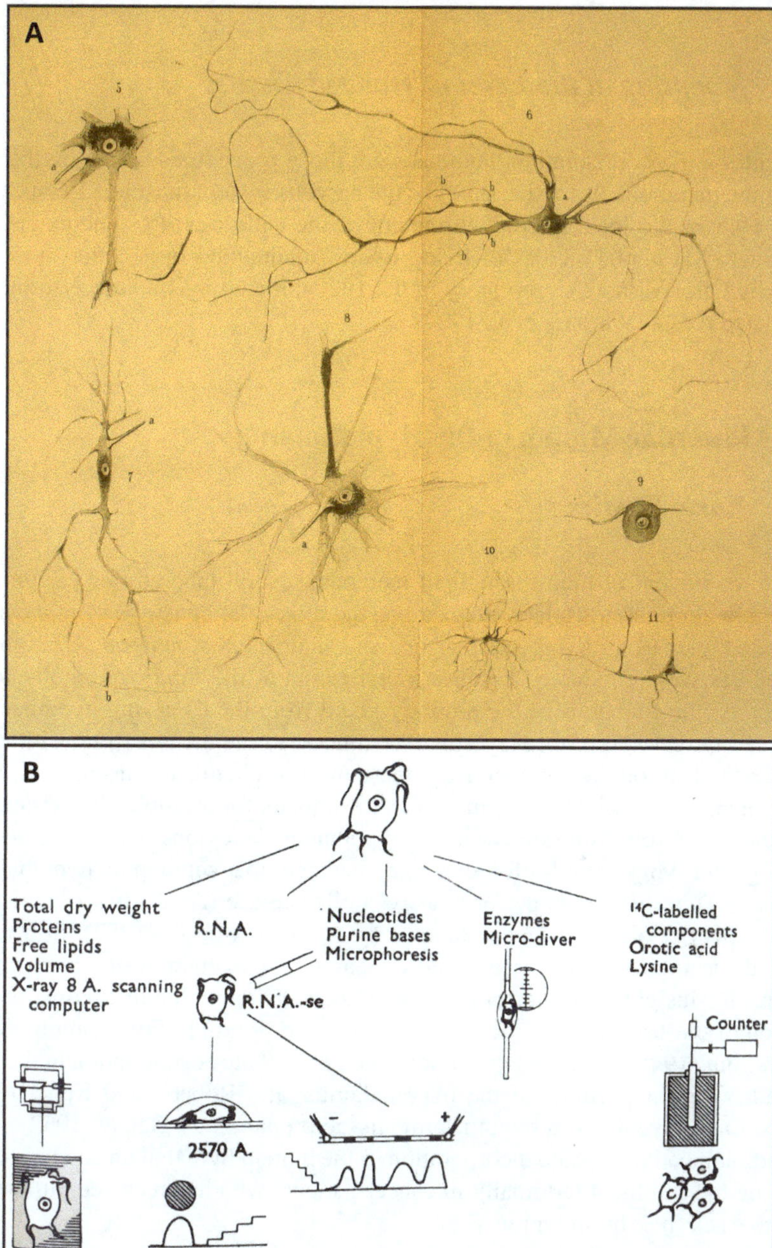

Fig. 6.1 Demonstrations of single-neuron isolation procedures, nearly a century apart, which anticipate isolation methods performed at present for studying single-cell transcriptomics, proteomics, and peptidomics. (**a**) *Plate II* of Deiters (1865), showing the morphologies of neurons collected from the central nervous system by hand microdissection. Note the recovery of many fine processes extended from each perikaryon, including axons and dendrites. These drawings are in the public domain. (**b**) *Figure 2* of Hydén (1959) showing a workflow schematic of possible assays that can be performed on single neurons isolated from freshly prepared tissue of the lateral vestibular nucleus. Neurons in this region are named "giant cells of Deiters" in honor of Deiters's initial description of these cells (see Sotelo and Palay 1968). Reprinted by permission from Macmillan Publishers, Limited: *Nature* 184(4684):433–435, 1959

6.3.3 *Sampling at the Level of Isolated Tissues*

Alongside single-cell isolation methods were those procedures driven by the need to examine metabolically active states of the nervous system in isolated tissue preparations where the local microenvironment of the cells was, to some extent, still maintained. Metabolic studies of living tissues maintained in isolation were pioneered by Otto Warburg's laboratory in the 1920s, including studies performed on the isolated retina (Warburg et al. 1924).

6.4 Molecular Mining of the Hypothalamus

6.4.1 *Early Studies*

Prior to the advent of high throughput methods, several laboratories performed a variety of techniques to isolate and examine the molecular constituents of the hypothalamus, either using living samples or fixed samples *post mortem*. A number of such studies were conducted because investigators at the time were motivated to differentiate the functions of the pituitary gland from the overlying hypothalamus (e.g., see Lisser 1927). Other investigators concerned themselves more with trying to understand, through histochemistry, the nature of chemical transmission in the hypothalamus (reviewed by Pilgrim 1974), to validate, for example, the existence of cholinergic neurotransmission within hypothalamic regions (see Sect. 6.3.1). Feldberg and Vogt (1948), for example, isolated the supraoptic hypothalamic nucleus (SO) in the dog to perform acetylcholinesterase (AchE) histochemistry, a method also performed in hypothalamus by Abrahams et al. (1957). Still others extended the tradition of Warburg and colleagues by examining the living hypothalamus for insights into metabolic processes occurring within this tissue, primarily through the use of radiolabeled phosphate incorporation. For example, Borell and Örström (1945) examined the radiolabeled phosphate accumulation in the anterior and posterior portions of the hypothalamus, and Roberts and Keller (1953, 1955) studied glycolysis in hypothalamic tissue preparations. Bakay (1952) examined radiolabeled phosphate incorporation in the human hypothalamus *post mortem* following the deaths of terminally ill cancer patients who had received intravenous tracer to track their brain tumors.

In what is perhaps the earliest demonstration of chemical analysis performed on an explicitly defined microdissected sub-region of the hypothalamus, Forssburg and Larsson (1954) sampled a portion of the hypothalamus from male and female rats that were either food-deprived for 24 h or ad libitum-fed and that received radioactive (^{14}C; $Na_2H^{32}PO_4$) tracer injections to track their carbon and phosphate metabolism. Brains were rapidly dissected and frozen, and 20–50 μm-thick sections were obtained of the brain, and examined carefully for the incorporation of ^{14}C and ^{32}P in chemically extracted fractions of the microdissected tissue. Importantly, the authors included a schematic to outline the areas they micropunched (Fig. 6.2A), including areas they sampled outside of the hypothalamus that served as a control. Their careful documentation of the sampled area and use of a custom-made micropunch tool (which they also illustrated in their study) anticipates the later use of similar instruments as developed by Palkovits and colleagues to sample discrete parts of the brain (see Palkovits 1973, 1975, 1986, 1989; Jacobowitz 1974; also see Jacobowitz 2006).

Using these micropunch methods, and leveraging refinements (O'Farrell 1975) of the original two-dimensional gel electrophoresis method (Smithies and Poulik 1956) that allowed proteins to be separated by their apparent molecular weights and isoelectric points (reviewed by Dunn 1987), Jacobowitz and colleagues pioneered the systematic study of proteins from discrete micropunched regions of the rat brain, including from within the hypothalamus (Heydorn et al. 1983). Importantly, their groundbreaking study included a schematic of atlas maps from the rat brain atlas of König and Klippel (1963) to identify the approximate locations and diameters of their tissue micropunches. Among the many brain regions sampled were the anterior, paraventricular, ventromedial and dorsomedial hypothalamic nuclei. Although the authors were able to obtain apparent molecular weights, isoelectric points and relative amounts of proteins from their tissue punches (see also Heydorn et al. 1986), their study does not specifically identify the proteins themselves except in a few cases. Methods to do so, involving annotated databases, had not yet been developed.

While micropunch methods continue to remain popular (e.g., see Atkins et al. 2011; Kasukawa et al. 2011), finer-grained studies that require more precise sampling of brain regions utilize LCM (Emmert-Buck et al. 1996), which is described in greater detail in Sect. 6.5, and a product of which is shown in Fig. 6.2B, C. This higher resolution sampling using LCM has now been performed at the level of single hypothalamic cells (e.g., see Fig. 6.2D–F).

6.4.2 *Studies of the Hypothalamus Using High Throughput Methods*

Tables 6.1, 6.2, and 6.3 provide a summary of selected studies performed to extract molecular data from the hypothalamus using high-throughput transcriptomic, proteomic, and peptidomic approaches; respectively. Transcriptomic approaches include microarray (Fodor et al. 1991; Maskos and Southern, 1992; also see Lenoir and Gianella, 2006; Pirrung and Southern, 2014) and next-generation sequencing (RNA-Seq; e.g., Mortazavi et al. 2008) technologies; proteomic and peptidomic approaches include protein separation methods such as electrophoresis and profiling technologies based on mass spectrometry (Gauss et al. 1999). A few of the tabulated studies are discussed below, beginning with studies which examined the hypothalamus as part of larger whole brain and/or multi-regional studies, and then on to studies in which the hypothalamus itself or its sub-regions were the main focus. Before these studies are examined in greater detail, it is useful to first

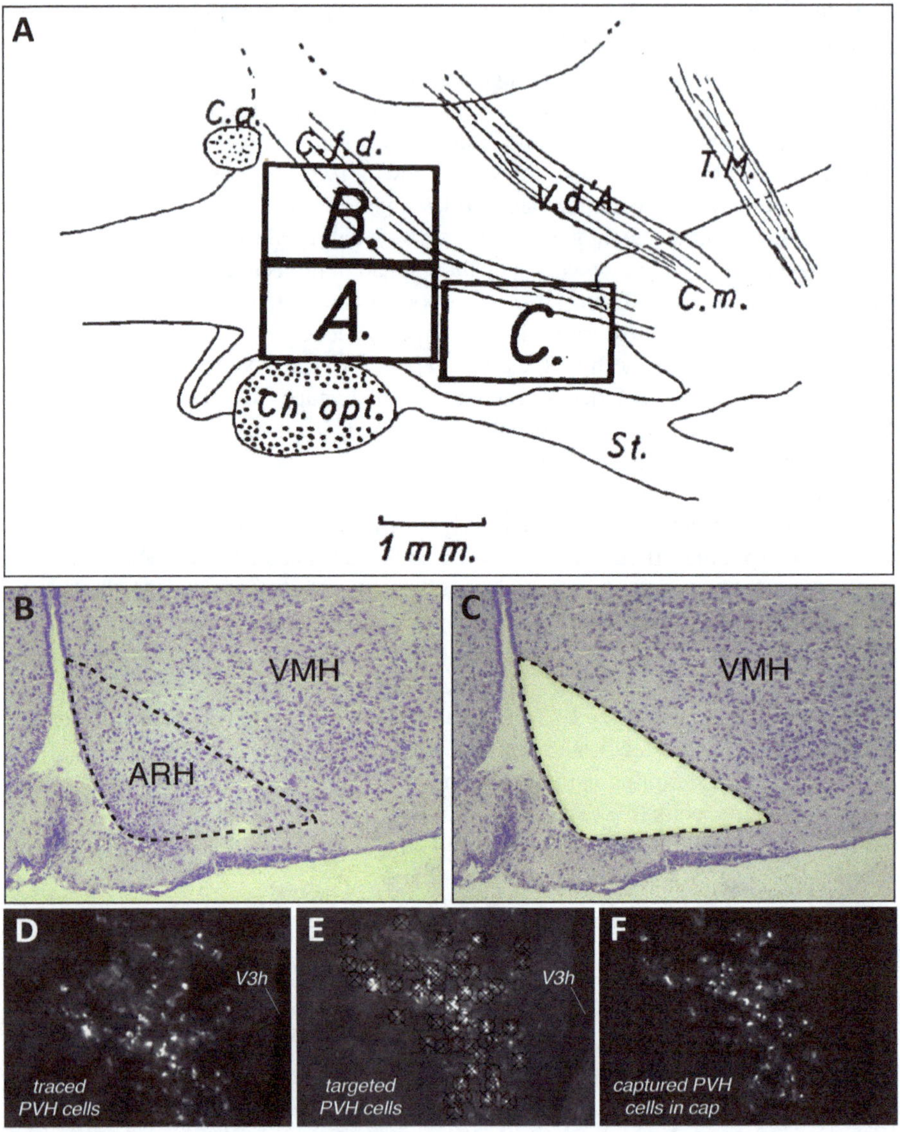

Fig. 6.2 Examples of documentation, past and more recent, of microdissected areas sampled from rat hypothalamus for chemical or molecular analyses—from gross micropunch (**A**), to region-level laser-capture microdissection (LCM) (**B, C**), to single-cell LCM (**D–F**). (**a**) *Figure 14* of Forssburg and Larsson (1954), showing a sagittal drawing of the hypothalamus, as defined by major fiber tract landmarks: Ch. Opt. = optic chiasm; C.a. = anterior commissure; C.f.d. = fornix; V.d'A. (*Tract of Vicq d'Azyr*) = mammillothalamic tract; T.M. (*Tractus Meynert*) = fasciculus retroflexus; St. = infundibular stalk. The boxes denoted by letters mark the regions micropunched from thin frozen sections at the locations indicated by the drawing, with *A* and *B* serving as control regions and *C* as the region of interest containing the lateral and ventromedial hypothalamus. Reproduced with permission from John Wiley & Sons, Ltd. (**B, C**). A Nissl-stained view of the arcuate

consider the "state of the field" as a whole in terms of how much sampling and coverage of the hypothalamus and its various regions have been undertaken thus far. Figure 6.3 provides a snapshot of the level of coverage reported by the studies listed in Tables 6.1, 6.2, and 6.3, organized by high throughput method and by spatial location within the hypothalamus. Specifically, a choropleth flatmap of the rodent brain, adapted from Swanson (2004), is utilized to highlight the degree to which either the whole hypothalamus (Fig. 6.3A), or individual sub-regions of the hypothalamus (Fig. 6.3B–D) have been sampled using transcriptomic, proteomic and peptidomic methods.

A number of observations can be made from an examination of the figure. First, of the total number of studies listed in Tables 6.1, 6.2, and 6.3, 45–83% of them (depending on which molecular analysis was performed) provided no sub-regional specificity for their sampling but rather sampled the whole hypothalamus (Fig. 6.3A). Second, of the studies performing high throughput extraction and molecular analysis of hypothalamic sub-regions, the greatest degree of coverage occurs for transcriptomic (Fig. 6.3B), followed by proteomic (Fig. 6.3C) and peptidomic (Fig. 6.3D) studies. Finally, across all methods, the overwhelming emphasis of sub-regional analyses of the hypothalamus has been on medially located nuclei, with little to no examination of sub-regions within the larger lateral hypothalamic area (LHA). Even for transcriptomic studies (Fig. 6.3B), the greater majority of studies of the LHA have focused mainly on a few key peptidergic cell types and not the whole region per se. Below, after describing a few studies that have focused on the hypothalamus in the context of whole-brain or multi-regional studies, we summarize a few key studies from among those listed in Tables 6.1, 6.2, and 6.3.

6.4.2.1 Whole Brain Extraction and Multi-Region Comparison Studies

There are many excellent reasons investigators opt to extract molecular information from the whole brain or large subdivisions of the brain without attending to where exactly in the brain the molecules are located. Such reasons include the need for investigators to survey the effects of factors that produce global, whole-organism or whole-subdivision effects that are poorly understood at a regional or cellular level.

Fig. 6.2 (continued) hypothalamic nucleus (ARH) and ventromedial hypothalamic nucleus (VMH) in rat brain tissue sectioned in the coronal plane, before (**B**) and after (**C**) the tissue was subjected to LCM. The dotted outline marks the region captured by the LCM instrument; note how the Nissl pattern helps to delineate the boundaries of the region to be sampled, and the remaining tissue after LCM can then be used to map the sampled region to a digital atlas. These images are provided courtesy of Dr. Rebecca Hull and Nishi Gill (*see Acknowledgments*). (**D–F**). Example of single-cell LCM of hypothalamic cells. These panels show photomicrographs adapted from *Figure 5* of Blevins et al. (2009), in which paraventricular hypothalamic (PVH) cells projecting to the hindbrain (as revealed in (**D**) by the presence of the retrograde tracer, cholera toxin subunit B (*white*), in PVH cells); have been targeted for LCM (see cross-patterns in (**E**)); and then have been collected into a microcentrifuge cap following LCM capture (**F**). The images in (**D–F**) are reproduced here in accordance with the policies of the *American Journal of Physiology*

Table 6.1 Selected transcriptomic studies in whole hypothalamus and by hypothalamic sub-region

Study	Animal	Extraction	Target(s) a priori?	Screen [S], Validation [V]	Map or schematic	Major findings
Whole hypothalamus						
Gautvik et al. (1996)	Rt	Dissection	N	Subtractive hybridization [S]; Southern and Northern blots; ISH [V]	N	Pioneering transcriptomic study of the hypothalamus; identified 53 hypothalamus-specific mRNAs
Jiang et al. (2001)	Ms	Dissection	N	μ-array [S]	N	Identified a few key genes that show differential expression in aged hypothalamus
Akhtar et al. (2002)	Ms	Dissection	N	μ-array [S] ISH [V]	N	Liver possesses cycling transcripts that are also in SCH but which do not cycle rhythmically there; Liver cycling dependent on intact SCH
Li et al. (2002)	Rt	Dissection	N	μ-array [S] Northern [V]	N	Fasting induced 96 mRNAs, and down-regulated 73 mRNAs
Yonehara et al. (2002)	Rt	Dissection	N	μ-array [S] RT-PCR [V]	N	12 genes display twofold greater increase in male vs. female neonates; 20 genes w twofold increase in female vs. male neonates
Mutsuga et al. 2004	Rt	Dissection	N	μ-array [S] ISH [V]	N	Found 1,385 genes expressed in SO at levels two times greater than in the hypothalamus as a whole
Prima et al. (2004)	Rt	NS	N	μ-array [S] Northern [V]	N	Found that ten weeks of cytokine exposure is associated with gene expression changes characteristic of chronic inflammation
Lachuer et al. (2005)	Ms	Dissection	N	μ-array [S] qRT-PCR [V]	N	Found NPY mRNA and AgRP mRNA to be down-regulated in *anx*/*anx* mice relative to wild-type mice
Lee et al. (2005)	Ms	Dissection	N	μ-array [S] RT-PCR [V]	N	108 of 6,016 genes identified were differentially expressed between control and immobilization-stressed mice

Zapala et al. (2005)	Ms	Dissection	N	μ-array	Y	Identified hypothalamus-enriched genes (see Tables 5 and 7 of Supplementary materials)
Shiue et al. (2006)	Ck	Dissection	N	qRT-PCR	N	16 mRNAs in high egg-yielding strain
Chen et al. (2007a)	Ck	Dissection	N	qRT-PCR	N	25 egg production-related mRNAs
Conti et al. 2007	Rt	Dissection	N	μ-array	Schematic	294/269 mRNAs up/down-regulated by fluoxetine treatment
Gao et al. (2007)	Rt	Dissection	N	μ-array [S] qRT-PCR [V]	N	Found differentially expressed genes between subject groups responsive and non-responsive to electroacupuncture analgesia
Kurrasch et al. (2007)	Ms, Fs	Dissection	N	μ-array [S] ISH, qPCR [V]	Ms: N, but photos; Fs: schematic, photos	Identified 200 genes enriched in neonatal VMH tissue; knockdown of some in zebrafish impairs development
Mennigen et al. (2008)	Fs	NS	N	μ-array	N	17 mRNAs induced, 70 mRNAs down-regulated by fluoxetine
Mercader et al. (2008)	Ms	Dissection	N	μ-array	N	In *anx*/*anx* mice, 141 mRNAs induced, 14 down-regulated, relative to wild-type
Xu et al. (2008)	Rt	Dissection	N	μ-array	N	27 mRNAs affected by high-fat diet: 14 induced, 13 down-regulated
Lee et al. (2009)	Ms	Dissection	N	μ-array [S] qRT-PCR [V]	N	Found caspase-1 up-regulated and μ-crystallin down-regulated in tubby mice.
Zhang et al. (2009)	Fs	NS	N	μ-array	N	873 genes differentially expressed among May, Aug, Dec seasonal periods
Byerly et al. (2010)	Ck	Dissection	N	μ-array [S] qRT-PCR [V]	N	Found differential expression of six genes in fat vs lean chickens involved in body fat control, and nine genes involved in glucose metabolism and glucose sensing
Ding et al. (2010)	Ms	Dissection	N	μ-array	N	Several genes differentially expressed after neonatal deprivation relative to adults

(continued)

Table 6.1 (continued)

Study	Animal	Extraction	Target(s) a priori?	Screen [S], Validation [V]	Map or schematic	Major findings
Higgins et al. (2010)	Ck	Dissection	N	μ-array	N	119 genes differentially expressed after fasting
Martyniuk et al. (2010a)	Fs	Dissection	N	μ-array	N	227 mRNAs differentially expressed after acute dieldrin exposure
Martyniuk et al. (2010b)	Fs	Dissection	N	μ-array	N	3,135 mRNAs differentially expressed after chronic dieldrin exposure
Orozco-Solís et al. (2010)	Rt	NS	N	μ-array	N	997 genes associated with nutritional deficiency during development
Popesku et al. 2010	Fs	Dissection	N	μ-array	N	3,088 ESTs were differentially regulated by dopamine receptor agonists
Poplawski et al. (2010)	Ms	NS	N	qRT-PCR	N	48-h fast shifts metabolism from glucose to lipid metabolism
Su et al. (2011)	Rt	Purchased from supplier	N	μ-array	N	Used an in-house fabricated microarray to analyze mitochondrial gene transcripts in hypothalamus, frontal cortex and hippocampus—proof of concept
Xu et al. (2011)	Fs	Dissection	N	μ-array	N	Nine genes differentially expressed
Zmora et al. (2012)	Fs	LCM	Y	qRT-PCR	N	Identified two kisspeptin systems
Chadwick et al. (2012)	Rt	Dissection	N	μ-array	N	GIT2 as aging-related molecule
González et al. (2012)	Rt	Dissection	N	qRT-PCR	N	Neuropeptide S and NPS-R both modulated by hyperthyroidism
Knight et al. (2012)	Ms	Dissection	Y	qRT-PCR, RNA-Seq, μ-array, IHC	N	Found various actively translating mRNAs in rats to be up-regulated, under various stimuli conditions
Mozhui et al. (2012)	Ms	Dissection	N	μ-array [S] qRT-PCR, ISH [V]	N	Found sexually divergent transcripts between males and females from recombinant inbred strains of mice, especially in certain hypothalamic nuclei

Paternain et al. (2012)	Rt	Dissection	N	qRT-PCR	N	A high fat/sucrose diet decreased expression of *Slc6a3*, *Npy*, and insulin receptor, and increased *Pomc* expression
Rabaglino et al. (2012)	Sh	Dissection	N	μ-array [S] qRT-PCR [V]	N	Estradiol-3-sulfate exposure altered fetal hypothalamic transcripts (NPY, AgRP, especially)
St-Amand et al. (2012)	Ms	Dissection	N	SAGE [S] qRT-PCR [V]	N	Found six unclassified and three novel transcripts enriched in hypothalamus
Farajzadeh et al. (2013)	Pg	Dissection	N	RNA-Seq	N	Transcriptional start site analysis revealed a proportionally greater number of sites for the hypothalamus relative to other regions sampled
Martyniuk et al. (2013)	Fs	Dissection	N	μ-array	N	Sexually dimorphic response to dieldrin
Nakazawa et al. (2013)	Rt	Dissection	N	μ-array	N	Found that relaxin administration was associated with expression of anxiety and fear-related genes, and feeding-related genes
Roy et al. (2013)	Dg	Dissection	N	RNA-Seq	N	Found significant differences in alternatively spliced genes in hypothalamus as compared to cerebral cortex
Sakakibara et al. (2013)	Ms	Dissection	N	μ-array [S] RT-PCR [V]	N	Found >100 genes downregulated by estradiol benzoate treatment underwent biphasic elevations in expression; validated a small subset of these genes by RT-PCR, including *Hcrt* and *Ptgds* (which encodes prostaglandin D2)
Schneeberger et al. (2013)	Ms	Dissection	N	μ-array	N	Observed down-regulation of genes associated with MAP kinase signaling, ubiquitin-proteasome signaling, autophagy and ribosome biosynthesis in subjects with targeted deletion of Dicer enzyme in *Pomc* neurons
Wood et al. (2013)	Sh	Dissection	N	μ-array [S] qRT-PCR [V]	N	Fetal hypoxia triggered changes in gene expression associated with reduced metabolism, mobilization of the immune and neuroendocrine response.

(continued)

Table 6.1 (continued)

Study	Animal	Extraction	Target(s) a priori?	Screen [S], Validation [V]	Map or schematic	Major findings
Zhang et al. (2013)	Pg	NS	N	μ-array [S] qRT-PCR [V]	N	Found 175 unique micro RNAs including 39 novel ones, in the hypothalamus
Balakrishnan et al. (2014)	Sp	NS	N	RNA-Seq [S] CZE [V]	N	Found transcripts with BLAST hits to 16,646 genes (93% of Ensembl annotated genes)
Fang et al. (2014)	Ck	Dissection	N	μ-array	N	Fasting up-regulated NPY and AgRP transcripts and those associated with fatty acid oxidation; and downregulated POMC, GHRH and other transcripts associated with fatty acid synthesis/transport
Luan et al. (2014)	Gs	Dissection	N	subtractive hybridization [S]; qRT-PCR [V]	N	Found 46 up-regulated and 49 down-regulated ESTs showing homology to known genes; identified GnRH-related regulatory genes to be expressed differentially during and after egg laying
Richter et al. (2014)	Fs	Dissection	N	μ-array	N	Methylmercury exposure triggers large-scale gene expression
Sangiao-Alvarellos et al. (2014)	Rt	Dissection	N	μ-array	N	Identified a number of microRNAs that displayed altered expression levels in response to caloric restriction and/or a high-fat diet
Fortes et al. (2015)	Cw	Dissection	N	RNA-Seq	N	Identified 978 genes expressed in hypothalamus
Gao et al. (2015)	Gs	NS	N	Illumina MiSeq [S] RT-PCR [V]	N	Found 48 hypothalamic transcripts up-regulated in the pre-egg laying period and 180 up-regulated during the laying period; found a few transcripts differentially expressed between the two periods
Kobayashi (2015)	Rt	Dissection	N	μ-array [S] semi-quant RT-PCR [V]	N	Showed a variety of gene expression changes in hypothalamic tissue following MK-801 exposure

Sun et al. (2015)	Ck	Dissection	N	μ-array [S] qRT-PCR [V]	N	Found heat shock proteins significantly altered in expression in response to thermal stress; identified 11 genes by qRT-PCR that were consistently expressed across samples, and 38 differentially expressing genes encoding growth-related functions and enzymatic activities.
Yelin-Bekerman et al. (2015)	Fs	Dissection/ digestion	Y	FACS, Illumina TruSeq [S]; RT-PCR, ISH [V]	N	Identified dozens of H/O-specific neuronal transcripts, and confirmed their expression and localization using imaging; identified Kcnh4a, which encodes a voltage-gated K+ channel, in H/O neurons; CRISPR-based silencing of this gene reduced sleep time in zebrafish
Fortes et al. (2016)	Cw	Dissection	N	μ-array	N	Identified five transcription factors with potential regulatory functions in hypothalamus that were expressed differentially pre- and post-pubertally
Klimov et al. (2016)	Rt	NS	N	RNA-Seq [S] qRT-PCR [V]	N	Found multiple differentially expressed genes in a hypertensive rat model
Rabaglino et al. (2016)	Sh	Dissection	N	μ-array [S] qRT-PCR [V]	N	Fetal hypothalamic transcripts for cell cycle, reproduction, and feeding were up-regulated after acute exposure to triclosan, whereas transcripts for steroid metabolism, lipoproteins, fatty acids and glucose were downregulated after exposure.
Tu et al. (2016)	Ck	NS	N	μ-array [S] qRT-PCR [V]	N	Found differentially expressed genes in hypothalamic samples as a result of heat stress, including genes encoding neuropeptides and heat shock proteins.
DiCarlo et al. (2017)	Ms	Dissection	N	RNA-Seq	photos of gross dissection	Found 63 differentially expressed genes in the hypothalamus across the estrous cycle, 12 of which encode oligodendrocyte- and myelin-specific proteins
Chen et al. (2017)	Ms	Dissection	N	Drop-Seq [S] ISH, IHC [V]	N	Identified 11 non-neuronal and 34 neuronal cell types, and the restricted expression of genes such as *Crabp1* and *Pax6*.

(continued)

Table 6.1 (continued)

Study	Animal	Extraction	Target(s) a priori?	Screen [S], Validation [V]	Map or schematic	Major findings
Cubuk et al. (2017)	Hm	Dissection	N	Illumina TruSeq [S]; qRT-PCR [V]	N	Identified 284 differentially expressed genes associated with entrance to torpor; 181 of which were up- and 103 of which were down-regulated
Johnson et al. (2017)	Ms	Dissection	N	Illumina TruSeq	N, but specify Bregma coordinates	Found bisphenol A and ethinyl estradiol exposure was associated with differential hypothalamic gene expression in California mice
Lee et al. (2017)	Fs	Dissection	N	RNA-Seq	N	Found differentially expressed genes in hypothalamus that correlated with lateralization of behavior. Many of these were unique to the hypothalamus as compared with other regions.
Nectow et al. (2017)	Ms	dissection	Y	vTRAP [S] ISH database, IHC, RNA-Seq [V]	N	Isolated translating mRNAs in MCH neurons using viral TRAP following injection of eGFP-L10a constructs into lateral hypothalamus; note that tissue isolation was at the level of the whole hypothalamus
Bochukova et al. (2018)	Hu	Dissection	N	RNA-Seq [S] qRT-PCR, FISH, IHC [V]	Photos of tissue furnished along with schematic	Identified up-regulated genes that are in common with genes that signal hunger encoded in the mouse AgRP neuron transcriptome; and down-regulated genes that are in common with POMC neuron expression profiles during feeding
Ivask et al. (2018)	Ms	Dissection	N	RNA-Seq [S]; qRT-PCR [V]	N	Found many differentially expressed genes in *WFS1* gene knockout mice relative to wild-type, including those that encode VP receptors.
Johnson et al. (2018)	Ms	μ-punch	N	qRT-PCR	N	Bisphenol A-exposed parenting California mice showed up-regulated hypothalamic expression of *Kiss1*, *Esr1* and *Esr2* genes relative to controls.
Lerner et al. (2018)	Ms	μ-punch	N	qPCR; LC/MS; MRM; MALDI MSI	N, but MSI images furnished	Found several lipid and transcriptomic changes in epileptic mice relative to controls

Qiu et al. (2018)	Fs	Dissection	N	RNA-Seq [S] qRT-PCR [V]	N	Found >30K unigenes mapping to known genes, 275 of which were expressed differentially in immature male and female adults, and 561 between mature male and female adults.
Sharma et al. (2018)	Bn	Dissection	N	RNA-Seq [S] qRT-PCR [V]	N	Found seasonal differences in gene expression in hypothalamic samples from black-headed buntings
Diencephalon						
Reyes et al. (2003)	Ms	Dissection	N	μ-array [S]; ISH, IHC [V]	Photo provided	Microdissected tissue comprising the full PVH, descending columns of the fornix, AHA, certain midline thalamic nuclei, and zona incerta displayed differential gene expression in animals receiving immune vs restraint stressors
Dalal et al. (2013)	Ms	Dissection	Y	TRAP [S]; μ-array/ISH [V]	N	Homogenized diencephalon to run TRAP assays from transgenic mice expressing eGFP-L10a fusion protein; confirmed identification of 15 transcripts expressed in H/O neurons
Hypothalamus (various sub-regions)						
Kasukawa et al. (2011)	Ms	μ-punch	N	μ-array [S]; qPCR; ISH [V]	Y	Micropunched several hypothalamic regions at various circadian times and analyzed transcriptomic content of each region; data available for each sub-region at http://brainstars.org
Medial hypothalamus (various sub-regions)						
Auger et al. (2006)	Rt	Dissection	N	μ-array [S] qRT-PCR [V]	Y	Sampled tissue containing preoptic area and mediobasal hypothalamus together; found expression pattern differences for 12 genes following progesterone treatment; four of which were confirmed by qRT-PCR

(continued)

Table 6.1 (continued)

Study	Animal	Extraction	Target(s) a priori?	Screen [S], Validation [V]	Map or schematic	Major findings
Romanov et al. (2017)	Ms	Dissection, manual dissociation	N	Single cell RNA-Seq [S]; Drop-Seq, IHC [V]	N	Sampled a large portion of the medial hypothalamus which included portions of the Preoptic nucleus, PVH, AHN, SCH, DMH and ARH; identified single phenotypes (62 in total) on the basis of clustering analysis, including novel subtypes of GABA, glutamate, and dopamine-containing neurons
AHA: Anterior hypothalamic area						
Sanna et al. (2005)	Rt	LCM	N	μ-array	Y	Established a working protocol for microarray analysis of LCM samples
AVPV: Anteroventral periventricular nucleus						
Del Pino et al. (2015)	Rt	dissection	N	μ-array [S] qPCR; ISH [V]	Y	Identified the RNA-binding protein, *Cugbp2*, as a gene enriched in AVPV and regulated by estradiol
ARH: Arcuate hypothalamic nucleus						
Middleton et al. (2004)	Rt	Dissection	N	μ-array	N	Observed fourfold changes in expression of ARH genes associated with diet-induced obesity
Li et al. (2005)	Rt	μ-punch	N	μ-array	N	118 mRNAs up-regulated and 203 mRNAs down-regulated after fasting
Segal et al. (2005)	Ms	LCM	N	μ-array [S] ISH [V]	N	Found genes for VMH enriched as compared to ARH
Xiao et al. (2005)	Rt	μ-punch	N	μ-array [S] RT-PCR [V]	N	In ARH tissue punches which also contained VMH, the authors found 12 genes differentially regulated during lactation.
Nilaweera et al. (2009)	Hm	LCM	N	μ-array	N	Found a number of genes in dorsomedial ARH that are regulated by photoperiod
Paulsen et al. (2009)	Rt	LCM	N	μ-array [S] qRT-PCR [V]	N	Fasting-induced changes in NPY and POMC expression; 3,480 other genes

Arai et al. (2010)	Ms	LCM	N	qRT-PCR	N	Increased NPY mRNA/peptide in neurogenin3 null mutants
Briski et al. (2010)	Rt	LCM	Y	qRT-PCR	N	Insulin-induced hypoglycemia is associated with alterations in approx. a half-dozen transcripts
Draper et al. (2010)	Ms	Dissection, FACS	N	μ-array [S]; RT-PCR, ISH, FISH; IHC [V]	Y	Found 20 genes differentially expressed between ARH and DMH NPY-GFP neurons; with ARH neurons expressing the leptin receptor and responding to leptin with pSTAT activation
Jovanovic et al. (2010)	Ms	LCM	N	μ-array	N	Fasting induces 639 genes and down-regulates 452 genes
Adler et al. (2012)	Rt	LCM	Y	multiplex, nested PCR	Y	Sex differences in WAT projection neuron neurochemistry
Amar et al. (2012)	Rt	μ-punch	N	RNA-Seq	schematic only	Found moderate to high expression for 20 miRNAs among 210 miRNA genes examined
Landmann et al. (2012)	Rt	LCM	Y	qRT-PCR	No, but atlas levels specified	Fasting induces AgRP but not POMC
Stocker et al. (2012)	Rt	LCM	N	qRT-PCR	N	Pups cross-fostered to dams fed low protein diet increase leptin and melanocortin-3 receptor expression
Zmora et al. (2012)	Fs	LCM	Y	qRT-PCR, ISH	N	Detected expression of kisspeptin genes and genes for their receptors in males and females
Henry et al. (2015)	Ms	Manual sorting	Y	RNA-Seq	N	Selective changes in AgRP neurons after food deprivation
Trivedi et al. (2015)	Rt	LCM	N	μ-array [S]; qPCR [V]	N	Identified tachykinin-1 as a gene down-regulated by ghrelin
Doubi-Kadmiri et al. (2016)	Rt	Dissection	Y	qRT-PCR	N	Analyzed >300 miRNAs from ARH/ME samples, and >30% of these underwent maternal diet-induced expression changes in progeny

(continued)

Table 6.1 (continued)

Study	Animal	Extraction	Target(s) a priori?	Screen [S], Validation [V]	Map or schematic	Major findings
Jeong et al. (2016)	Ms	Dissection; aspiration	Y	Single-cell qRT-PCR	N	Characterized transcripts in single cells captured in ARH that had a cholinergic phenotype; found that the cells diverged in the types of transcripts each expressed
Kabra et al. (2016)	Ms	LCM	N	qRT-PCR	N	HDAC5 is an important component of leptin signaling and food intake control
Campbell et al. (2017)	Ms	Dissection	N	Drop-Seq, single-cell RNA-Seq [S]; ISH database; IHC [V]	Y	Catalogued and identified 34 distinct neuronal populations and 36 non-neuronal populations in ARH-ME (arcuate hypothalamus-median eminence) samples from >20K individual profiles of ARH cells.
DMH: Dorsomedial hypothalamic nucleus						
Segal et al. (2005)	Ms	LCM	N	μ-array [S] ISH [V]	N	Found genes for VMH enriched as compared to DMH
Draper et al. (2010)	Ms	Dissection, FACS	N	μ-array [S]; RT-PCR, ISH, FISH; IHC [V]	Y	Found 20 genes differentially expressed between ARH and DMH NPY-GFP neurons; with DMH neurons showing a conspicuous absence of leptin receptor expression
Lee et al. (2012)	Ms	LCM	N	μ-array	camera lucida	Highly expressed DMH genes: *Gpr50*, *Pcsk5*, *Sulf1*, *Rorb*, others
GnRH population/preoptic: GnRH motor neuron pool of the preoptic area, and preoptic area						
Vasilache et al. (2007)	Ms	LCM	Y	qRT-PCR	N	Distinct EP3 receptor isoform profiles
Soga et al. (2012)	Ms	LCM	Y	qRT-PCR	Y	Neonatal dexamethasone exposure up-regulates GnIH-GnRH pathway
Vasilache et al. (2013)	Ms	LCM	N	μ-array	N	Prostaglandin E synthase 1 KO and inflammation induce some gene expression changes
Eberwine and Bartfai (2011)	Ms	Patch pipette	Y	μ-array	N	Unique receptor on warm-sensitive neurons

LHA: lateral hypothalamic area						
Volgin et al. (2004)	Rt	Acute dissociation	Y	ICC [S], RT-PCR [V]	N	Demonstrated single-cell isolation, immunocytochemical identification, and mRNA recovery for H/O and MCH peptidergic neurons of the LHA
Ahmed et al. (2005)	Rt	Dissection	N	μ-array	N	75–100 mRNAs up-/down-regulated with cocaine escalation
Harthoorn et al. (2005)	Rt	LCM	Y	ICC [S], RT-PCR [V]	N	Identified mRNAs for MCH, H/O, CART, dynorphin, various receptors, and GABA/Glu markers in H/O and MCH neurons
Sanna et al. (2005)	Rt	LCM	N	μ-array	Y	Established a working protocol for microarray analysis of LCM samples
Honda et al. (2009)	Hu, Ms	Dissection	N	μ-array [S], RT-PCR, IHC, ISH [V]	N	Compared transcriptomes of control and narcoleptic *post mortem* human brains, and control vs. transgenic mice lacking H/O neurons; found insulin-like growth factor binding protein (IGFBP3) downregulated in both transgenic mouse and narcoleptic human brains
Chen et al. (2013)	Ms	NS	Y	μ-array	N	Syndecan-3 mRNA was up-regulated in LHA after cocaine self-administration
Mickelsen et al. (2017)	Ms	Dissection, FACS	Y	single-cell qPCR [S]; dual FISH, IHC [V]	Y	Found H/O and MCH neurons express 48 key genes encoding multiple neuropeptides and markers for fast neurotransmission; found, strikingly, that virtually all MCH neurons, and about half of the H/O neurons, express markers for glutamate release and GABA synthesis, but not GABA release
Preoptic area						
Akbari et al. (2013)	Rt	Dissection	N	μ-array	N, but did specify atlas	Maternal behavior was associated with changes in expression for dopamine-related genes, neurotransmitter and neuropeptide receptors, and especially glucocorticoid gene family

(continued)

Table 6.1 (continued)

Study	Animal	Extraction	Target(s) a priori?	Screen [S], Validation [V]	Map or schematic	Major findings
Aubert et al. (2013)	Mk	LCM	N	μ-array [S]; qRT-PCR [V]	specified atlas and coordinates	Found that serotonin receptor agonist administration was associated with altered expression of various transcripts in marmoset tissue samples
Chung et al. (2017)	Ms	Dissection, FACS	N	TRAP; single-cell RNA-Seq	N	Identified GABAergic preoptic neurons projecting to the tuberomammillary nucleus that are sleep-active, including biomarkers within these neurons
PVH: paraventricular hypothalamic nucleus						
Bonaventure et al. (2002)	Rt	LCM	N	μ-array	N	Found gene-relatedness based correlations in brain sub-regions in PVH
Sanna et al. (2005)	Rt	LCM	N	μ-array	Y	Established a working protocol for microarray analysis of LCM samples
Hindmarch et al. (2006)	Rt	Dissection	N	μ-array	N	Found mRNAs regulated by dehydration, enriched in PVH and SO
Heisler et al. (2007)	Ms	LCM	N	μ-array	Fos map	Found 5-HT2CR and 5-HT1DR mRNAs
Hindmarch et al. (2007)	Rt	NS	N	μ-array	N	mRNA expression differences between strains for the neurohypophysial system
Tung et al. (2008)	Ms	LCM	N	μ-array [S] qRT-PCR [V]	Y	Profiled transcripts from ad libitum-fed vs 48 h-fasted mice with or without leptin treatment. Found 527 transcripts with altered expression by fasting that could at least be partially reversed by leptin
Blevins et al. (2009)	Rt	LCM	Y	qRT-PCR	injections	Found MC4R mRNAs in NTS-projecting PVH neurons
Atkins et al. (2011)	Ms	Dissection	N	RNA-Seq	N	Established protocol
Amar et al. (2012)	Rt	μ-punch	N	RNA-Seq	Y	Found moderate to high expression for 20 miRNAs among 210 miRNA genes examined

Kohno et al. (2014)	Ms	Dissection	N	μ-array [S] qRT-PCR, IHC [V]	N	TH and galanin up-regulated in Sim1-specific Dnmt3a deletion mice, who displayed hyperphagia, decreased energy expenditure, glucose intolerance, and increased serum insulin and leptin levels
Nedungadi and Cunningham (2014)	Rt	LCM	Y	qRT-PCR	N, but atlas levels specified	Found TRPC4 channel expression, but hepatic cirrhosis is not associated with changes in its expression in PVH
Romanov et al. (2014)	Ms	Dissection, dissociation	Y	RNA-Seq	Y	Phenotyped 151 neurons from the mouse PVH, including neuropeptide phenotypes in cells with excess of 100 mRNA copy numbers per cell: somatostatin, galanin, cholecystokinin, neurotensin S, and CART
Novoselova et al. (2016)	Ms	LCM	N	μ-array [S]; qRT-PCR, WB [V]	LCM image	Found *Mrap2* deficient mice displayed down-regulated expression of *Sim1*, *Trh*, *Oxt* and *Crh* relative to wild-type subjects
SCH: suprachiasmatic hypothalamic nucleus						
Panda et al. (2002)	Ms	dissection	N	μ-array [S]; RT-PCR, ISH [V]	N	Found approx. 650 cycling transcripts in the SCH
Porterfield et al. (2007)	Ms	LCM	N	μ-array [S]; RT-PCR [V]	LCM image	Identified a number of genes differentially up-regulated following light pulse exposure
Winrow et al. (2009)	Rt	LCM	N	μ-array	N	Differential profiles across circadian cycle
Porterfield and Mintz (2009)	Ms	LCM	N	qRT-PCR	N	Induction of genes in early dark phase to light pulse
Boone et al. (2012)	Rt	LCM	N	qRT-PCR	LCM image	TBI model shows altered circadian gene expression patterns
Zhu et al. (2012)	Ms	LCM	N	qRT-PCR	N	Transcript differences in core and shell at time points in and out of phase of light reset
Boone et al. (2013)	Rt	LCM	Y	qRT-PCR, μ-array	LCM image	TBI model shows altered gene expression patterns in SCH and hippocampus

(continued)

Table 6.1 (continued)

Study	Animal	Extraction	Target(s) a priori?	Screen [S], Validation [V]	Map or schematic	Major findings
Pembroke et al. (2015)	Ms	LCM	N	RNA-Seq [S]; ISH [V]	N	Identified 146 genes highly enriched in the SCH; four of these were confirmed using ISH; also identified twin-peaking genes in the SCH and novel transcripts with circadian profiles
Park et al. (2016)	Ms	LCM	Y	qRT-PCR	N	Identified transcriptional changes in dark-adapted mice and those dark-adapted and then exposed to a brief light pulse; identified distinct expression profiles across groups, but no specific spatial organization of expression patterns
SFO: Subfornical organ						
Hindmarch et al. (2008) (also see Hindmarch and Ferguson, 2016)	Rt	dissection	N	μ-array	N	Found 46 genes with altered expression in association with dehydration, including BDNF, calcium-sensing receptors, and apelin receptors
Walch et al. (2014)	Rt	LCM	N	qRT-PCR	N	Detected AT1aR expression in SFO that was markedly reduced by virally mediated RNA interference
SO: supraoptic hypothalamic nucleus						
Ghorbel et al. (2003)	Rt	dissection	N	μ-array [S]; IHC, ISH, WB [V]	N	Identified nine candidate genes, four of which were up-regulated by dehydration (including interleukin-6) and five were down-regulated
Mutsuga et al. (2004)	Rt	LCM	N	μ-array [S] ISH [V]	N	Found 1,385 genes expressed in SO at levels two times greater than in the hypothalamus as a whole
Hindmarch et al. (2006)	Rt	Dissection	N	μ-array	N	Found mRNAs regulated by dehydration, enriched in PVH and SO
Yue et al. (2006)	Rt	LCM	Y	μ-array	N	40 mRNAs greater in hypo-osmotic vs. normo-osmotic conditions

Gouraud et al. (2007)	Rt	Dissection	N	µ-array [S]; RT-PCR [V]	N	Confirmed up-regulation of 14-3-3 family of proteins in dehydrated SO and also identified a novel 14-3-3 binding partner protein
Hindmarch et al. (2007)	Rt	NS	N	µ-array	N	mRNA expression differences between strains for the neurohypophysial system
Qiu et al. (2011)	Rt	Dissection	N	µ-array [S] ISH [V]	N	Found 567 genes commonly regulated by dehydration in the male and by lactation and euhydration in the female.
Stewart et al. (2011)	Ms	LCM	N	µ-array [S] ISH [V]	N	Identified 69 genes that have altered gene expression under conditions of dehydration in mice (and in rats compared from a previous data set); four of these genes were validated by ISH and were found to be up-regulated as a result of dehydration
Nedungadi et al. (2012b)	Rt	LCM	Y	qRT-PCR	N, but Bregma-based ranges specified	TRPV2 mRNA detected
Humerick et al. (2013)	Rt	LCM	Y	qRT-PCR	N	Transcription factors differentially expressed in OT and VP neurons
Nedungadi and Cunningham (2014)	Rt	LCM	Y	qRT-PCR	N, but atlas levels specified	Found TRPC4 channel expression, and its up-regulation in association with hepatic cirrhosis
Qiu et al. (2014)	Rt	Dissection	N	µ-array [S] EMSA, ELISA, qPCR [V]	N	Found changes in binding for 26 consensus elements in dehydrated relative to control rats
Greenwood et al. (2015)	Rt	µ-punch	N	µ-array [S] qPCR [V]	N	Compared salt loading vs water deprivation on transcript expression in SO; identified and validated five new genes and confirmed nine others
Johnson et al. (2015)	Rt	LCM	N	RNA-Seq, µ-array [S]; IHC, qPCR [V]	N	Detected 9,709 genes by RNA-Seq, 552 of which altered their expression in SO as a result of salt-loading

(continued)

Table 6.1 (continued)

Study	Animal	Extraction	Target(s) a priori?	Screen [S], Validation [V]	Map or schematic	Major findings
VMH: ventromedial hypothalamic nucleus						
Segal et al. (2005)	Ms	LCM	N	qRT-PCR	N	Four of twelve mRNAs reduced in steroidogenic factor 1 knockouts
Xiao et al. (2005)	Rt	μ-punch	N	μ-array [S] RT-PCR [V]	N	In ARH tissue punches which also contained VMH, the authors found 12 genes differentially regulated during lactation.
Kurrasch et al. (2007)	Ms	Dissection	N	μ-array [S]; qRT-PCR, ISH [V]	N	Identified approx. 200 mRNAs enriched in neonatal VMH, including several transcriptional regulators
Kim et al. (2012)	Ms	NS	N	μ-array	N	Found several differentially expressed genes in SF-1-specific FOXO deletion mice relative to wild-type mice
Trivedi et al. (2015)	Rt	LCM	N	μ-array [S]; qPCR [V]	N	Identified tachykinin-1 as a gene down-regulated by ghrelin

5-HT1DR serotonin (5-HT) 1d receptor, *5-HT2CR* serotonin (5-HT) 2c receptor, *μ-array* microarray, *μ-punch* micropunch, *AT1aR* angiotensin 1a receptor, *BDNF* brain-derived neurotrophic factor, *Bn* bunting, *CART* cocaine- and amphetamine-related transcript, *Ck* chicken, *Cw* cow, *CZE* capillary zone electrophoresis, *Dg* dog, *Drop-Seq* droplet encapsulated single-cell transcriptional profiling, *eGFP* enhanced green fluorescent protein, *ELISA* enzyme-linked immunosorbent assay, *EMSA* electrophoretic mobility shift assay, *ESTs* expressed sequence tags, *FACS* fluorescence activated cell sorting, *FISH* fluorescence in situ hybridization, *Fs* fish, *GABA* gamma-amino butyric acid, *Glu* glutamate, *Gs* goose, *Hm* hamster, *H/O* hypocretin/orexin, *Hu* human, *ICC* immunocytochemistry, *IHC* immunohistochemistry, *ISH* in situ hybridization, *LCM* laser-capture microdissection, *MC4R* melanocortin 4 receptor, *MCH* melanin concentrating hormone, *miRNA* microRNA, *MiSeq* next-generation sequencing, *Mk* monkey, *Ms* mouse, *MSI* mass spectrometric imaging, *NPY* neuropeptide Y, *NS* not stated, *NTS* nucleus of the solitary tract, *OT* oxytocin, *Pg* pig, *POMC* pro-opiomelanocortin, *qRT-PCR* quantitative real-time polymerase chain reaction, *RNA-Seq* next-generation RNA sequencing, *Rt* rat, *SAGE* serial analysis of gene expression, *Sh* sheep, *Sp* sparrow, *TBI* traumatic brain injury, *TRAP* translating ribosome affinity purification, *TRPV2* transient receptor potential cation channel subfamily V member 2, *TruSeq* next-generation sequencing, *VP* vasopressin, *vTRAP* viral translating ribosome affinity purification, *WAT* white adipose tissue, *WB* western blotting

Table 6.2 Selected proteomic studies in whole hypothalamus and by hypothalamic sub-region

Study	Animal	Extraction	Target(s) a priori?	Screen [S], Validation [V]	Map or Schematic	Major findings
Whole hypothalamus						
Sung et al. (2004)	Rt	Dissection	N	MALDI-TOF MS	N	36 proteins expressed in a neuropathic pain model relative to controls
Kuo et al. (2005)	Ck	Dissection	N	2D-GE [S] LC-MS/MS, qRT-PCR [V]	N	6 proteins associated with high egg production
Roth et al. (2006)	Rt	Dissection	N	2D-LC-MS/MS & clCAT	N	Found differential expression of five proteins involved in glutamate metabolism in juvenile versus peri-pubertal females
Skynner et al. (2006)	Ms	Dissection	N	MALDI-TOF MS	N	Chronic corticosterone altered markers of glycolysis, gluconeogenesis and nitrogen metabolism
Kuhla et al. (2007)	Cw	Dissection	N	MALDI-TOF MS [S]	N	Found nine proteins differentially expressed in ad libitum fed vs. energy restricted cows
Ropp et al. (2008)	Ms	Dissection	N	SELDI-TOF	N	Distinct protein profiles following sub-acute pyridostigmine treatment
Sarkar et al. (2008)	Ms	Dissection	N	2D-GE, WB, MALDI-TOF-MS	N	Found seven proteins that were significantly different in hypothalamus in control versus microgravity-treated animals
Lee et al. (2009)	Rt	Dissection	N	2D-GE; MALDI-TOF/MS	N	Found several proteins up-regulated following lithium treatment
Mishra et al. (2009)	Rt	Dissection	N	2D-GE, LC-MS/MS [S]; WB [V]	N	Light-dark shifts in circadian cycle resulted in increased food intake, body weight gain, retroperitoneal fat mass, and expression levels of 4 of 5 hypothalamic 2D-GE spots; these were identified by MS and included glycolytic and citric acid cycle enzymes.

(continued)

Table 6.2 (continued)

Study	Animal	Extraction	Target(s) a priori?	Screen [S], Validation [V]	Map or Schematic	Major findings
Kim et al. (2010)	Rt	Dissection	N	2D-GE, MALDI-TOF MS	N	Maternal separation was associated with down-regulation of 14 proteins from hypothalamus; maternal separation with acupuncture was associated with five down- and nine up-regulated proteins relative to maternal separation alone.
Popesku et al. (2010)	Fs	Dissection	N	iTRAQ labeling & MS	N	42 proteins differentially regulated by treatment with DA receptor agonists
Argüelles et al. (2011)	Rt	NS	N	MALDI-TOF/TOF-MS	N	Oxidative stress could be involved in the alterations of eEF-2 and several other proteins
Wang et al. (2011)	Rt	Dissection	N	MALDI-TOF/TOF-MS	N	Ubiquitin was significantly decreased in diet-resistant rats but not changed in diet-induced obese rats
Alexandre-Gouabau et al. (2012)	Rt	Dissection	N	LC-MS/MS	N	Protein restriction in utero alters numerous pathways
Gasperini et al. (2012)	Rt	Dissection	N	2D-GE, MALDI-TOF MS/MS	N	Found 26 of 28 protein spots on 2D gels for hypothalamus show significant expression after i.c.v. PACAP; including cytoskeletal, signaling and synaptic proteins
Guest et al. (2012)	Rt	Dissection	N	LC-MS	N	Identified hypothalamic proteins that differ in expression in rats subjected to a low-protein diet as compared to wild-type controls
Pedroso et al. (2012)	Rt	Dissection	N	MALDI-TOF MS [S]; 2D-GE, WB [V]	N	Identified 86 hypothalamic proteins in Wistar rats
Stelzhammer et al. (2012)	Rt	Dissection	N	LC-MS/MS [S]	N	Found 21,455 peptides that corresponded to 622 unique proteins
Zhang et al. (2012)	Ms	Dissection	N	LC-FT-MS/MS, SIEVE™ software-based analysis, spectral analysis	N	Identified 367 peptides from neuropeptide precursors
Ihnatko et al. (2013)	Ms	Dissection	N	LC-MS/MS	N	Differential up- and down-regulation of proteins in tumor-bearing mice and caloric-restricted pair fed mice

Iqbal et al. (2013)	Rt	Dissection	N	HPLC/ESI-ion trap; HPLC/ESI-Q-TOF MS	N	Identified 198 proteins, 78 of which were common to both sets of methods; 58 unique proteins identified by Q-TOF and 62 by HPLC/ESI-ion trap.
Kefaloyianni et al. (2013)	Rt	Dissection	N	LC-MS^E and LC-MS/MS	N	KATP channels in different tissues assemble with proteins having common functions
Taraslia et al. (2013)	Ms	Dissection	N	2D-GE, MALDI-TOF MS [S]	N	515 different single-gene products were identified, eight of which were unique to hypothalamus
Iqbal et al. (2014a)	Rt	Dissection	N	HPLC/ESI-TOF & HPLC-Q-TOF	N	35 and 97 significantly differentially expressed proteins by two analyses in simulated microgravity model
Iqbal et al. (2014b)	Rt	Dissection	N	HPLC/ESI-TOF & HPLC-Q-TOF	N	Differential expression of 17 specific cellular defense proteins in simulated microgravity model
Kim et al. (2014)	Rt	Dissection	N	LC-ESI-MS/MS [S]; WB, IHC [V]	N	Following chronic partial sleep deprivation in rats for 7 d, 89 and 50 proteins were up- and down-regulated, respectively
Chao et al. (2015)	Rt	Dissection	N	2D-GE, LC-MS/MS [S]; WB [V]	N	Found a few proteins induced by heatstroke that had their levels normalized by cooling
Zhong et al. (2015)	Ms	Dissection	N	ESI-LC-MS/MS	N	Found 31 overexpressed proteins in wild-type group compared to *EPHX2* KO group
Manousopoulou et al. (2016)	Ms	Dissection	N	ESI-LC-MS/MS [S]; qPCR [V]	N	Quantitative profiling yielded 9,249 protein groups, with 7,718 groups profiled with a minimum of two unique peptides each; high-fat diet or lipopolysaccharide challenge produced unique proteomic profiles
Azzam et al. (2017)	Ms	Dissection	N	RPLC-MS/MS [S] SRM MS [V]	N	Found 39 proteins showing differences in expression in mouse models of narcolepsy
Pedroso et al. (2017)	Rt	Dissection	N	Q-TOF MS [S]; 2D-GE [V]	N	1,356 proteins were identified and 348 were quantified, along with 127 metabolites. Intrauterine growth restriction resulted in down-regulation of 36 proteins and 5 metabolites, and up-regulation of 21 proteins and 9 metabolites in the hypothalamus.

(continued)

Table 6.2 (continued)

Study	Animal	Extraction	Target(s) a priori?	Screen [S], Validation [V]	Map or Schematic	Major findings
Udvari et al. (2017)	Rt	Dissection	N	2D-DIGE [S] LC/MS-MS; WB; ISH; IHC; EM [V]	Y	Identified 26 proteins Isolated synaptosome fractions from maternal rats, 7 of which up-regulated and 19 were down-regulated. Identified a complement cascade protein by WB, ISH, IHC and EM to be present within ARH and VMH.
Zettergren et al. (2017)	Ms	Dissection	N	MS and MS/MS; iTRAQ [S]; PRM [V]	N	Identified 2,998 proteins in hypothalamus and amygdala of neonatal male, female and androgen receptor knockout male mice; of which 173 proteins were expressed differentially in males and females. Verified expression of seven genes using targeted proteomics.
Cao et al. (2018)	Gs	Dissection	N	iTRAQ; LC-MS/MS [S]; qRT-PCR; WB [V]	N	Found 18 proteins up-regulated and 16 down-regulated in association with conditions of periods before and during egg laying
Firmino et al. (2018)	Rt	NS	N	LC-MS/MS	N	Found 7,021 proteins, many of which exhibited changes in relative abundance in immune-activated rats relative to controls
Govindaraj et al. (2018)	Rt	Dissection	N	2D-GE; MALDI-TOF/TOF MS [S]; semi-quant RT-PCR; WB [V]	N	Found 21 protein spots differentially expressed in preoptic, whole hypothalamic, hippocampal and pituitary tissues of females exposed neonatally to estradiol
Nobis et al. (2018)	Ms	Dissection	N	2D-GE; HPLC; LC/Q-TOF [S]; LC-ESI-MS/MS; WB [V]	N	Identified 22 proteins that dismayed alterations in levels in the hypothalamus among three groups: activity-based anorexic, limited-food access, and ad libitum-fed
Zhang et al. (2018)	Ck	Dissection	N	iTRAQ; LC-MS/MS	N	Found 235 differentially expressed proteins between L-arginine-fed and control subjects
ARH: arcuate hypothalamic nucleus						
Amigó-Correig et al. (2012)	Ms	Dissection	N	MALDI-TOF/TOF	N	Adult lean and high fat diet-induced obese mice orally treated with sodium tungstate had modified levels of proteins involved in cell morphology, axonal growth and tissue remodeling

Preoptic area						
Govindaraj et al. (2018)	Rt	Dissection	N	2D-GE; MALDI-TOF/TOF MS [S]; semi-quant RT-PCR; WB [V]	N	Found 21 protein spots differentially expressed in preoptic, whole hypothalamic, hippocampal and pituitary tissues of females exposed neonatally to estradiol
PVH: paraventricular hypothalamic nucleus						
Romanov et al. (2014)	Rt	Dissection	Y	Illumina HiSeq2000 sequencer, MALDI-TOF	Y	Profiled secretagogin neurons as a distinct CRH-releasing neuron population
SCH: suprachiasmatic hypothalamic nucleus						
Deery et al. (2009)	Ms	NS	N	2D-DIGE & MS	N	13% of soluble proteins were found to be subject to circadian regulation
SO: supraoptic hypothalamic nucleus						
Gouraud et al. (2007)	Rt	Dissection	N	2D-GE, MALDI-TOF MS [S]; WB, IHC [V]	N	Identified 14-3-3 proteins that are up-regulated as a consequence of chronic dehydration
VLPO: ventrolateral preoptic nucleus						
Dooley et al. (2010)	Rt	Dissection	N	MALDI-TOF/TOF MS	N	Identified diaminochlorotriazine (DACT) protein adducts formed in Atrazine-exposed rats
VMH: ventromedial hypothalamic nucleus						
Mo et al. (2006)	Rt	µ-punch	N	2D-GE [S] LC-ESI-MS/MS [V]	N[1]	Identified 99 unique proteins based on data from 2D-GE experiments, which comprise a "primary proteome database" for the VMH
Mo et al. (2008)	Rt	µ-punch	N	RPLC-nanoESI-MS/MS	N[a]	Up-regulation of 29 identified proteins with estradiol treatment

2D-DIGE two-dimensional difference gel electrophoresis, *2D-GE* two-dimensional gel electrophoresis, *2D-LC* two-dimensional liquid chromatography, *µ-punch* micropunch, *cICAT* cleavable isotope-coded affinity tags, *Ck* chicken, *CRH* corticotropin-releasing hormone, *Cw* cow, *EM* electron microscopy, *ESI* electrospray ionization, *Fs* fish, *FT* Fourier transformation, *FTICR* Fourier-transform ion cyclotron resonance, *Gs* goose, *HCD and ETD-based MS/MS* high-energy collisional dissociation and electron-transfer dissociation-based tandem mass spectrometry, *IHC* immunohistochemistry, *iTRAQ* isobaric tag for relative and absolute quantitation, *KO* knockout, *LC* liquid chromatography, *LC-MS[E]* liquid chromatography-label-free mass spectrometry, *LTQ* linear trap quadrupole, *MALDI-TOF* matrix-assisted laser desorption/ionization-time of flight, *Ms* mouse, *MS/MS* tandem mass spectrometry, *MSPD* matrix solid-phase dispersion, *nanoESI* nano-electrospray ionization, *NS* not stated, *PACAP* pituitary adenylate cyclase-activating polypeptide, *PRM* parallel reaction monitoring, *qRT-PCR* quantitative real-time polymerase chain reaction, *Q-TOF* quadrupole time of flight, *RPLC* reversed-phase liquid chromatography, *SELDI-TOF* surface-enhanced laser desorption/ionization-time of flight, *semi-quant* semi-quantitative, *SRM* selected reaction monitoring, *Triple Quad MS* triple quadrupole mass spectrometry, *Rt* rat, *WB* western blotting

[a]The authors reference specific atlas plates and locations for their micropunches

Table 6.3 Selected peptidomic studies in whole hypothalamus and by hypothalamic sub-region*

Study	Animal	Extraction	Target(s) a priori?	Screen [S], Validation [V]	Map or schematic	Major findings
Whole hypothalamus						
Bures et al. (2001)	Ms	Dissection	N	LC-MS [S]; LC-MS/MS [S]	N	Identified 27 peptides derived from known neuropeptides as well as 25 additional peptides not known to be in the neuropeptide processing pathway; all up-regulated in carboxypeptidase E mutant mice
Svensson et al. (2003)	Rt, Ms	Dissection	N	nanoLC-ESI-Q-TOF-MS	N	Detected 550 endogenous peptides
Che et al. (2005)	Ms	Dissection	N	LC-MS/MS	N	Knockdown of carboxypeptidase E activity in two paradigms for decreasing body mass show different peptide profiles
Décaillot et al. (2006)	Ms	Dissection	N	Isotopic labeling & Nano-LC MS/MS	N	Detection 27 distinct peptides from hypothalamus and striatum in *Cpe*$^{fat/fat}$ mice, with some showing changes in levels in mice chronically treated with morphine
Pan et al. (2006)	Ms	Dissection	N	Isotopic labeling & LC-ESI-MS	N	Approx. one-third of the peptides found in wild-type mice were not found in prohormone convertase KO mice
Che et al. (2007)	Ms	Dissection	N	Nano-LC MS/MS	N	Identified 95 peptides from samples, 64 of which were neuropeptides or other peptides derived from proteins in the secretory pathway; found OT to be preferentially abundant in hot-acid extracts over hot-water extracts
Sköld et al. (2007)	Ms	Dissection	N	Nano-LC MS [S]; Q-TOF LTQ MS/MS [V]	N	Identified 23 neuropeptides, hormones and potentially biologically active peptides; all were primarily up-regulated in control mouse brain relative to brains processed with longer *post mortem* times
Mihailova et al. (2008)	Rt	Dissection	N	Capillary 2-D LC/MS	N	Identified 107 peptides, 26 of which displayed differences in concentration under hypoxic stress conditions

Cai et al. (2011)	Pg	Dissection	N	MSPD & Nano-LC MS/MS	N	14 potential endogenous peptides were identified using MSPD extracts versus to peptides using acid extracts
Colgrave et al. (2011)	Cw	Dissection	N	LC-MS/MS	N	Used thermal stabilization methods to refine the yield of neuropeptides isolated from hypothalamus
Nilsson et al. (2012)	Ms	Dissection	N	nano-LC-ESI-LTQ MS/MS or nano-LC-ESI-LTQ-FTICR-MS/MS	N	14 peptides were significantly regulated by imipramine treatment
Zhang et al. (2012)	Ms	Dissection	N	LC-FT-MS/MS	N	Identified 367 peptides from neuropeptide precursors from hypothalamic samples.
Fouillen et al. (2013)	Sw	Dissection	N	LC-FT-MS/MS, SIEVE™ software-based analysis,	N	12 hypothalamic peptides were up-regulated following prolonged general anesthesia
Frese et al. (2013)	Rt	Dissection	N	HCD and ETD-based MS/MS [S]; LC-MS [V]	N	Identified 1,292 unique peptides from hypothalamus in rats fed on a regular diet, HFHS diet, restricted chow diet, or chocolate diet. HFHS diet produced the greatest increases in peptides as determined by label-free quantification.
Gao et al. (2013)	Rt	Dissection	N	2D-GE, MALDI-TOF MS [S]; qRT-PCR, WB [V]	N	Identified 17 hypothalamic proteins with twofold or greater expression after electroacupuncture intervention of sciatic pain
Nakazawa et al. (2013)	Rt	Dissection	N	RP HPLC; Nano-LC-MS/MS [S]; WB [V]	N	Found hundreds of peptides in relaxin- and saline-treated rats; two of which exhibited signatures both in microarray experiments and peptidomic experiments: OT and CART; with OT markedly up-regulated after relaxin exposure
Schmidlin et al. (2015)	Rt	Dissection	N	LC-MS/MS [S]; SRM/Triple-Quad MS [V]	N	Demonstrated the feasibility of using SRM to evaluate a priori selected transitions of key neuropeptide fragments from the hypothalamus
Secher et al. (2016)	Rt	Dissection	N	LC-MS/MS [S]	N	Identified 14,416 peptides in 786 protein families; sorted these by LPVs to isolate 2,835 peptides derived from 356 prohormone precursors; of these, 105 LPVs were not previously described

(continued)

Table 6.3 (continued)

Study	Animal	Extraction	Target(s) a priori?	Screen [S], Validation [V]	Map or schematic	Major findings
Yang et al. (2017)	Rt	Dissection	N	nanoESI; nanoLC-MS/MS; MRM	N	Evaluated the efficacy of a rapid conductive sample heating system in stabilizing proteins from whole hypothalamic extracts
DeAtley et al. (2018)	Cw	Dissection	N	LC-MS/MS; MRM	N	Observed 143 peptides in hypothalamus of pre- and post-pubertal heifers that were assigned neuropeptide status; three of which differed between the conditions
SCH: Suprachiasmatic hypothalamic nucleus						
Hatcher et al. (2008)	Rt	Acute tissue slice prep; μ-punch	N	HPLC, LC, SPE beads [S]; MALDI-TOF MS/MS ; LTQ FTMS [V]	N	Identified peptides released from acute slice preparations containing SCH, including after electrical stimulation of the retinohypothalamic input to the SCH; found peptide content in releasates to be stimulation-specific
Lee et al. (2010)	Rt	μ-punch	N	LC-FTMS/MS	N	list of 102 endogenous peptides, including 33 that were previously unidentified; also identified novel post-translational modifications
Lee et al. (2013)	Rt	μ-punch	N	LC-FTMS/MS, SIEVE™ software-based analysis	N	list of 190 endogenous peptides from 310 identified
Chiang et al. (2014)	Ms	Dissection	N	fractionation, LC MS/MS [S]; RT-PCR, WB, IHC [V]	N	quantified 2,112 proteins, 20% of which exhibited a time-of-day-dependent profile; found 48 proteins exhibiting circadian rhythms of expression from this time-of-day proteome
Southey et al. (2014)	Rt	μ-punch	N	Spectral count, spectra index, SIEVE™ software-based analysis	N	differential peptide abundances between day and night conditions
Yang et al. (2017)	Rt	Dissection	N	nanoESI; nanoLC-MS/MS; MRM	N	Evaluated the efficacy of a rapid conductive sample heating system in stabilizing proteins from SCH extracts

Study	Animal	Extraction	Target(s) a priori?	Screen [S], Validation [V]	Map or schematic	Major findings
SO: Supraoptic hypothalamic nucleus						
Bora et al. (2008) [also see: Perkel (2008)]	Rt	μ-punch	Y	LC/MS & tandem mass spectrometry	N	20 unique peptides identified

2D-GE two-dimensional gel electrophoresis, *μ-punch* micropunch, *CART* cocaine- and amphetamine-related transcript, *Cw* cow, *ESI* electrospray ionization, *Fs* fish, *FT* Fourier transformation, *FTICR* Fourier-transform ion cyclotron resonance, *HCD and ETD-based MS/MS* high-energy collisional dissociation and electron-transfer dissociation-based tandem mass spectrometry, *HFHS* high fat and high sucrose, *Hu* human, *KO* knockout, *LC* liquid chromatography, *LPVs* longest peptide variants, *LTQ* linear trap quadrupole, *MALDI-TOF* matrix-assisted laser desorption/ionization-time of flight, *MRM* multiple reaction monitoring, *Ms* mouse, *MS/MS* tandem mass spectrometry, *MSPD* matrix solid-phase dispersion, *nanoESI* nanoscale electrospray ionization, *nanoLC-MS/MS* nanoscale liquid chromatography coupled to tandem mass spectrometry, *OT* oxytocin, *Pg* pig, *qRT-PCR* quantitative real-time polymerase chain reaction, *Q-TOF* quadrupole time of flight, *SRM* selected reaction monitoring, *Triple Quad MS* triple quadrupole mass spectrometry, *Rt* rat, *SPE* solid-phase extraction, *Sw* shrew

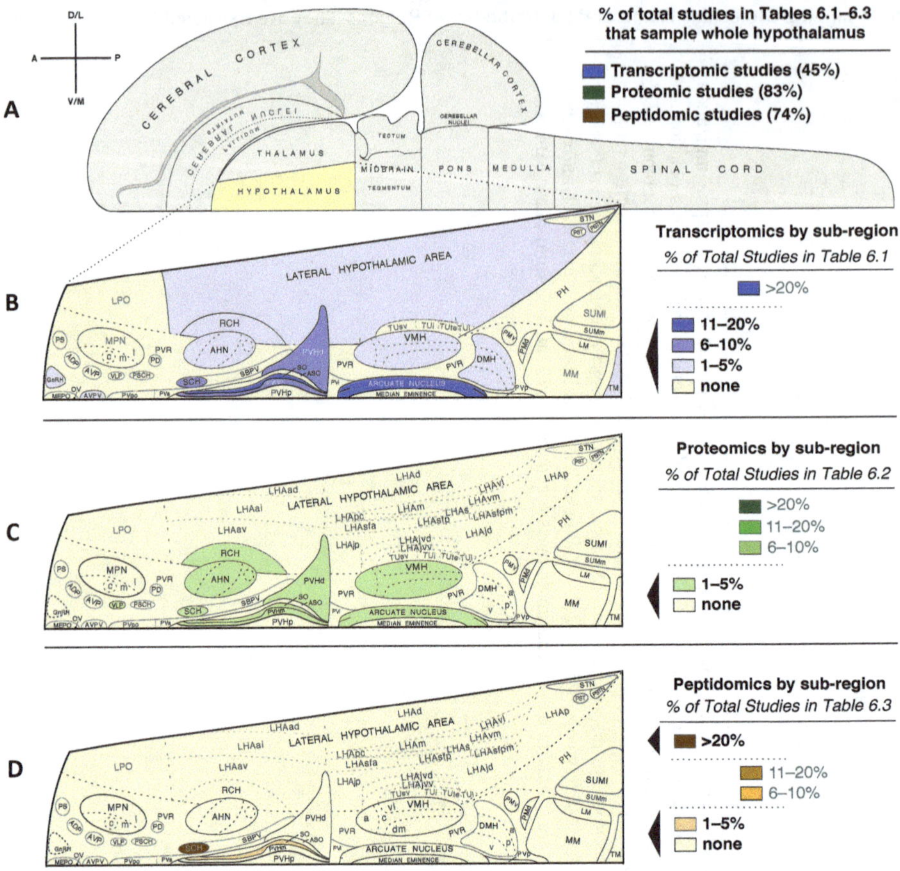

Fig. 6.3 Survey of coverage for the hypothalamus or its various regions by published transcriptomic, proteomic and peptidomic studies listed in Tables 6.1, 6.2, and 6.3. (**A**) Choropleth flatmap of the rat central nervous system (CNS), modified from Swanson (2004), illustrating the various major CNS subdivisions, including the hypothalamus. Note the *legend* at the upper left, which indicates directions of orientation (A, anterior; P, posterior; D/L, dorsal/lateral; V/M, ventral/medial). The *chart* to the right of the flatmap in (**A**) lists the % of total studies reported in each of the tables in this review (Tables 6.1, 6.2, and 6.3) that conducted transcriptomic, proteomic, or peptidomic studies of the *whole* hypothalamus, respectively. (**B–D**) A breakdown, by *hypothalamic region*, of the percentage of studies in which a particular region was sampled for analysis, with choropleth flatmaps in (**B**) showing all hypothalamic regions analyzed by transcriptomic analyses, in (**C**) showing all proteomic studies, and in (**D**) all peptidomic studies. The percentage ranges are coded by colors and reflect percentages of the total number of studies listed in each table. Note that although rat brain flatmaps are used here, the studies are across many different taxonomic groups, including fish, chicken, goose, cow, pig, sheep, shrew, mouse, rat, guinea pig, hamster, dog and human. Therefore, the maps are meant to be convenient vehicles to convey a sense of the amount of coverage in the literature for any particular region, differences in their neuroanatomy or cytoarchitectural boundaries notwithstanding. Note also that for many shaded regions, the smaller abbreviations have been removed for sub-regions, to emphasize that studies did not sample at that level of resolution. Thus, the lateral hypothalamic area (LHA) may have been sampled, but the LHAjvv was not. Conversely, although large areas are shaded, in certain cases only a few cell types were specifically mined from the region rather than the region sampled as a whole, but this is not reflected in the diagrams. For explanation of all abbreviations, please see abbreviations list. The flatmaps from Swanson (2004) (and available at https://larrywswanson.com) are reproduced here under the conditions of a Creative Commons BY-NC 4.0 license (https://creativecommons.org/licenses/by-nc/4.0/legalcode)

These include environmental agents (Li et al. 2014), pharmacological interventions (Jin et al. 2016), ontogenic state (e.g., see introductory remarks in Balivada et al. 2017), or physiological processes. Miller et al. (2014) examined various hypothalamic sub-regions within the context of a hemispheric tissue analysis in prenatal human brain using high throughput transcriptomic methods. Zapala et al. (2005) contextualized regional specificity with embryonic development, taking care to provide supplementary information that includes photographic documentation of the tissue they dissected for their hypothalamic sample. In contrast, it is disappointing that in their "in-depth analysis of the mouse brain and its major regions and cell types" for the proteome, K. Sharma et al. (2015) neglected to sample the hypothalamus in what is otherwise a detailed and interesting study.

6.4.2.2 Molecular Extraction from Whole Hypothalamus

In non-mammalian vertebrates, the hypothalamus has been studied for transcriptomics, proteomics, and peptidomics in fishes and birds; in some cases, in the context of animal husbandry. For example, hypothalamic and pituitary molecules associated with high egg production in chickens have been analyzed at the transcriptomic (Shiue et al. 2006; Chen et al. 2007a; Table 6.1) and proteomic (Kuo et al. 2005) levels. Egg-laying traits have also been compared alongside transcripts identified to be associated with high egg production (Chen et al. 2007b). The hypothalamic transcriptome and proteome of the Huoyan goose (Luan et al. 2014; Cao et al. 2018) and the hypothalamic transcriptome of the Sichuan white goose (G. Gao et al. 2015) have been profiled before, during, or after their egg laying periods in the interests of finding clues to improve the reproductive performance of these economically valuable domestic animals (also see Fig. 1 of Li et al. 2011). In the interests of optimizing feed intake in chickens or to understand how they cope with environmentally-induced pressures, many studies have also examined the role of body composition, fasting, diet, or heat stress on gene expression in chicken hypothalamus (e.g., Byerly et al. 2010; Fang et al. 2014; Sun et al. 2015; Tu et al. 2016; see also Kuenzel et al. 1999). Despite the intensive investigations of chicken hypothalamus for molecular mining and extraction, these studies have not contextualized sub-regional changes in expression for molecules in relation to published stereotaxic atlases of the chicken that include illustrations, maps and drawings of the hypothalamus with stereotaxic coordinates (van Tienhoven and Juhász 1962; Yasuda and Lepkovsky 1969; Feldman et al. 1973). Seasonal changes in hypothalamic gene expression have also been documented in the black-headed bunting, a migratory songbird (A. Trivedi et al. 2014; A. Sharma et al. 2018).

In mammals, whole hypothalamus has been mined for gene transcripts in mouse, rat, hamster, guinea pig, shrew, pig, cow, sheep, dog and human (Table 6.1). Recently, human induced pluripotent stem cells differentiated into "hypothalamic-like" neurons have also been profiled for their transcriptomes (Rajamani et al. 2018). The first large-scale in situ hybridization-based study of hypothalamus-enriched transcripts was provided by Gautvik et al. (1996) in the rat by using directional tag PCR subtraction, which led to the discovery of the hypocretin neuropeptides (de Lecea et al. 1998; also see Sutcliffe, 2001; Sutcliffe and de Lecea, 2002). Friedman and colleagues utilized a novel molecular technique that extends the principles underlying an earlier approach (Heiman et al. 2008), to isolate and extract activated transcriptional systems in the hypothalamus under conditions of salt-loading, fasting, light exposure, or various other stimulus paradigms (Knight et al. 2012). Specifically, they immunoprecipitated the phosphorylated form of the ribosomal protein, S6, to isolate and enrich mRNAs that are actively being translated (i.e., in transcriptionally activated neurons) in mouse hypothalamic samples. Using TaqMan® technology (Holland et al. 1991), RNA-Seq and microarrays, they isolated several mRNAs, many of which displayed expression in pS6-immunoreactive neurons in various sub-regions of the hypothalamus.

Using Drop-Seq, a method that allows for single-cell transcriptomics to be performed in a manner that preserves the cell provenance of the RNA that is extracted (Macosko et al. 2015), Chen et al. (2017) reported single-cell RNA sequencing results from the adult mouse hypothalamus. They used clustering analysis to identify 11 non-neuronal (including oligodendrocytes, astrocytes, ependymocytes, tanycytes, microglia, and macrophages) and 34 neuronal cell types (including 15 glutamatergic and 18 GABAergic clusters, and one histaminergic neuron cluster) from tissue dissociated from manually dissected hypothalamus, and confirmed some of their key findings by performing immunohistochemistry for neuropeptides or comparing their results with those found in the publicly available Allen Brain Atlas. Importantly, their workflow revealed the spatially restricted expression of novel molecules in the hypothalamus, including retinoic acid binding protein (*Crabp1*) in the ARH. They also found restricted expression of the neurodevelopmental factor, *Pax6*, in the zona incerta, which the authors assign as a hypothalamic structure but which is considered as a thalamic structure by others (e.g., see Swanson 2018). Importantly, their datasets indicate that all hypothalamic peptidergic neurons can also be classified by the small neurotransmitter they synthesize (glutamate or GABA). Recently, Romanov et al. (2017) provided evidence of numerous novel neuronal phenotypes of hypothalamic cells using single cell RNA-Seq and DropSeq technologies, but the only provenance that could be attributed to these cells was from within the large heterogeneous group of hypothalamic sub-regions partially sampled within their microdissected tissue sample, which include large portions of the medial, but not lateral hypothalamus. In contrast, Yelin-Bekerman et al. (2015) sampled from the whole hypothalamus of zebrafish to identify transcripts specific to neurons—isolated by fluorescence-activated cell sorting (FACS)—that expressed the neuropeptide hypocretin/orexin (H/O); these neurons are typically enriched in the lateral hypothalamus in most species (Table 6.1).

Very few studies have examined proteomic or peptidomic profiles of whole hypothalamic samples. Extending the protocol they developed for peptidomic analysis of small microdissected brain regions such as the motor cortex, thalamus and striatum (Sköld et al. 2002), the Andrén laboratory reported identifying novel peptides from hypothalamic extracts (Svensson et al. 2003; Sköld et al. 2007). Fälth et al. (2006) developed a database for endogenous peptides identified by mass spectrometry, into which they have incorporated their hypothalamic datasets. Nakazawa et al. (2013) took the rather novel approach of performing both transcriptomics and peptidomics on separate sets of whole hypothalamic extracts (a "cross-omics" approach), and reported consensus results from both methods for oxytocin up-regulation in association with intracerebroventricular relaxin administration in rats. Recently, "cross-omics" approaches have been extended to combined transcriptomics/lipidomics of hypothalamus (Lerner et al. 2018).

6.4.2.3 Molecular Extraction from the Hypothalamic Circadian System

The suprachiasmatic hypothalamic nucleus (SCH), a well-defined compact nucleus within the hypothalamus that is amenable to precise sampling or molecular studies (e.g., see Fig. 1 of Porterfield et al. 2007; Fig. 1 of Boone et al. 2013), is the primary neural substrate for the master circadian clock in the body, which receives signals that allow organisms to respond to shifts in light during the day-night cycle. Often, circadian rhythms are characterized by changes in gene expression within the SCH; studies using microarray analysis demonstrated, for example, that approximately 650 transcripts undergo cyclic changes in expression in the SCH and the liver of mice, with many of these specific to the SCH (Panda et al. 2002). After certain stimuli, immediate early genes in the SCH peak and return to baseline, while a few others maintain their expression levels to protect the nuclei from excitotoxicity (Porterfield et al. 2007; Porterfield and Mintz 2009). Similar to contrasts between light and dark cycles, the transcriptome of the SCH is also distinct during wake and sleep cycles (Winrow et al. 2009), and there is recent transcriptomic evidence that certain classes of genes in the SCH peak twice in their expression levels across the circadian cycle (Pembroke et al. 2015). Single-cell transcriptomic analyses of mouse SCH neurons isolated by LCM have also revealed novel transcripts expressed in correlation with phase shifts in the circadian cycle (J. Park et al. 2016).

During shifts in circadian time, gene expression is not the only mechanism affected, but protein levels as well. Certain studies have examined proteomic changes in the whole hypothalamus after experimental disruptions in circadian rhythms (Mishra et al. 2009). Moreover, analysis of the proteome has revealed that 13% of soluble proteins expressed in the SCH undergo circadian regulation (Deery et al. 2009), and that a "time of day proteome" exists in this structure, with several proteins exhibiting marked fluctuations specifically during the transitions from light to dark and vice versa (Chiang et al. 2014). Interestingly, the SCH has become something of a model system for peptidomic studies, in that most of the peptidomic studies to date for a hypothalamic sub-region have been focused mainly on this structure (Fig. 6.3D). Peptidomic studies have revealed differential peptide abundances that correlate with changes in the time of day, including vasoactive-intestinal polypeptide (VIP) and pituitary adenylate cyclase-activating polypeptide (PACAP) (Lee et al. 2010; Southey et al. 2014). However, peptidomic signatures of the SCH do not necessarily mark peptides designated for release, and an analysis of releasates has made it possible to detect peptides designated for cell-to-cell communication (Hatcher et al. 2008; see review by Mitchell et al. 2011). Future work along these lines could help to determine differential peptide release from SCH sub-regions (e.g., the core and shell), which are known to have distinct physiological characteristics (reviewed by Moore et al. 2002). For example, neurons have a firing rhythm that need to be reset after responding to stimuli and the dynamics in gene expression patterns associated with phase resetting are different between the core and shell (Zhu et al. 2012).

6.4.2.4 Molecular Extraction from the Hypothalamo-Neurohypophysial System

The supraoptic nucleus of the hypothalamus (SO) is a well-studied structure known for its role in mediating fluid homeostasis and regulating parturition, and exhibits structural and functional plasticity in association with these physiological processes that signal underlying alterations in molecular expression. These hallmarks of plasticity include changes in nucleolar numbers (Hatton et al. 1972) that signify changes in ribosomal RNA synthesis; i.e., protein synthetic machinery levels (Pederson 2011). Studies on the SO have been conducted to profile the transcriptome under normal, physiological conditions or after the effects of hypo-osmolality and/or dehydration. The main neuronal phenotypes of the SO are oxytocin (OT)- and vasopressin (VP)- expressing magnocellular neurons (MNs), which have been found to express 1,385 genes at levels that are more than twice those found in the rest of the hypothalamus, when sampled as a whole (Mutsuga et al. 2004). Taking advantage of the two types of MNs, Humerick et al. (2013) isolated SO MNs by their expression of OT or VP and found differential expression patterns; most notably in their transcription factors. However different these neuronal subtypes are, many studies have also examined global effects on MNs. For example, hypo-osmolality inhibits both OT and VP MNs and alters their transcriptome in comparison to the whole hypothalamus (Yue et al. 2006). Single MNs have also been isolated from rat SO and analyzed for neuropeptide phenotype markers (Xi et al. 1999; Glasgow et al. 1999; Yamashita et al. 2002; reviewed by Mutsuga and Gainer 2006).

Together with the MNs of the paraventricular hypothalamic nucleus (PVH), the SO makes up the hypothalamo-neurohypophysial system (HNS) that, along with several other functions, mediates fluid homeostasis. Dehydration/salt-loading can alter the HNS transcriptome, with certain genes enriched in the PVH and SO being especially sensitive to this physiological condition (Hindmarch et al. 2006; J. Qiu et al. 2011, 2014; Stewart et al. 2011; Greenwood et al. 2015; see also Hindmarch et al. 2013).

Similarly, the HNS proteome is also altered by dehydration, where 25 and 45 proteins have been reported to be affected in the SO and neurointermediate lobe (NIL), respectively (Gouraud et al. 2007). K. Johnson et al. (2015) have employed next generation sequencing technology (RNA-Seq) to examine the effects of salt loading on gene expression in the SO of rats, and found that nearly 6% of the genes alter their expression levels following this intervention.

Given the role of OT and VP in the HNS system, there is also a rich interest in other peptides MNs may express. For example, Bora et al. (2008) identified 85 peptides from isolated MNs of the SO. Moreover, Hazell et al. (2012) provide an overview of their studies concerning the presence of various G-protein coupled receptors in the PVH and SO using high-throughput methods, along with other techniques.

Along with MNs, the PVH also harbors distinct parvicellular neurons (PNs), although their similarity is highlighted by their comparable gene expression profiles (Bonaventure et al. 2002). Of the 2,145 profiled genes within these cell types, 65% were validated via in situ hybridization. The PNs of the PVH that express corticotropin-releasing hormone (CRH) are involved in the stress response as part of the hypothalamic–pituitary–adrenal (HPA) axis, and distinct stressors can produce differential gene expression in the PVH (Reyes et al., 2003). Some studies on the PVH have been conducted to examine a handful of genes in PNs without technically resorting to "high-throughput methods", such as focused studies of certain genes using real-time PCR. For example, S. Wang et al. (2008) examined LCM-captured human hypothalamic tissue collected *post mortem*, and identified an up-regulation of corticotropin-releasing hormone (CRH) and other gene products in associated with patients who suffered from clinical depression. Other studies have used modern "-omics" technologies to either profile the transcriptome alone (Atkins et al. 2011) or to investigate a mechanistic role for PVH genes within the HPA axis. For example, transcriptomic analysis, combined with morphometric and immunohistochemical evidence, demonstrated that select neurons, likely to be true PNs, express the gene encoding the molecule secretagogin, which is functionally linked to CRH release from these neurons (Romanov et al. 2014).

6.4.2.5 Molecular Extraction from the Arcuate Hypothalamic Nucleus (ARH)

The arcuate hypothalamic nucleus (ARH) is a structure involved in the maintenance of energy homeostasis (see Andermann and Lowell (2017) for a recent review of ARH function within feeding control networks). Transcriptomic analyses have been conducted across multiple studies examining the effects of diet, peripheral signals, and environment on gene expression in ARH neurons. For example, Paulsen et al. (2009) identified changes in neuropeptide Y (NPY) and pro-opiomelanocortin (POMC) mRNAs and an additional 3,480 transcripts in fasted, diet-induced obese rats. Similarly, Jovanovic et al. (2010) showed changes in hundreds of genes in the ARH after leptin treatment in 48-h fasted animals. Using cell sorting methods, Draper et al. (2010) isolated NPY-expressing neurons in the mouse ARH and ran microarray analysis to identify novel genes in this specific cell population in comparison to NPY-expressing neurons elsewhere in the hypothalamus (DMH), including the gene encoding the leptin receptor. At a more detailed level, Landmann et al. (2012) used LCM to sample the ARH in fasted rats, fed rats, and rats refed with a glucose load and found an up-regulation of Agouti-Related Peptide (AgRP) mRNA under fasted conditions that was greater in magnitude within single, LCM-captured neurons compared to what the authors term "ARH cell layers", which essentially meant a complete LCM of the full ARH expanse along its cytoarchitectonic boundaries in each sampled coronal section (note the investigators performed single-cell LCM on one hemisphere, and full ARH LCM on the opposite hemisphere). In response to the refed condition, AgRP was conversely found to be down-regulated and POMC mRNA up-regulated. Importantly, the authors specified a brain atlas they used and the specific atlas levels from which they sampled the ARH, setting this study apart from most others in its more careful delineation of anatomical boundaries.

Conducting cell type-specific transcriptomics, Henry et al. (2015) identified molecular pathways specific to AgRP neurons that were differentially affected in fed and food-deprived animals. Similarly, Campbell et al. (2017) found using Drop-Seq methodology (see Sect. 6.4.2.2) that thousands of genes coding for non-neuronal and neuronal cell types displayed altered expression in association with changes in feeding conditions and energy states. They found that the transcriptional response to fasting was generally stronger than that produced by a high-fat diet, with neuronal types responsive to fasting also responsive to high-fat feeding.

Transcriptomics has also addressed questions about the relationship between the ARH and the peripheral nervous system. For example, Adler et al. (2012) characterized the transcriptome of retrogradely-labeled neurons within the ARH projecting to white adipose tissue. Neurogenin 3, a transcription factor that helps differentiate pancreatic endocrine cells also comprised a portion of the transcriptomic profile of NPY neurons of the ARH (Arai et al. 2010). Other cell types in the ARH that have been targets of molecular profiling include cholinergic neurons, many of which were found to also express tyrosine hydroxylase and markers for GABAergic neurotransmission (Jeong et al. 2016).

Transcriptomic analyses have also been used to address how the environment can affect the ARH. For example, low protein diet during postnatal development reduces body fat, and increases leptin and melanocortin receptors (Stocker et al. 2012). The ARH also maintains stability in its expression patterns under certain changes within the internal environment, such as during pregnancy. Specifically, Phillipps et al. (2013) showed that despite higher shifts in plasma leptin and insulin and low blood glucose induced by pregnancy, there are no changes in the ARH transcriptome. These studies have provided understanding as to what extent the ARH transcriptome is affected by environment. Finally, transcriptomics traditionally provides information about the expression levels of mRNA but can also provide valuable information expression concerning microRNA levels (Taouis 2016). A set of more than 210 microRNA genes was profiled in both the ARH and the PVH as potential regulators of mRNA (Amar et al. 2012).

In contrast to transcriptomics, only a few investigators have investigated the ARH proteome. For example, proteomic analysis of protein markers in the ARH after exposure of the organism to an inorganic compound demonstrated a few proteins that are altered in their levels of expression that are related to cell morphology, axonal growth and tissue remodeling (Amigó-Correig et al. 2012).

6.4.2.6 Molecular Extraction from Other Hypothalamic Sub-Regions (LHA, VMH)

A number of studies have performed molecular analyses of peptidergic neurons known to be enriched in the LHA, a relatively large expanse of the hypothalamus that harbors a diversity of cell types (Bonnavion et al. 2016). For example, Volgin et al. (2004) reported isolating individual slices of brain containing portions of the

LHA and creating suspensions of dissociated cells from that region. They then identified the peptidergic phenotypes of the cells using antibodies raised against the precursor peptide encoding hypocretin/orexin (H/O), pre-pro-H/O, or melanin-concentrating hormone (MCH); and performed RT-PCR on each cell for the respective mRNAs for these neuropeptides, providing a proof of concept for their delicate methods. Harthoorn et al. (2005) reported using single-cell LCM to generate transcriptional profiles of neurons expressing MCH and H/O, and found that these neurons express transcripts for several other neuropeptides, such as dynorphin and cocaine- and amphetamine-related transcript (CART).

Using a translational profiling technique called TRAP (Translating Ribosome Affinity Purification; Doyle et al. 2008; Heiman et al. 2008; Dougherty et al. 2010), which involves affinity purification of polysomal mRNAs in defined cell populations, Dalal et al. (2013) generated mouse transgenic lines that expressed a fusion protein encoding enhanced green fluorescent protein and the large-subunit ribosomal protein L10a (eGFP-L10a) in hypothalamic neurons that express H/O. The expression of this fusion protein allows for the isolation of those mRNAs within H/O-expressing neurons that are undergoing translation at the site of polyribosomes, effectively allowing a translational profiling of a chemically identified neuron. Using this approach, the investigators identified >6000 transcripts with signal above background levels; 188 of these were highly enriched in H/O neurons (Dalal et al. 2013). Fifteen of these transcripts were confirmed to be present within intact H/O neurons by dual-label in situ hybridization, including the transcription factor Lhx9, which the authors showed, using gene ablation experiments, that it contributes to maintaining wakefulness in mice. Using an extension of the TRAP approach on the same problem, which they dubbed "vTRAP" ("viral TRAP"), Nectow et al. (2017) engineered a Cre-dependent adeno-associated virus to harbor a construct encoding eGFP-L10a, to translationally profile a specific variety of cell types in layer 5 of the cerebral cortex, the dorsal thalamus, ventral tegmental area, dorsal raphe nucleus, and LHA. Within the latter region, they focused on targeting their viral construct to MCH-expressing neurons.

The Jackson laboratory has recently reported single-cell transcriptomic data obtained from LHA H/O-expressing neurons and MCH-expressing neurons in mouse transgenic lines (Mickelsen et al. 2017). Importantly, in their study, they show specific delineations of the regions they dissected using atlas-based coordinates and drawings of the estimated areas they micropunched. A surprising finding from their careful analyses was that virtually all MCH neurons and approximately half of H/O neurons express markers for glutamate release and GABA synthesis (but not GABA vesicular release), underscoring the importance of fast-acting, small neurotransmitters within these peptidergic neurons and highlighting potentially interesting roles for GABA metabolism with glutamatergic neurons.

Studies have also been conducted to analyze the molecular expression patterns within the ventromedial hypothalamic nucleus (VMH). The Elmquist laboratory performed LCM to isolate and analyze the VMH from mice and used microarrays to detect genes enriched in this region of the hypothalamus (Segal et al. 2005). They compared the genes they obtained with those obtained from nearby regions (the

ARH and dorsomedial hypothalamic nucleus; DMH). They used real-time PCR to validate nine of the twelve most robustly expressed genes, and went on to confirm the expression of three of these genes using in situ hybridization. Their work complements that conducted by the Ingraham laboratory, which furnished a transcriptome from manually microdissected tissue samples obtained from the developing mouse (Kurrasch et al. 2007), in which they identified and confirmed the expression of six different VMH-enriched markers from their initial screens. At the protein level, the Renner laboratory conducted studies in which they micropunched the VMH from female rats in an atlas-guided fashion, and identified several proteins that could be reproducibly resolved via 2-D gel electrophoresis from the micropunches, including several sensitive to estradiol regulation (Mo et al. 2006; 2008).

6.4.3 A Note About "Hypothalamic-Derived" Molecules

Before moving on to discuss LCM, it is worth ending this portion of the narrative with a brief note regarding molecular provenance from the perspective of evolution. In this section, we have focused on molecular extraction of molecules from the hypothalamus, including, to name a few, neuropeptides of the hypothalamo-neurohypophysial system (OT and VP), the circadian system (VIP), and wakefulness and energy balance (H/O, MCH, AgRP). However, it is important to bear in mind that these "hypothalamic-derived" molecules are not strictly linked to the vertebrate hypothalamus per se, since large-scale molecular phylogenetic studies have identified precursors and analogs of these molecules in animal taxa that have evolved nervous systems lacking a hypothalamus (Jékely 2013; Mirabeau and Joly 2013; E. Williams et al. 2017). For example, Semmens et al. (2016) performed transcriptomic studies of the radial nerve cords of the European starfish, *Asterias rubens*, and identified >40 neuropeptide precursors in this echinoderm, many of which have homologs in the vertebrate hypothalamus. Indeed, precursors to neuropeptides found in the mammalian hypothalamus can be found in many phylogenetically ancient animal taxa (see Supplemental Data of Elphick et al. 2018). Thus, in our quest to preserve the provenance of molecular data from the hypothalamic regions from which they are extracted, we must bear in mind the ironic fact that many of the molecules, from an evolutionary standpoint, never "belonged" to the hypothalamus in the first place.

6.5 Laser-Capture Microdissection Studies: Methodological Considerations

In this section, we describe how laser-capture microdissection (LCM) techniques are a useful step for precisely delineating regions of interest within the hypothalamus for subsequent high throughput molecular analyses. We describe a few approaches involving this technique and their advantages and disadvantages, followed in Sect. 6.6 with how such samples can be traced back to their regions of extraction using digital atlas-based mapping techniques.

Since its development in the late 1990s (Emmert-Buck et al. 1996), LCM has been a useful procedure for obtaining RNA from single cells or whole regions of tissue (for selected reviews of techniques, see Espina et al. 2006; Baskin and Bastian 2010; Datta et al. 2015). LCM has been widely used to collect individual cells (Backholer et al. 2010; J. Park et al. 2016) or groups of cells from tissue slices or cultured cells that have been identified using immunocytochemistry (termed immuno-LCM; Baskin and Bastian 2010) or specific fluorescent tags (e.g., GFP) or fluorescent dyes such as Alexa Fluor™ 488). These approaches have enabled users to examine the expression of anywhere from a few genes of interest upwards to several hundred genes in specific cell types for various applications including genomics, transcriptomics (next-generation sequencing, microarrays; Paulsen et al. 2009), and proteomics (Moulédous et al. 2003; for a review of applications, see Datta et al. 2015). In this section, we describe findings and/or present data from our use of two different LCM systems: 1) the Arcturus AutoPix Fluorescent LCM System (Thermo Fisher Scientific, Waltham, MA) and 2) Leica LMD 7000 Microscope (Leica Microsystems Inc., Buffalo Grove, IL). In contrast to the Arcturus AutoPix LCM model in which the dissected tissue was collected onto a plastic cap (CapSure LCM Caps) *above the slide*, both dissected tissue and membrane surrounding the tissue was collected *below the slide* containing a UV-absorbing membrane (MembraneSlide) into a microcentrifuge tube cap using the Leica LMD7000 Microscope. Both LCM instruments have now been replaced by more recent models, including the ArcturusXT™ LCM System (now distributed through Thermo Fisher Scientific) and the Leica LMD6/LMD7. Here, we present ways in which LCM has been used to collect: (1) regions of tissue from anatomically distinct areas of the brain (Sect. 6.5.1); and (2) targeted populations of cells that have been identified using immunocytochemistry (Sect. 6.5.2) or fluorescent conjugates (Sect. 6.5.3). We discuss the advantages and pitfalls to using these approaches.

6.5.1 LCM for General Sampling of Brain Regions

This is the most common application of LCM for collection of brain tissue involves collecting anatomically matched regions of tissue across several rostrocaudal levels of a particular brain site. For example, we have used the Arcturus AutoPix Fluorescent LCM System to confirm sufficient knockdown of OT receptor mRNA following hindbrain nucleus of the solitary tract (NTS) injection of OT-saporin toxin relative to control saporin toxin (Baskin et al. 2010). We collected bilateral samples of NTS tissue from slide-mounted cryostat sections (10 μm) at the level of the area postrema (AP) and rostral to the AP at 200 μm intervals (n = 8 slides/brain). Following LCM collection, sections were dehydrated in ethanol and xylene, and then air-dried. We have found that this approach was suitable for measuring differences in NTS expression of OT receptor mRNA. In addition, we have used the Arcturus AutoPix

Fluorescent LCM System to confirm the "expected" reduction in cholecystokinin 1 receptor (CCK1R) mRNA in both the ARH (−3.48 mm to −2.04 mm from Bregma; Paxinos and Watson 2007) and dorsomedial hypothalamic nucleus (DMH) (−3.60 mm to −2.80 mm from Bregma; Paxinos and Watson 2007) in rats that lack CCK1Rs relative to wild-type rats (Blevins et al. 2012). As before, slide-mounted cryostat sections (10 μm) of ARH and DMH were selected at 200 μm intervals, dehydrated in ethanol and xylene, and then air-dried (n = 6 slides/brain). Bilateral samples were collected from brain sites that normally express CCK1R (i.e., ARH and DMH). Lastly, we have used the Arcturus AutoPix Fluorescent LCM System (Fig. 6.4, *left panel*) and Leica LMD 7000 Microscope (Fig. 6.4, *right panel*) to confirm the increase of NPY/AgRP in the ARH from 48-h fasted rats relative to ad libitum fed rats (T. Hahn et al. 1998; Mizuno and Mobbs 1999; Korner et al. 2001; Swart et al. 2002; Bi et al. 2003). We collected bilateral samples of ARH (−3.48 mm to −2.04 mm from Bregma; Paxinos and Watson 2007) from slide-mounted cryostat sections (10 μm) at 200 μm intervals (n = 6 slides/brain). In all cases, sections from adjacent slides were stained with cresyl violet (Blevins et al. 2000a, b; Baskin et al. 2010) to enable anatomical matching. As noted earlier, Fig. 6.2B, C show an example of LCM of the ARH from cresyl violet-stained tissue sections. Landmann et al. (2012) have extended these findings by using LCM to demonstrate that fasting results in increased AgRP mRNA expression from the ARH (both when collected as a region or as single neuron pools consisting of 100 neurons). LCM has been used by other labs to profile the molecular composition of various hypothalamic regions (Tables 6.1, 6.2, and 6.3). For example, to highlight just a few studies by way of illustration, LCM has been used to confirm: (1) the effectiveness of adeno-associated viral knockdown of angiotensin II receptor subtype 1a in the subfornical organ (SFO) of rat brain (Walch et al. 2014); (2) reductions in gene expression in brains of steroidogenic factor 1 (SF-1) in the VMH knock-out mice (Segal et al. 2005); and (3) fasting-elicited changes in gene expression in the PVH and the impact of leptin replacement on these genes (Tung et al. 2008).

6.5.2 *Immuno-LCM*

6.5.2.1 Advantages

Immuno-LCM (Fassunke et al. 2004; Fink et al. 2000; Waller et al. 2012) is the approach of using immunocytochemistry to identify cells to be collected by LCM. One of the primary advantages of immuno-LCM is that it enables the user to phenotype specific cells of interest that could not be as easily identified using anatomical landmarks alone. This may be a particularly useful approach given that tissue sections collected by LCM cannot be coverslipped and, as a result, may not allow sufficient resolution to identify anatomical landmarks readily. One of our laboratories (JEB) has used this approach to identify (following a rapid immunostaining procedure for tyrosine hydroxylase (TH); a marker of catecholamine neurons) and collect catecholamine immunopositive neurons from the A2/C2 catecholamine cell groups in the hindbrain NTS. We used the Arcturus AutoPix Fluorescent LCM System to confirm the specificity of this approach by measuring TH mRNA from TH+ neurons relative to TH– neurons in order to confirm its presence for qPCR analysis (D. Williams et al. 2008). There are a number of protocols for rapid immunostaining that have already been published (Fink et al. 2000; Uz et al. 2005; D. Williams et al. 2008; Baskin and Bastian 2010; Briski et al. 2010; Carreño et al. 2011; Nedungadi and Cunningham 2014; J. Park et al. 2016).

6.5.2.2 Challenges and Pitfalls

There are a number of challenges when using the rapid immunostaining approach that must be considered prior to incorporating immuno-LCM. For example, as reviewed by Baskin and Bastian (2010), the process of immunostaining can introduce the potential for RNA extraction and degradation. In an effort to minimize loss and degradation of RNA, common strategies are to implement rapid immunostaining

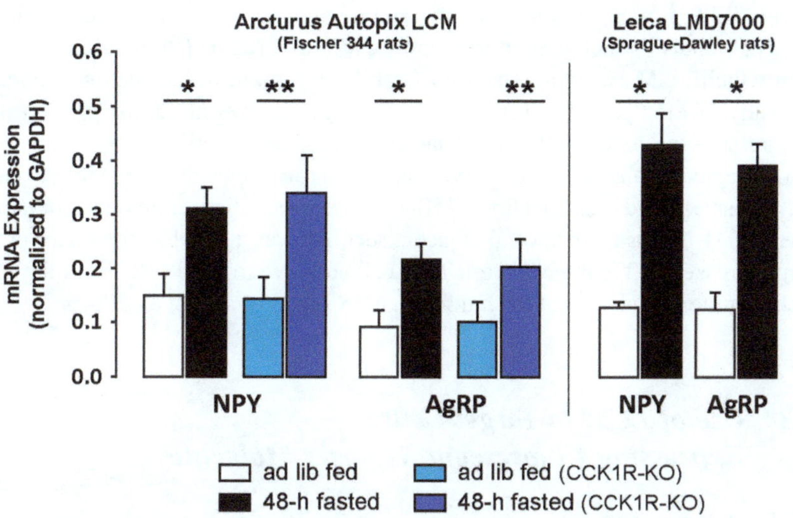

Fig. 6.4 Unpublished data furnished here to illustrate the efficacy of two LCM systems (Arcturus and Leica), along with qPCR, to confirm fasting-elicited increases in NPY and AgRP mRNA expression in the ARH in rats. Slide-mounted cryostat sections (10 μm-thick) were generated through the ARH from ad libitum fed or 48-h fasted wild-type or cholecystokinin-1 receptor (CCK1R) knockout (KO) Fischer 344 rats (*left panel*) or ad libitum fed or 48-h fasted wild-type Sprague-Dawley rats (*right panel*). Sections were thawed briefly prior to dehydration in ethanol and xylene and allowed to air dry as described previously (Blevins et al. 2000a, 2012; Baskin et al. 2010). LCM was used to collect bilateral sections (10 μm-thick) from the ARH (200 μm intervals) from six anatomically matched levels and sections were mounted onto slides. All microdissected samples from a single brain were pooled for RNA extraction. NPY and AgRP mRNA expression were measured using qPCR. Data are expressed as mean ± SEM. $*P < 0.05$ or $**P = 0.1$ fed vs. fasted

protocols and the use of alcohol fixation (methanol or ethanol) in place of formaldehyde-based fixatives (which can result in much of the RNA being fragmented and degraded by formalin) (Gillespie et al. 2002; Soukup et al. 2003; J. Su et al. 2004; Hu et al. 2005; Baskin and Bastian 2010; Carreño et al. 2011; Nedungadi and Cunningham, 2014; J. Park et al. 2016). We have previously shown that brief thawing (~30–60 s) of cryostat-cut sections of frozen rat brain in combination with quick immunostaining after methanol fixation (~3 min) works well for immuno-LCM and qPCR for mRNA (D. Williams et al. 2008). Other challenges to the use of the rapid immunostaining approach include antibodies that require a low titer or are relatively nonspecific as well as antigens that are found in low-abundance (Baskin and Bastian 2010). As Baskin and Bastian (2010) indicate, adjustments in staining times, incubation temperatures or more sensitive fluorochromes, may increase the specificity to acceptable levels. Rapid immunostaining approaches may be less suitable for targeting and collection of cells with low gene expression. One other challenge when selecting specific cells is that contamination from neighboring cells may also be included in the sample. For example, Okaty et al. (2011) reported in their meta-analysis of various cell isolation methods conducted by certain laboratories that LCM produced higher contamination from spurious signals, as compared to other cell isolation methods, such as TRAP, FACS, immunopanning, and manual sorting of fluorescently labeled cells. Their analysis included an immuno-LCM study (C.-Y. Chung et al. 2005) and one in which LCM was performed on fresh-frozen brain tissue sections containing genetically labeled cells from transgenic mice (Rossner et al. 2006). One means to address this issue is to collect an equal number of neighboring cells outside of the intended region of analysis as negative controls to run alongside the positively labeled cells. We have found that selecting ~150–200 TH+ and adjacent non-catecholaminergic cells (TH–) cells from several adjacent sections was a suitable approach for measuring increases in TH mRNA from TH+ cells relative to TH– cells from the A2/C2 catecholamine cell groups in the hindbrain NTS (D. Williams et al. 2008).

6.5.3 *Use of LCM to Target Cells Expressing Fluorescent Reporter Molecules*

6.5.3.1 Advantages

Similar to immuno-LCM, this approach enables the phenotyping of specific cells of interest that could not be as readily identified using anatomical landmarks. In contrast to immuno-LCM, there is no need for rapid immunostaining as the fluorescent tag is already present. We have used this approach previously (Blevins et al. 2009; see Figure 6.2D–F) to identify those neurons in the PVH that project to the hindbrain NTS using Alexa Fluor™ 488-conjugated retrograde tracer, cholera toxin subunit B (CTB). We have found that brief thawing (~30–60 s) of cryostat-cut sections of frozen rat brain, in combination with selecting ~250 CTB+ cells from three or four anatomically matched coronal sections from PVH, was a suitable approach for measuring OT, CRH, and melanocortin-4 (MC4-R) receptor mRNAs (Blevins et al. 2009). We also collected the same number of neurons from the SCH as a negative control, as this site expresses relatively low levels of each of these transcripts (Mountjoy et al. 1994; Jing et al. 1998). In addition, unlabeled cells from the PVH were collected and screened for OT mRNA, CRH, and MC4-R mRNAs.

6.5.3.2 Challenges and Pitfalls

One potential limitation of using LCM to collect GFP-labeled cells is that free GFP is soluble and can leak out from cryostat-cut sections in the absence of fixation (Jockusch et al. 2003), thus necessitating perfusion and/or post-fixation of the tissue. Soluble eGFP is preserved in paraformaldehyde (PFA)-fixed tissues that are post-fixed in 50% ethanol and 100% *n*-butanol (Khodosevich et al. 2007). The authors noted that while PFA fixation of mouse tissue is sufficient in preserving the EGFP signal for up to 30–60 min, it was not sufficient in preserving EGFP signal for longer periods of time (Khodosevich et al. 2007). They also indicated that post-fixation in alcohol is "necessary not only to remove the water to prevent RNA degradation, but also to render the aldehyde-crosslinks more stable, thus preserving the fluorescence" (p. 2). They added that "alcohol fixation alone also was not sufficient to preserve fluorescence of the soluble EGFP and prevent it from leaching out and diffusing to neighboring tissue making it impossible to specifically identify green fluorescent cells" (p. 2). Although some groups have reported relative disadvantages of using formaldehyde-based fixatives to retrieve PCR product from LCM-sampled non-neural (Goldsworthy et al. 1999) and neural tissues (J. Su et al. 2004), there are instances where LCM has been shown to work successfully on formaldehyde-fixed tissues (e.g., Kabra et al. 2016). Recent papers indicate that EGFP+ (or EYFP+) cells can also be harvested from fresh frozen mouse (Rossner et al. 2006; Liu et al. 2011) and rat brain tissue (Liberini et al. 2016), but the extent to which the fluorescent signal may have diffused or faded beyond 30–60 min were not addressed in these studies. It is worth noting that Leica has produced a protocol designed to optimize visualization of GFP from post-fixed tissue to be used for LCM.

6.5.4 RNA Integrity

The RNA Integrity Number (RIN) value is a tool developed by Agilent Technologies to assess RNA integrity using the Agilent 2100 Bioanalyzer and RNA LabChip® kits. The RNA integrity is based on the electrophoretic trace of the sample and allows the user to assess the amount of degradation products in the sample and to determine integrity of the sample. It is an important consideration when assessing gene expression data from samples generated by LCM. The RIN algorithm assesses RIN values that range from 1–10 with 1 representing completely degraded RNA, 5 representing partially degraded RNA, and 10 representing completely intact

RNA. We have used the 2100 Electrophoresis Bioanalyzer (Agilent Technologies) to obtain RIN values from ARH tissue samples that had been stored for ~3 months at −80 °C. We obtained RIN values ranging from 7.6–8.2 (7.92 ± 0.09). These RIN values are comparable to those we obtained from ARH tissue (7.8 and 8.5) that had been stored ~7–8 months at −80 °C. While these values are in the higher range it does indicate some degree of degradation. These findings are also consistent to the RIN values (6.2) reported from tissue collected from patients with oral cancer that was stored for ~48 h at −80 °C (Xu et al. 2008b), as well as RIN values (6–7) reported from pancreatic tissue collected from rats and humans (Butler et al. 2016). They are also consistent with RIN values (6.6–7.6) reported for hypothalamic tissue sampled using LCM from the supraoptic (SO) nucleus; the LCM was performed within one month following tissue sectioning and storage of the slide-mounted sections at −80 °C (K. Johnson et al. 2015).

6.6 Anchoring Molecular Information to Their Native Regions Using Digital Atlas Maps

Having reviewed in the preceding sections the importance of location information in the brain (Sect. 6.2), the historical antecedents of current high throughput work concerning molecular extraction of the brain (Sect. 6.3) and the hypothalamus (Sect. 6.4), and the methodology of LCM (Sect. 6.5); we now turn to the topic that constitutes the principal thesis of this review; namely, the mapping of datasets to standardized atlases of the brain. Using the backdrop of LCM procedures described in the preceding section, we discuss first how documenting the location of the native substrate from where tissue is extracted is critical for the subsequent mapping of that location, and then describe the mapping steps themselves.

6.6.1 Documenting the Native Substrate Before Extraction

Applying LCM to a tissue section to capture and sample a particular region of interest can be performed in a number of ways, a few of which were described in Sect. 6.5. Unstained tissue sections can be viewed under a dark field microscope to observe the region of interest in relation to white matter tracts that might be nearby. Such landmarks can aid greatly in the accurate and repeated sampling of a region, especially for large sub-regions of the hypothalamus, a part of the brain replete with white matter landmarks (e.g., anterior commissure, optic chiasm, optic tract, fornix, mammillothalamic tract). Indeed, what is perhaps the first documented sampling of the hypothalamus was reported diagrammatically in relation to many of these fiber systems (see Fig. 6.2A). Micropunch methods, first developed before the establishment of LCM, involve procedures where tissue punches are harvested from unstained frozen or fresh tissue sections; in such cases, white matter tracts also serve as important landmarks to orient the experimentalist as to where a particular region of interest was located and how much tissue to collect from that region (Palkovits 1973; Jacobowitz 1974).

Apart from unstained tissue, the most common method for identifying regions of interest in sectioned brain tissue is through the use of Nissl stains (stains that label basophilic substrates—'Nissl substance'—in the cell, including rough endoplasmic reticulum and the nucleus, sites where nucleic acid molecules are concentrated). The use of such stains on brain tissue sections prior to LCM-based sampling from those sections is a common way of accurately delineating regions of interest for LCM-targeting (e.g., Ginsberg and Che 2004; Boone et al. 2013). Nissl-based stains such as cresyl violet (Fig. 6.2B), thionin, and hematoxylin have been used to guide sampling of hypothalamic sub-regions and cells, including the preoptic region (Vasilache et al. 2007; Aubert et al. 2013), ARH (Segal et al. 2005; Jovanovic et al. 2010), SCH (Porterfield et al. 2007; Boone et al. 2013; Pembroke et al. 2015), SO (Xi et al. 1999; Glasgow et al. 1999; Yamashita et al. 2002; S. Wang et al. 2008), VMH (Segal et al. 2005; Kurrasch et al. 2007), DMH (Segal et al. 2005), and PVH (S. Wang et al. 2008; Blevins et al. 2009; Novoselova et al. 2016). Importantly, investigators have performed LCM on the Nissl-stained tissue itself (Segal et al. 2005; Porterfield et al. 2007; Jovanovic et al. 2010; Boone et al. 2013), but in principle, one can also use adjacent sections stained for Nissl substance to help delineate regions of interest on unstained companion sections sampled by LCM, as has been done for human tissue samples collected post mortem for the PVH and SO (S. Wang et al. 2008). In addition to using Nissl staining as a tool to help delineate LCM-captured tissue sample boundary conditions, other stains and labeling strategies have also been used in conjunction with LCM, including FluoroJade for delimiting tissue pathology (Boone et al. 2013), Cy3-conjugated secondary antibody to identify antibody-labeled peptidergic neurons (Nedungadi et al. 2012a, b; Nedungadi and Cunningham, 2014), immunoperoxidase-based detection of peptidergic neurons (Briski et al. 2010), NeuroTrace staining for visualizing fluorescent Nissl-like profiles (Benner et al. 2015), and in situ hybridization in human *postmortem* tissue (Bernard et al. 2009). Finally, it is worth noting that although LCM procedures themselves do not appear to result in significant losses of protein as compared to manually dissected samples of comparably located regions, Nissl staining itself can be detrimental to the full retention of some proteins for subsequent proteomic analyses (Moulédous et al. 2002), and the use of Nissl stains such as neutral red, cresyl violet, or NeuroTrace reportedly contributes to lower yields of transcripts from LCM-captured brain tissue (Kerman et al. 2006; Benner et al. 2015).

6.6.2 Mapping to Standardized Atlases

Using aids such as the Nissl stain to identify a region of interest to be sampled by LCM not only helps ensure accurate sampling of that region, but also provides an opportunity to document the location of the excised tissue itself using standardized atlases of the brain. Such atlases have existed for several decades, and many have been created for a variety of animal models, including—to name but a few—toads (Hoffmann 1973), frogs (Wada et al. 1980), lizards (Greenberg 1982), guinea pigs (Tindal 1965), rabbits (Sawyer et al. 1954), mice (Dong 2008; Paxinos and Franklin

2012), and rats (Swanson 2004; Paxinos and Watson 2007) (for a detailed listing, see Ten Donkelaar and Nicholson 1998). As detailed in Khan (2013), there are many advantages of using standardized atlases to map experimental data, not least of which is to be able to spatially align different datasets from diverse studies and contextualize them with some rigor and precision (also see Khan et al. 2018).

How is mapping experimental data to a reference atlas of the brain performed? Simmons and Swanson (2009) describe many aspects of how mapping experimental data to a standardized reference atlas is undertaken. A critical factor is reconciling the plane of section of the experimental tissue with the plane of the atlas map that will be used to contain the mapped dataset. Differences in plane of section, determined by the angle of cutting on the microtome or cryostat instrument used to section blocks of brain tissue, can potentially constitute a significant source of mapping errors, especially in the absence of any global (e.g., Nissl) stain to mark the cytoarchitecture of the tissue being sectioned. It is surprising to us how few investigators explicitly discuss how they have dealt with plane of section issues when analyzing the results of expression and distribution studies.

For example, many investigators have utilized immunohistochemistry of the transcription factor and immediate-early gene product, Fos, to identify regions of the brain post mortem that may be associated with patterns of activation or with particular behaviors that the organism was involved in during the life history immediately preceding death. However, to our knowledge, none of these studies presents a comparison of Fos expression patterns between groups of animals while providing an explicit discussion of how planes of section were taken into account in their determination of regional comparisons. Thus, as part of a collaborative study (Zséli et al. 2016), a few of us (AM, AMK) performed a plane of section analysis to map patterns of Fos expression in rats who had fasted for 40 h (but had ad libitum access to water) versus rats who fasted for 40 h but then were allowed to re-feed for 2 h. Figure 6.5 shows a portion of these data. Compared to the plane of section of the Swanson (2004) reference atlas (Fig. 6.5A, *top panel*), the planes of section for the subjects examined for Fos expression were markedly different (Fig. 6.5A, *middle and bottom panels*), and any accurate comparison of the same regions at similar rostrocaudal levels between fasted and re-fed cohorts required reconciling the planes of section for tissues sectioned from both cohorts with the plane of section of the reference atlas. This was not only important for representing the patterns of expression on the atlas, but also to ensure that we were not comparing levels of expression between regions that did not correspond with one another in terms of spatial positioning within the brain. As detailed in our study (Zséli et al. 2016), we utilized the digital atlas maps for Swanson (2004), which are now also available online (larrywswanson.com). These were manipulated in Adobe Illustrator (AI) software. Nissl counterstain in the Fos-immunoreacted tissue sections was used as a guide to identify cytoarchitectural boundaries for each section. The photomicrographs were imported into separate layers of AI, scaled, and compared to the atlas plates to determine whether there were differences in plane across the mediolateral and dorsoventral axes. In some cases, as delineated by Simmons and Swanson (2009), patterns on a tissue section require mapping to more than one level of a reference atlas, and the differences in plane of section are often in more than one plane simultaneously, necessitating a segment-by-segment translation of the region of interest to the relevant location on a map or set of maps.

Another important point to note about deriving information for mapping on the basis of Nissl-stained tissues is that often Nissl stains do not fully reveal distinct patterns of cytoarchitecture within tissue; in such cases, it is at times difficult to discern a particular sub-region within a tissue section and determine precisely the boundaries of a region. In such cases, we have opted to report the uncertainty in our mapping that results from such ambiguous staining patterns, by noting within the reference atlas those portions of the map that are based on inferred positions of cytoarchitectonic boundaries as opposed to those that were directly observed (and distinct) within the stained tissue section. As shown in Fig. 6.5C, we found certain sub-regions of the LHA to display Nissl patterns that were indistinct, which permitted us to only infer positions of the Fos-immunoreactive cells we were mapping. This uncertainty was represented in the form of a pale yellow color for the dotted line boundaries for those regions (Fig. 6.5C).

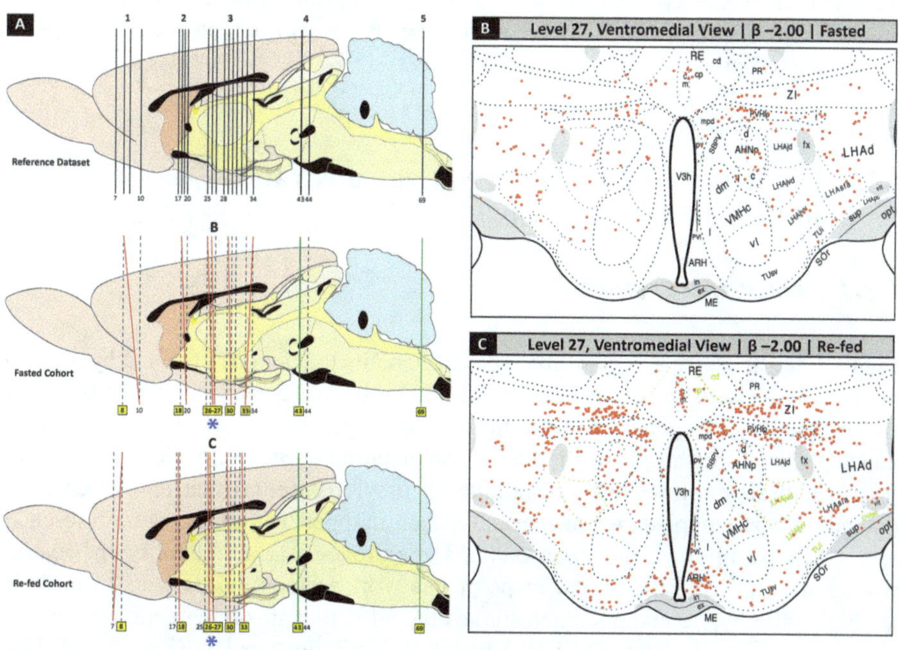

Fig. 6.5 Example of a plane-of-section analysis conducted on Fos transcription factor expression maps of hypothalamic tissue sections obtained from fasted versus refed adult male Wistar rats, conducted by the UTEP Systems Neuroscience Laboratory as part of a published collaborative study with the Hungarian Academy of Sciences and Tufts University (Zséli et al. 2016). (**A**) Sagittal views of the rat brain, modified from Swanson (2004), showing the plane of section of the Swanson atlas (*top panel*: 'Reference Dataset'), followed by the planes of section for a subject from the fasted cohort (*middle panel*) and the refed cohort (*bottom panel*). (**B, C**) Locations of Fos-immunoreactive neurons plotted onto coronal-plane maps from Swanson (2004) showing the ventromedial views of atlas level 27 for each cohort (which are denoted by *asterisks* in **A**). Figures in (**A–C**) are reproduced from Zséli et al. (2016) with permission from John Wiley & Sons, Ltd

For LCM-captured brain tissue, an outline of basic steps for mapping the sampled tissue to a reference atlas is shown in Fig. 6.6. First, as described in Sects. 6.5.2.2 and 6.5.3.2, investigators have to decide whether to employ fixatives such as methanol, alcohol or formaldehyde to preserve their tissues of interest before sectioning them, or instead opt to use freshly frozen, unfixed tissue sections (Fig. 6.6A, *Step 1*). Once sectioned and mounted onto slides (Fig. 6.6A, *Step 2*), a given tissue series can be Nissl-stained (Fig. 6.6A, *Step 3*), and then placed within an LCM instrument to excise a region of interest (ROI; Fig. 6.6A, *Step 4*). Apart from the sequestration and processing of the LCM-captured tissue of interest for further analyses using transcriptomics, proteomics, or peptidomics, etc., the remaining tissue section (i.e., the rest of the section that remains *after* the region of interest has been excised) can now be used as a key to unlock the precise location of the sampled area within a standardized reference space of the brain. Similar to the example of a plane-of-section analysis furnished in Fig. 6.5, the section can be examined in relation to the Nissl-based landmarks of photographs within the reference atlas to be used, and the tissue's plane of section assigned to appropriate levels of the reference atlas (Fig. 6.6B, *Step 5*). The ROI within the tissue section can then be mapped using a digital atlas map of that reference level.

6.7 The Benefits of Mapping Native Substrates and Anchoring Datasets

6.7.1 Data Integration

Figure 6.6C provides a view of the types of benefits that can be obtained by assiduously mapping a tissue sample obtained by LCM to a reference map of the brain. In addition to the data generated from the high throughput "-omics"-based extraction and analysis of the sample itself, the precise mapping of the sample in relation to its native landscape allows one to examine all previous studies that have been conducted on that sampled region that have been mapped within the same reference space. For example, for the Swanson (2004) reference atlas of the rat brain, several studies have utilized the digital maps of this work to map the datasets from their studies of the hypothalamus. Such studies include those involving central microinjections of molecules into the PVH (Khan et al. 2007; reviewed in Khan et al. 2017), protein expression in the LHA in response to water deprivation (S. Yao et al. 2005), transcription factor activation in several hypothalamic regions in response to fasting or re-feeding (Zséli et al. 2016), deposits of neuroanatomical tract tracer molecules into any of several hypothalamic regions (e.g., Hahn and Swanson 2010), and mapping of key neuropeptides within distinct subdivisions of the LHA (Swanson et al. 2005; J. Hahn 2010). In the hypothetical scenario furnished in Fig. 6.6C, the location of the portion of the ARH sampled by LCM maps to Level 28 of Swanson (2004). Specifically, this region—at this same rostrocaudal level—could also have been the focus of investigations concerning anterograde tracing, central drug

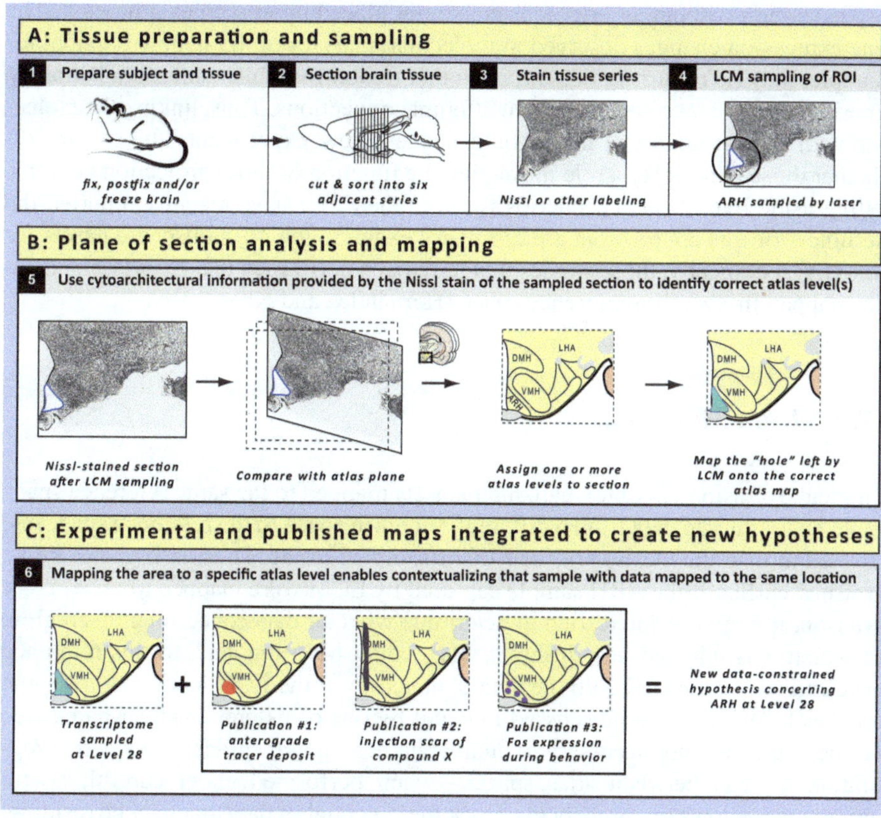

Fig. 6.6 Steps for mapping tissue samples collected using LCM, and the advantages of such mapping. Following steps to process brain tissue through the tissue fixation (**A**, *Step 1*) and sectioning/staining/LCM-based sampling steps (**A**, *Steps 2–4*), the remaining mounted and Nissl-stained tissue section from which the sample was excised can be mapped and analyzed for the location of the 'hole' left from sampling in relation to a standardized reference atlas of the brain (**B**, *Step 5*). The hypothetical example shown here is for LCM-based capture of the arcuate hypothalamic nucleus (ARH) (**A**, *Step 4*, *circled region* on the Nissl photomicrograph; the white ROI outlined in *dark blue* shows the region that was 'sampled'). (**C**) Once mapped, the location of the sampled tissue can be examined in relation to any other dataset that has been mapped to the same atlas level (*Step 6*). In this example, three hypothetical published datasets (a tract-tracing deposit, a drug injection site, and the distribution of activated immediate-early gene products) have all been mapped from different brains to the same reference location, thereby allowing a contextualization of the data in a spatially rigorous manner to generate new hypotheses concerning the ARH at that rostrocaudal level

injection, and Fos transcription factor expression. Therefore, all of those published datasets could be considered in conjunction with the molecular analyses performed on the LCM-sampled ARH, and new hypotheses can be constructed that are constrained by the spatial patterns of data from these maps, when they are considered as a collective (Fig. 6.6C). Together, therefore, the maps constitute a powerful way to help investigators see relationships among datasets, for the same region mapped in the same reference space, that they otherwise may not have seen or which they may have seen without any rigorous constraint placed upon such examination. The gene expression changes observed in molecular analyses of the sampled region, for example, may be occurring in neurons in that region for which anterograde tracing experiments have revealed prominent efferent connections. Thus, linking the molecular with neuroanatomical data would suggest new experiments that could test whether those genes play a role in shaping the function of those projection neurons. Setting aside hypothetical scenarios for a moment, we have recently reported the usefulness of this approach in a preliminary examination of published datasets for the LHA; specifically, those studies that have been performed that report LHA datasets mapped in Swanson reference space (Hernandez and Khan 2016).

6.7.2 *Data Migration*

A logical extension of contextualizing datasets mapped to the same reference space would be to migrate data from a different reference space to the reference space one has used to map the location of their LCM-sampled tissue. Thus, as in the example furnished earlier, if the ARH sample captured by LCM were mapped to Level 28 of Swanson, it would be interesting to determine whether data concerning this region, but which was mapped in a different atlas, could be "migrated" to this reference space and contextualized with the data obtained from the LCM sample at the same atlas level. This has been discussed in detail by one of us previously (Khan 2013) and the details are not necessary to enumerate here again; suffice it to say that registration of data between atlas spaces—when performed under careful, lawful parameters—allows researchers to unlock the potential of data that may be residing, unattended, in a different reference space. This is important because many researchers use different atlases to map their datasets; this is true for the hypothalamus as much as any other brain region. For example, the locations of recording electrodes used to perform electrophysiological recordings of neurons in the PVH have been mapped to the atlas of Paxinos and Watson, along with inferred stereotaxic coordinates for the locations of the maps (Aramakis et al. 1996). The recordings are for responses these PVH neurons have to application of NPY or its receptor agonists, and understanding the locations of the neurons displaying these responses could be better contextualized in relation to other datasets mapped in Swanson reference space if the data were migrated to that space. Fortunately, the alignment and registration between these atlas spaces appear to constitute a tractable problem (Wells and Khan 2013; Khan 2013; Hernandez and Khan 2016; Perez et al. 2017), the mature, fully fledged solution for which may help to bring together datasets that would otherwise be separated in time and space. As a step towards such a solution, we have recently developed and implemented a computer vision algorithm that matches features detected in photomicrographs of the Nissl-stained sections of the Paxinos and Watson and Swanson reference atlases to provide independent support of alignments we performed separately between the reference atlases based on craniometric measures in relation to the skull landmark, Bregma (Khan et al. 2018).

The algorithm produces matches between atlas levels that are in close agreement with matches produced on the basis of craniometric alignments, providing support for the feasibility of data migration between the two reference spaces.

Other, older datasets could also be potentially migrated between atlas reference spaces, provided that the reference spaces can be aligned and registered in a fashion similar to that described above for the Paxinos and Watson/Swanson reference atlases. For example, Jacobowitz and colleagues combined micropunch methods with two-dimensional gel electrophoretic separation methods to generate protein profiles from discrete sub-regions of the hypothalamus and other brain regions (Heydorn et al. 1983), mapping their data using coordinates derived from the König and Klippel (1963) rat brain atlas. In principle, such data can be contextualized more broadly if they were migrated to other extant reference spaces.

6.7.3 Data Refinement

Another benefit of mapping the location of LCM-captured brain tissue is the ability to improve our understanding of hypothalamic organization by refining the data generated from previously published studies. Prior to the advent of LCM (Emmert-Buck et al. 1996), the ability to sample brain tissue with high spatial resolution found perhaps its most precise expression in the micropunch methods mentioned above (reviewed by Palkovits 1975, 1986, 1989). Notwithstanding notable examples using these and other methods (e.g., Heydorn et al. 1983), LCM offers investigators the ability for an even greater precision of sampling of brain tissue within a given region's 3-D expanse, thereby allowing more careful examination of subregions to detect possible differences in molecular expression patterns within a defined neural substrate.

This level of spatial resolution is important, as data has emerged that suggest heterogeneous neuronal constituents along the rostrocaudal extent of hypothalamic nuclei and areas. For example, within the ARH, data from the mouse model demonstrate a segregation of the effects of acutely administered leptin and insulin on populations of ARH POMC neurons (K. Williams et al. 2010). Specifically leptin-induced excitation of neurons was found throughout the rostrocaudal extent of the retrochiasmatic area (RCH) and ARH, but most of the POMC neuronal excitation was recorded from neurons in the lateral RCH and medial POMC group in the ARH. In contrast, insulin-induced hyperpolarization of POMC neurons was restricted to medial RCH and rostromedial ARH (K. Williams et al. 2010). More recently, Lam et al. (2017) used single-cell RNA sequencing to determine that the POMC neuronal population in the mouse ARH consists of heterogeneous populations that differ on the basis of their cell surface receptor expression. Clustering analysis resulted in the investigators identifying four different classes of POMC neuron. Similarly, an elegant study by Foster et al. (2016) has demonstrated the presence of distinct subsets of neurons in the VMH in the rat model that show a selective *absence* of Fos immunoreactivity in association with the hypoglycemia produced by systemic insulin injections. In particular, they found that the VMHdm (dorsomedial part of the VMH) and the smaller VMHc (central part) show marked reductions in Fos immunoreactive neurons from hypoglycemic animals as compared to their euglycemic controls, and that these reductions were proportional to the reductions in terminal plasma glucose concentrations. In contrast, sub-regions such as the VMHvl (ventrolateral part), which are believed to be involved mainly in social and reproductive behaviors, do not exhibit such reductions. Clearly, then, sampling from these smaller sub-regions of the ARH and VMH warrants careful documentation and mapping.

Our own preliminary data (Martinez et al. 2015, 2016) on ARH connectivity underscores this point as well. Specifically, initial experiments in which the retrogradely transportable tracers, Fluorogold or CTb, were injected into the rostral and caudal portions of the ARH have yielded results showing subtle differences in the distribution and density of retrogradely labeled neurons throughout the forebrain that project to these portions of the ARH. A summary of these unpublished data is furnished in Fig. 6.7, simply to emphasize the point that it is no longer tenable to sample only one tissue section of a large expansive brain region such as the ARH, and make claims about its function as a whole without taking into consideration the possibility of heterogeneous properties for neurons along its full extent. A difference in afferent input implies different qualities for incoming signals to ARH neurons in the rostral end of the structure versus its caudal end; this in turn, implies that perhaps the neuronal populations receiving these differential signals may also be heterogeneous. Therefore, their molecular expression patterns, in terms of either phenotype or intrinsic state (or both) will likely also be non-uniform. Moreover, sex-specific differences in gene expression have also been reported for the ARH (Mozhui et al. 2012). Thus, the greater spatial resolution afforded by LCM sampling methods—together with careful digital atlas mapping of those locations of those samples—allows us as a community to continually refine our coarse datasets, rendering them sharper and more information-rich.

6.8 Concluding Remarks and Future Directions

In this article, we have surveyed the historical antecedents of high throughput technologies to extract molecular information from the brain, focusing on studies of the hypothalamus. After surveying selected articles reporting high throughput transcriptomic, proteomic and peptidomic studies of the hypothalamus or its subregions, we discussed the importance of LCM and digital atlasing methods in facilitating the anchoring (mapping) of such information to a tractable spatial model of the brain. In doing so, we build upon earlier efforts to link molecular information with spatial locations in the brain in a large-scale manner, such as grid-based mapping based on voxelation methods (Chin et al. 2007; C. Park et al. 2009; Petyuk et al. 2007; 2010) and analysis of gene expression patterns in the hypothalamus from the rich repository of in situ hybridization data within the Allen Brain Atlas (Olszewski et al. 2008).

6.8.1 Future Directions in Data Management

We also apply the topics presented here to previous discussions we have raised concerning automated, informatics-based management of neuroscientific data; for example, using electronic laboratory notebooks to perform digital mapping and documentation of analyzed datasets (see Fig. 4 of Burns et al. 2003; Khan et al. 2006). As greater sophistication is brought to bear using methods that combine neuroanatomical tract tracing with molecular analysis of the traced projection neurons (e.g., see Pomeranz et al. 2017), both the mapping and management of data concerning such projection systems will become even more streamlined. An important aspect of developing informatics tools and methodologies is that much of the information used by expert biologists is technically specified, but informally defined. Naturally, expert neuroscientists are trained to understand the spatial structure of a published brain atlas, a flatmap, or a stained histological slide without requiring an explicitly defined logical representation. Informatics systems, however, must be grounded in a well-defined ontological model, which inevitably leads to some disagreement concerning the optimal design of such ontologies for neuroanatomical data. There are number of approaches put forward within the neuroinformatics community to represent mapped neuroanatomical data in ontologies. These include representations with a neuroimaging focus (Nichols et al. 2010); philosophically grounded approaches to neuroanatomy (Osumi-Sutherland et al. 2012); or comprehensive, cross-species methods for neuroanatomical phenotype (Haendel et al. 2014; Mungall et al. 2017; also see Deans et al. 2015). It is important to note that selecting an appropriate formalization can have a deep impact on how a neuroinformatics system functions, and we feel that any formalization used to represent the data described in this chapter should reflect the expertise and practices of experimental scientists working in this field. Thus, we recommend lightweight, data-centric formalizations that mirror scientists' use of standard atlases, such as the Allen Brain Atlas portal (Sunkin et al. 2013).

Neuroscientific knowledge carries a structured context that is inherited from the experimental design that ultimately generates the data. One methodology for representing this context in a general way is based on the relationships between independent and dependent variables within studies. This may serve as a convenient framework for describing neuroanatomically grounded data by treating the location of the phenomena of interest in the brain simply as one of several independent variables that describe the context of a particular datum (Russ et al. 2011; Tallis et al. 2011).

6.8.2 Future Directions in Imaging

Though it has not yet achieved the mesoscale resolution required to permit detailed mapping of most molecules, mass spectrometry imaging (MSI)—including imaging based on matrix assisted laser desorption technology (Karas et al. 1987), known as MALDI (Spengler et al. 1994; Caprioli et al. 1997)—will hopefully provide

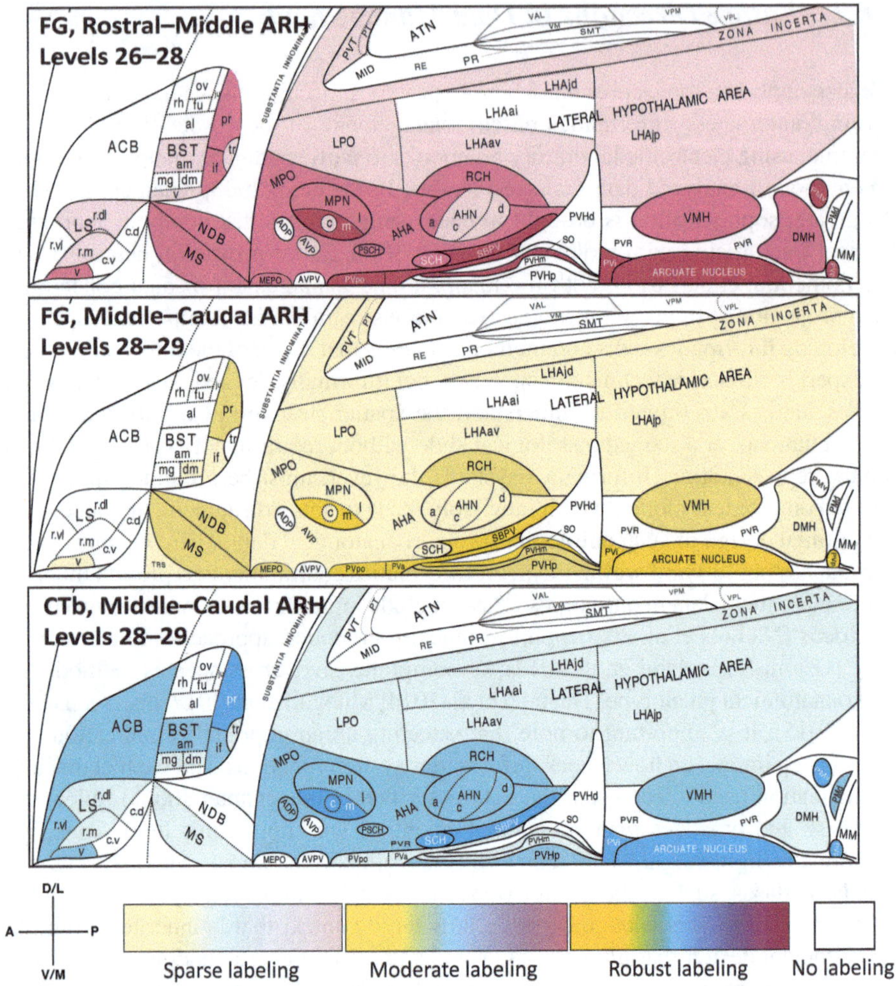

Fig. 6.7 Provisional choropleth flatmaps of unpublished data (Martinez et al. 2015, 2016) are furnished here to illustrate differences in the distributions and densities of retrogradely labeled neurons traced from rostral versus caudal locations within the rat ARH using two different retrograde tracers across three test subjects (adult male Sprague-Dawley rats). The upper two panels show distributions and densities for neurons traced retrogradely from ARH injection sites containing deposits of the retrograde tracer, Fluorogold (FG). The bottom (*third from top*) panel is from a test subject receiving the retrogradely transportable tracer, cholera toxin subunit B (CtB). Note the locations within the ARH of the deposits, which were mapped to specific atlas levels of Swanson (2004) that contain the ARH. The deposit in the case shown in the upper panel was localized to the anterior half (rostral–middle) of the ARH, whereas the lower two panels were from subjects receiving deposits mapped to the posterior third (middle–caudal) of the ARH. Note that at atlas level 28 there was overlap of injection deposits for all three cases. Light to heavy shading for each panel corresponds to sparse to robust numbers of retrogradely labeled cells for each sub-region, with no shading indicating an absence of cells in that region. While many regions show similar densities and distributions of afferent input as would be expected from an overlapping series of injection deposits, there are also clear significant differences in such input to the rostral versus caudal portions of the ARH; most notably, the dorsomedial subdivision of the bed nuclei of the stria terminalis (BSTdm), the

investigators the ability to rapidly sample the molecular landscape of the brain while simultaneously facilitating the preservation of the provenance of this molecular information at a resolution comparable to our proposed methods to map such information (for reviews on MALDI, see Cornett et al. 2007; Römpp et al. 2010; Shariatgorji et al. 2014b). MALDI has now been performed for single neurons (Schwartz et al. 2003) and tissue sections (e.g., Heijs et al. 2015), including sections containing hypothalamus (Altelaar et al. 2006; Groseclose et al. 2007; I. Yao et al. 2008; Shariatgorji et al. 2014a) and pituitary (Altelaar et al. 2007). It is now being applied for metabolomics studies of the brain as well (Esteve et al. 2016). Modifications of the original method, including MALDI Fourier Transform Ion Cyclotron Resonance (MALDI FTICR; Spraggins et al. 2016), offer greater mass resolution and accuracy. A promising future direction for MALDI with respect to mapping of molecular information to canonical atlases is the recently reported strategy of combining MALDI with LCM and LC-MS/MS on the same brain section (Dilillo et al. 2017), which would facilitate the retention of provenance information for the molecular datasets mined from the section. Similarly, image fusion strategies that create one image of a tissue section from two registerable source images produced by two separate imaging modalities (MALDI, optical microcopy) also hold great promise for mapping molecular information (van de Plas et al. 2015). Other modalities, such as Raman spectroscopic imaging (Manciu et al. 2013), may offer additional opportunities for high spatial resolution analysis of molecular datasets in the brain.

6.8.3 Future Directions in Molecular Analysis

Alongside developments in imaging technologies are enhanced technologies that allow for spatially resolved molecular sampling of tissue (see Crosetto et al. 2015, for a review). For example, fluorescent in situ sequencing (FISSEQ) of RNA has been developed for intact tissue samples (Lee et al. 2014, 2015). Similarly, Ståhl et al. (2016) have reported the novel use of arrayed reverse transcription primers accompanied by unique positional barcodes, which can be used to generate RNA-sequencing data directly on tissue slides in a manner that preserves the location of the information (also see Navarro et al. 2017). Additionally, single-cell transcriptomic analysis can be performed on individual nuclei obtained from fixed tissue;

Fig. 6.7 (continued) central part of the medial preoptic nucleus (MPOc), the periventricular part of the paraventricular hypothalamic nucleus (PVHp), and the dorsal premammillary nucleus (PMd). These differences in afferent distribution and density suggest that ARH recipient neurons are heterogeneous in at least certain functions (and therefore may differ in cellular state or phenotype), underscoring the need for accurate ARH sampling along the rostrocaudal axis of the nucleus for transcriptomic, proteomic or peptidomic studies. The flatmaps from Swanson (2004) (and available at https://larrywswanson.com) are reproduced and modified here under the conditions of a Creative Commons BY-NC 4.0 license (https://creativecommons.org/licenses/by-nc/4.0/legalcode)

these nuclei are sorted after tissue dissociation procedures via fluorescence-activated cell sorting (FACS) or nucleic acid barcoding. For example, Lake et al. (2016) characterized the single nuclear transcriptomes of cerebral cortical neurons from fixed post mortem human brain. Habib et al. (2017) used barcoded beads to sort individual nuclei taken from fresh or frozen brain samples from mouse and human, and developed a microfluidic device that enables the sorting process. A key future direction would be to integrate spatially resolved transcriptomics procedures and single-cell sequencing efforts into a pipeline that allows for the retention and mapping of the locations from where the samples originate with respect to canonical brain atlases. Along these lines, the Retro-TRAP technology developed by Jeffrey Friedman and colleagues (Ekstrand et al. 2014; Nectow et al. 2015; Pomeranz et al. 2017; derived from the original TRAP technology to identify activated neurons: Knight et al. 2012) to retrogradely label neurons with GFP constructs and then capture translating mRNAs from these neurons using anti-GFP nanobodies (i.e., single-domain antibodies), could potentially allow for projections being mapped for neurons from which single-cell molecular information can also be harvested in a spatially documentable manner. A current limitation of the method for such purposes is that fresh not frozen tissue needs to be harvested to generate sufficient RNA yields, precluding the freezing of tissue sections in preparation for LCM.

6.8.4 *Future Directions in Mapping*

A larger issue concerning the mapping of molecular information is the need to change the scientific culture so that best practices of reporting molecular information in the brain include procedures to map the information to standardized atlases. At present, this is not a common practice by most investigators in neuroscience. Such changes in culture would greatly accelerate the integration of datasets among researchers, and the need to do so is now more critical than ever, given the deluge of spatial, molecular data that has already been shared in repositories such as the Allen Brain Atlas (http://www.brain-map.org) and the GENSAT Project (http://www.gensat.org) (also see Gray et al. 2004; Magdaleno et al. 2006; Lein et al. 2007; Shimogori et al. 2010).

6.8.5 *Final Remarks*

For all of these and future advancements, it will remain critical to preserve information about the native lands from which so many molecules become expatriated, lest the information provided by these molecular datasets fails to link up with the larger neuronal information networks from which they came. Mapping the sampled tissue will provide the critical information that will bridge the gap between the systems biology of molecular information networks on the one hand, and the systems neuroscience of cellular information networks on the other. Without such a bridge, these domains of inquiry may never converge to form a unifying model of a dynamic brain, replete with diverse molecular citizens hailing from different but interconnecting cells, and communicating across local and regional boundaries to signal their neighbors, both near and far.

Acknowledgments We thank Dr. Sabiha Khan (UTEP) for thoughtful discussion on the organization of the manuscript, and Dr. Harold Gainer (National Institute of Neurological Disorders and Stroke) for his timely feedback. We would like to thank the anonymous reviewer who provided critical and constructive feedback on an earlier draft of this manuscript. We also acknowledge our debt to the late Dr. Claude F. Baxter, who served as Emeritus Professor of Psychiatry and Biobehavioral Sciences at the UCLA Brain Research Institute and past historian of the American Society for Neurochemistry, for having generously provided AMK access to his personal library of seminal works in neurochemistry. His kindness and hospitality are treasured memories. We would also like to acknowledge the contributions of Dr. Rebecca Hull and Nishi Gill for the images provided in Fig. 6.2B, C. Finally, we thank Dr. Alexander C. Jackson (University of Connecticut) for providing us with access to unpublished data from his single-cell transcriptomic studies of neuron populations in the mouse lateral hypothalamic area. This chapter is dedicated to the memory of Dr. John H. Ashe (University of California, Riverside), whose instruction and mentorship have deeply informed this narrative.

Funding Work in the UTEP Systems Neuroscience Laboratory is supported by grants awarded to AMK from the National Institutes of Health (NIH; SC3GM109817 and SC1GM127251), the Howard Hughes Medical Institute (UTEP PERSIST Education Grant; PI: S. Aley), and the UTEP Office of Research and Sponsored Projects (Grand Challenges Award). This work is also supported by funds awarded to the Border Biomedical Research Center by the National Institute of Minority Health and Health Disparities of the NIH (5G12MD007592). AHG is supported by the Research Initiative for Scientific Enhancement (RISE) Graduate Fellowship program of the NIH (R25GM069621). AM has been supported by UTEP PERSIST funds and an NSF GK–12 fellowship. Some data in this study were also based upon work supported by the Office of Research and Development, Medical Research Service, Department of Veterans Affairs (VA); specifically, by Merit Review Awards 1I01BX001213-01A1 and BX004102-01 from the United States (U.S.) Department of Veterans Affairs Biomedical Laboratory Research and Development Service to JEB as well as NIH R01DK115976 to JEB. The contents do not represent the views of the U.S. Department of Veterans Affairs or the U.S. Government. This study was also supported by the University of Washington Diabetes Research Center Cellular and Molecular Imaging Core, which is supported by NIH grant P30DK017047. The contribution by GAPCB to this work was funded by the Defense Advanced Research Projects Agency (DARPA) Big Mechanism program under Army Research Office (ARO) contract W911NF-1-0436 and by NIH grant R01LM012592.

References

Abrahams VC, Koelle GB, Smart P. Histochemical demonstration of cholinesterases in the hypothalamus of the dog. J Physiol (London). 1957;139:137–44.

Adler ES, Hollis JH, Clarke IJ, Grattan DR, Oldfield BJ. Neurochemical characterization and sexual dimorphism of projections from the brain to abdominal and subcutaneous white

adipose tissue in the rat. J Neurosci. 2012;32(45):15913–21. https://doi.org/10.1523/JNEUROSCI.2591-12.2012.
Adrian ED. Ferrier Lecture: the localization of activity in the brain. Proc R Soc London B: Biol Sci. 1939;126:433–49. https://doi.org/10.1098/rspb.1939.0001.
Ahmed SH, Lutjens R, van der Stap LD, Lekic D, Romano-Spica V, Morales M, et al. Gene expression evidence for remodeling of lateral hypothalamic circuitry in cocaine addiction. Proc Natl Acad Sci U S A. 2005;102(32):11533–8. https://doi.org/10.1073/pnas.0504438102.
Akbari EM, Shams S, Belay HT, Kaiguo M, Razak Z, Kent CF, Westwood T, Sokolowski MB, Fleming AS. The effects of parity and maternal behavior on gene expression in the medial preoptic area and the medial amygdala in postpartum and virgin female rats: a microarray study. Behav Neurosci. 2013;127(6):913–22.
Akhtar RA, Reddy AB, Maywood ES, Clayton JD, King VM, et al. Circadian cycling of the mouse liver transcriptome, as revealed by cDNA microarray, is driven by the suprachiasmatic nucleus. Curr Biol. 2002;12:540–50. https://doi.org/10.1016/S0960-9822(02)00759-5.
Alexandre-Gouabau M-CF, Bailly E, Moyon TL, Grit IC, Coupé B, et al. Postnatal growth velocity modulates alterations of proteins involved in metabolism and neuronal plasticity in neonatal hypothalamus in rats born with intrauterine growth restriction. J Nutr Biochem. 2012;23:140–52. https://doi.org/10.1016/j.jnutbio.2010.11.008.
Altelaar AFM, Klinkert I, Jalink K, de Lange RPJ, Adan RAH, Heeren RMA, Piersma SR. Gold-enhanced biomolecular surface imaging of cells and tissue by SIMS and MALDI mass spectrometry. Anal Chem. 2006;78:734–42. https://doi.org/10.1021/ac0513111.
Altelaar AFM, Taban IM, McDonnell LA, Verhaert PDEM, de Lange RPJ, Adan RAH, et al. High-resolution MALDI imaging mass spectrometry allows localization of peptide distributions at cellular length scales in pituitary tissue sections. Int J Mass Spectr. 2007;260:203–11. https://doi.org/10.1016/j.ijms.2006.09.028.
Alzate O. Neuroproteomics. Boca Raton: CRC Press/Taylor & Francis; 2010. https://www.ncbi.nlm.nih.gov/books/NBK56022/.
Amar L, Benoit C, Beaumont G, Vacher CM, Crepin D, Taouis M, et al. MicroRNA expression profiling of hypothalamic arcuate and paraventricular nuclei from single rats using Illumina sequencing technology. J Neurosci Methods. 2012;209(1):134–43. https://doi.org/10.1016/j.jneumeth.2012.05.033.
Amigó-Correig M, Barceló-Batllori S, Soria G, Krezymon A, Benani A, Pénicaud L, et al. Anti-obesity sodium tungstate treatment triggers axonal and glial plasticity in hypothalamic feeding centers. PLoS One. 2012;7(7):e39087. https://doi.org/10.1371/journal.pone.0039087.
Andermann ML, Lowell BB. Toward a wiring diagram understanding of appetite control. Neuron. 2017;95:757–78. https://doi.org/10.1016/j.neuron.2017.06.014.
Anglade P, Larabi-Godinot Y. Historical landmarks in the histochemistry of the cholinergic synapse: perspectives for future researches. Biomed Res. 2010;31:1–12. https://doi.org/10.2220/biomedres.31.1.
Arai Y, Gradwohl G, Kameda Y. Expression of neuropeptide Y and agouti-related peptide in the hypothalamic arcuate nucleus of newborn neurogenin3 null mutant mice. Cell Tissue Res. 2010;340:137–45. https://doi.org/10.1007/s00441-009-0925-4.
Aramakis VB, Stanley BG, Ashe JH. Neuropeptide Y receptor agonists: multiple effects on spontaneous activity in the paraventricular hypothalamus. Peptides. 1996;17:1349–57. https://doi.org/10.1016/S0196-9781(96)00222-7.
Argüelles S, Cano M, Machado A, Ayala A. Effect of aging and oxidative stress on elongation factor-2 in hypothalamus and hypophysis. Mech Ageing Dev. 2011;132:55–64. https://doi.org/10.1016/j.mad.2010.12.002.
Atkins N, Miller CM, Owens JR, Turek FW. Non-laser capture microscopy approach for the microdissection of discrete mouse brain regions for total RNA isolation and downstream next-generation sequencing and gene expression profiling. J Vis Exp. 2011;57:e3125. https://doi.org/10.3791/3125.
Aubert Y, Allers KA, Bernd S, Ronald de Kloet E, Abbott DH, Datson NA. Brain region specific transcriptomic markers of Serotonin1A receptor agonist action mediating sexual rejection and aggression in female marmoset monkeys. J Sex Med. 2013;10(6):1461–75.

Auger CJ, Jessen HM, Auger AP. Microarray profiling of gene expression patterns in adult male rat brain following acute progesterone treatment. Brain Res. 2006;1067(1):58–66.

Azzam S, Schlatzer D, Nethery D, Saleh D, Li X, Akladious A, et al. Proteomic profiling of the hypothalamus in two mouse models of narcolepsy. Proteomics. 2017;17(13–14):1600478. https://doi.org/10.1002/pmic.201600478.

Backholer K, Smith JT, Rao A, Pereira A, Iqbal J, Ogawa S, et al. Kisspeptin cells in the ewe brain respond to leptin and communicate with neuropeptide Y and proopiomelanocortin cells. Endocrinology. 2010;151:2233–43. https://doi.org/10.1210/en.2009-1190.

Badea A, Johnson GA, Williams RW. Genetic dissection of the mouse brain using high-field magnetic resonance microscopy. NeuroImage. 2009;45(4):1067–79. https://doi.org/10.1016/j.neuroimage.2009.01.021.

Bakay L. Studies on blood-brain barrier with radioactive phosphorus: II. Hypophysis and hypothalamus in man. AMA Arch Neurol Psychiatry. 1952;68(5):629–40.

Balakrishnan CN, Mukai M, Gonser RA, Wingfield JC, London SE, Tuttle EM, Clayton DF. Brain transcriptome sequencing and assembly of three songbird model systems for the study of social behavior. PeerJ. 2014;2:e396. https://doi.org/10.7717/peerj.396.

Balázs R, Patel AJ, Richter D. Metabolic compartments in the brain: their properties and relation to morphological structures. In: Balázs R, Cremer JE, editors. Metabolic compartmentation in the brain. New York: John Wiley & Sons; 1972. p. 167–84.

Balivada S, Ganta CK, Zhang Y, Pawar HN, Ortiz RJ, Khan AM, Kenney MJ. Microarray analysis of aging-associated immune system alterations in the rostral ventrolateral medulla of F344 rats. Physiol Genomics. 2017;49(8):400–15. https://doi.org/10.1152/physiolgenomics.00131.2016.

Baskin DG, Bastian LS. Immuno-laser capture microdissection of rat brain neurons for real time quantitative PCR. Methods Mol Biol. 2010;588:219–30. https://doi.org/10.1007/978-1-59745-324-0_23.

Baskin DG, Kim F, Gelling RW, Russell BJ, Schwartz MW, Morton GJ, et al. A new oxytocin-saporin cytotoxin for lesioning oxytocin-receptive neurons in the rat hindbrain. Endocrinology. 2010;151(9):4207–13. https://doi.org/10.1210/en.2010-0295.

Bayliss WM, Starling EH. The mechanism of pancreatic secretion. J Physiol (London). 1902;28(5):325–53. https://doi.org/10.1113/jphysiol.1902.sp000920.

Benner S, Kakeyama M, Endo T, Yoshioka W, Tohyama C. Application of NeuroTrace staining in the fresh frozen brain samples to laser microdissection combined with quantitative RT-PCR analysis. BMC Res Notes. 2015;8:252. https://doi.org/10.1186/s13104-015-1222-9.

Bennett MR. History of the synapse. Amsterdam: Harwood Academic Press; 2001.

Bernard R, Kerman IA, Meng F, Evans SJ, Amrein I, Jones EG, et al. Gene expression profiling of neurochemically defined regions of the human brain by in situ hybridization-guided laser capture microdissection. J Neurosci Meth. 2009;178(1):46–54. https://doi.org/10.1016/j.jneumeth.2008.11.012.

Bi S, Robinson BM, Moran TH. Acute food deprivation and chronic food restriction differentially affect hypothalamic NPY mRNA expression. Am J Physiol Regul Integr Comp Physiol. 2003;285:R1030–6. https://doi.org/10.1152/ajpregu.00734.2002.

Blevins JE, Hamel FG, Fairbairn E, Stanley BG, Reidelberger RD. Effects of paraventricular nucleus injection of CCK-8 on plasma CCK-8 levels in rats. Brain Res. 2000a;860:11–20. https://doi.org/10.1016/S0006-8993(99)02478-6.

Blevins JE, Stanley BG, Reidelberger RD. Brain regions where cholecystokinin suppresses feeding in rats. Brain Res. 2000b;860:1–10. https://doi.org/10.1016/S0006-8993(99)02477-4.

Blevins JE, Morton GJ, Williams DL, Caldwell DW, Bastian LS, Wisse BE, et al. Forebrain melanocortin signaling enhances the hindbrain satiety response to CCK-8. Am J Physiol Regul Integr Comp Physiol. 2009;296:R476–84. https://doi.org/10.1152/ajpregu.90544.2008.

Blevins JE, Moralejo DH, Wolden-Hanson TH, Thatcher BS, Ho JM, Kaiyala KJ, et al. Alterations in activity and expenditure contribute to lean phenotype in Fischer 344 rats lacking the cholecystokinin-1 receptor gene. Am J Physiol Regul Integr Comp Physiol. 2012;303:R1231–40. https://doi.org/10.1152/ajpregu.00393.2012.

Bochukova EG, Lawler K, Croizier S, Keogh JM, Patel N, Strohbehn G, Lo KK, Humphrey J, Hokken-Koelega A, Damen L, Donze S, Bouret SG, Vincent P, Sadaf Farooqi I. A transcriptomic signature of the hypothalamic response to fasting and BDNF deficiency in Prader-Willi syndrome. Cell Rep. 2018;22(13):3401–8.

Bonaventure P, Guo H, Tian B, Liu X, Bittner A, Roland B, et al. Nuclei and subnuclei gene expression profiling in mammalian brain. Brain Res. 2002;943:38–47. https://doi.org/10.1016/S0006-8993(02)02504-0.

Bonnavion P, Mickelsen LE, Fujita A, de Lecea L, Jackson AC. Hubs and spokes of the lateral hypothalamus: cell types, circuits and behaviour. J Physiol. 2016;594(22):6443–62. https://doi.org/10.1113/JP271946.

Boone DR, Sell SL, Micci M-A, Crookshanks JM, Parsley M, Uchida T, et al. Traumatic brain injury-induced dysregulation osf the circadian clock. PLoS One. 2012;7(10):e46204. https://doi.org/10.1371/journal.pone.0046204.

Boone DR, Sell SL, Hellmich HL. Laser capture microdissection of enriched populations of neurons or single neurons for gene expression analysis after traumatic brain injury. J Vis Exp. 2013;74:e50308. https://doi.org/10.3791/50308.

Bora A, Annangudi SP, Millet LJ, Rubakhin SS, Forbes AJ, Kelleher NL, et al. Neuropeptidomics of the supraoptic rat nucleus. J Proteome Res. 2008;7(11):4992–5003. https://doi.org/10.1021/pr800394e.

Borell U, Örström Å. Metabolism in different parts of the brain, especially in the epiphysis, measured with radioactive phosphorus. Acta Physiol Scand. 1945;10(3–4):231–42.

Boring EG. Sensation and perception in the history of experimental psychology. New York: Appleton Century Crofts; 1942.

Briski KP, Nedungadi TP, Koshy Cherian A. Effects of hypoglycaemia on neurotransmitter and hormone receptor gene expression in laser- dissected arcuate neuropeptide Y/agouti-related peptide neurones. J Neuroendocrinol. 2010;22(6):599–607.

Bures EJ, Courchesne PL, Douglass J, Chen K, Davis MT, Jones MD, et al. Identification of incompletely processed potential Carboxypeptidase E substrates from CpE^{fat}/CpE^{fat} mice. Proteomics. 2001;1:79–92. https://doi.org/10.1002/1615-9861(200101)1:1<79::AID-PROT79>3.0.CO;2-8.

Burns GAPC, Khan AM, Ghandeharizadeh S, O'Neill MA, Chen Y-S. Tools and approaches for the construction of knowledge models from the neuroscientific literature. Neuroinformatics. 2003;1(1):81–109.

Butler AE, Matveyenko AV, Kirakossian D, Park J, Gurlo T, Butler PC. Recovery of high-quality RNA from laser capture microdissected human and rodent pancreas. J Histotechnol. 2016;39:59–65. https://doi.org/10.1080/01478885.2015.1106073.

Byerly MS, Simon J, Cogburn LA, Le Bihan-Duval E, Duclos MJ, Aggrey SE, et al. Transcriptional profiling of hypothalamus during development of adiposity in genetically selected fat and lean chickens. Physiol Genomics. 2010;42:157–67. https://doi.org/10.1152/physiolgenomics.00029.2010.

Cai X, Wang C, Xu J, Xue X, Zhang X, et al. Application of matrix solid-phase dispersion methodology to the extraction of endogenous peptides from porcine hypothalamus samples for MS and LC-MS analysis. J Chromatogr B. 2011;879:657–61. https://doi.org/10.1016/j.jchromb.2011.01.038.

Campbell JN, Macosko EZ, Fenselau H, Pers TH, Lyubetskaya A, Tenen D, et al. A molecular census of arcuate hypothalamus and median eminence cell types. Nat Neurosci. 2017;20(3):484–96. https://doi.org/10.1038/nn.4495.

Cao Z, Fan R, Meng B, Xing Z, Liu M, Gao M, Luan X. Comparative proteomic analysis of hypothalamus tissue from Huoyan geese between pre-laying period and laying period using an iTRAQ-based approach. Anim Sci J. 2018;89(7):946–55.

Caprioli RM, Farmer TB, Gile J. Molecular imaging of biological samples: localization of peptides and proteins using MALDI-TOF MS. Anal Chem. 1997;69:4751–60. https://doi.org/10.1021/ac970888i.

Carreño FR, Walch JD, Dutta M, Nedungadi TP, Cunningham JT. Brain-derived neurotrophic factor-tyrosine kinase B pathway mediates NMDA receptor NR2B phosphorylation in the supraoptic nuclei following progressive dehydration. J Neuroendocrinol. 2011;23:894–905. https://doi.org/10.1111/j.1365-2826.2011.02209.x.

Chadwick W, Martin B, Chapter MC, Park S-S, Wang L, Daimon CM, et al. GIT2 acts as a potential keystone protein in functional hypothalamic networks associated with age-related phenotypic changes in rats. PLoS One. 2012;7(5):e36975. https://doi.org/10.1371/journal.pone.0036975.

Chao C-M, Cheng B-C, Chen C-Y, Lin M-T, Chang C-P, Yang S-T. Proteomic analysis of hypothalamic injury in heatstroke rats. Proteomics. 2015;15:1921–34. https://doi.org/10.1002/pmic.201400492.

Che F-Y, Yuan Q, Kalinina E, Fricker LD. Peptidomics of $Cpe^{fat/fat}$ mouse hypothalamus. J Biol Chem. 2005;280:4451–61. https://doi.org/10.1074/jbc.M411178200.

Che F-Y, Zhang X, Berezniuk I, Callaway M, Lim J, Fricker LD. Optimization of neuropeptide extraction from the mouse hypothalamus. J Proteome Res. 2007;6:4667–76. https://doi.org/10.1021/pr060690r.

Chen C-F, Shiue Y-L, Yen C-J, Tang P-C, Chang H-C, Lee Y-P. Laying traits and underlying transcripts, expressed in the hypothalamus and pituitary gland, that were associated with egg production variability in chickens. Theriogenology. 2007a;68:1305–15. https://doi.org/10.1016/j.theriogenology.2007.08.32

Chen J, Repunte-Canonigo V, Kawamura T, Lefebvre C, Shin W, Howell LL, et al. Hypothalamic proteoglycan syndecan-3 is a novel cocaine addiction resilience factor. Nat Commun. 2013;4:1955. https://doi.org/10.1038/ncomms2955.

Chen L-R, Chao C-H, Chen C-F, Lee Y-P, Chen Y-L, Shiue Y-L. Expression of 25 high egg production related transcripts that identified from hypothalamus and pituitary gland in red-feather Taiwan country chickens. Animal Reproduction Sci. 2007b;100:172–85. https://doi.org/10.1016/j.anireprosci.2006.07.005

Chen R, Wu X, Jiang L, Zhang Y. Single-cell RNA-Seq reveals hypothalamic cell diversity. Cell Rep. 2017;18:3227–41. https://doi.org/10.1016/j.celrep.2017.03.004.

Chiang C-K, Mehta N, Patel A, Zhang P, Ning Z, Mayne J, et al. The proteomic landscape of the suprachiasmatic nucleus clock reveals large-scale coordination of key biological processes. PLoS Genet. 2014;10(10):e1004695. https://doi.org/10.1371/journal.pgen.1004695.

Chin MH, Geng AB, Khan AH, Qian W-J, Petyuk VA, Boline J, et al. A genome-scale map of expression for a mouse brain section obtained using voxelation. Physiol Genomics. 2007;30:313–21. https://doi.org/10.1152/physiolgenomics.00287.2006.

Chu L-Y. A cytological study of anterior horn cells isolated from human spinal cord. J Comp Neurol. 1954;100(2):381–413.

Chung C-Y, Seo H, Sonntag KC, Brooks A, Lin L, Isacson O. Cell type-specific gene expression of midbrain dopaminergic neurons reveals molecules involved in their vulnerability and protection. Hum Mol Genet. 2005;14:1709–25. https://doi.org/10.1093/hmg/ddi178.

Chung S, Weber F, Zhong P, Tan CL, Nguyen TN, Beier KT, Hörmann N, Chang W-C, Zhang Z, Do JP, Yao S, Krashes MJ, Tasic B, Cetin A, Zeng H, Knight ZA, Luo L, Yang D. Identification of preoptic sleep neurons using retrograde labelling and gene profiling. Nature. 2017;545(7655):477–81. https://doi.org/10.1038/nature22350.

Colgrave ML, Xi L, Lehnert SA, Flatscher-Bader T, Wadensten H, et al. Neuropeptide profiling of the bovine hypothalamus: thermal stabilization is an effective tool in inhibiting post-mortem degradation. Proteomics. 2011;11:1264–76. https://doi.org/10.1002/pmic.201000423.

Conti B, Maier R, Barr AM, Morale MC, Lu X, Sanna PP, et al. Region-specific transcriptional changes following the three antidepressant treatments electro convulsive therapy, sleep deprivation and fluoxetine. Mol Psychiatry. 2007;12:167–89. https://doi.org/10.1038/sj.mp.4001897.

Cornett DS, Reyzer ML, Chaurand P, Caprioli RM. MALDI imaging mass spectrometry: molecular snapshots of biochemical systems. Nat Methods. 2007;4:828–33. https://doi.org/10.1038/NMETH1094.

Crosetto N, Bienko M, van Oudenaarden A. Spatially resolved transcriptomics and beyond. Nat Rev Genet. 2015;16:57–66. https://doi.org/10.1038/nrg3832.

Cubuk C, Kemmling J, Fabrizius A, Herwig A. Transcriptome analysis of hypothalamic gene expression during daily torpor in Djungarian hamsters (Phodopus sungorus). Front Neurosci. 2017;11:122.

Dalal J, Roh JH, Maloney SE, Akuffo A, Shah S, Yuan H, et al. Translational profiling of hypocretin neurons identifies candidate molecules for sleep regulation. Genes Dev. 2013;27:565–78. https://doi.org/10.1101/gad207654.112.

Datta S, Malhotra L, Dickerson R, Chaffee S, Sen CK, Roy S. Laser capture microdissection: big data from small samples. Histol Histopathol. 2015;30:1255–69. https://doi.org/10.14670/HH-11-622.

de Lecea L, Kilduff TS, Peyron C, Gao X-B, Foye PE, Danielson PE, Fukuhara C, Battenberg ELF, Gautvik VT, Bartlett FS, Frankel WN, van den Pol AN, Bloom FE, Gautvik KM, Sutcliffe JG. The hypocretins: hypothalamus-specific peptides with neuroexcitatory activity. Proc Natl Acad Sci. 1998;95(1):322–7.

De Palma A, Pareti G. Bernstein's long path to membrane theory: radical change and conservation in nineteenth-century German electrophysiology. J Hist Neurosci. 2011;20(4):306–37. https://doi.org/10.1080/0964704X.2010.532024.

Deans AR, Lewis SE, Huala E, Anzaldo SS, Ashburner M, Balhoff JP, et al. Finding our way through phenotypes. PLoS Biol. 2015;13:e1002033. https://doi.org/10.1371/journal.pbio.1002033.

DeAtley KL, Colgrave ML, Cánovas A, Wijffels G, Ashley RL, Silver GA, et al. Neuropeptidome of the hypothalamus and pituitary gland of Indicine × Taurine heifers: evidence of differential neuropeptide processing in the pituitary gland before and after puberty. J Proteome Res. 2018;17:1852–65. https://doi.org/10.1021/acs.jproteome.7b00875.

Décaillot FM, Che F-Y, Fricker LD, Devi LA. Peptidomics of $Cpe^{fat/fat}$ mouse hypothalamus and striatum: effect of chronic morphine administration. J Mol Neurosci. 2006;28(3):277–84.

Deery MJ, Maywood ES, Chesham JE, Sládek M, Karp NA, Green EW, Charles PD, Reddy AB, Kyriacou CP, Lilley KS, Hastings MH. Proteomic analysis reveals the role of synaptic vesicle cycling in sustaining the suprachiasmatic circadian clock. Curr Biol. 2009;19(23):2031–6.

Deiters O. Untersuchungen über Gehirn und Rückenmark des Menschen und der Säugethiere. Brunswick: Von Friedrich & Son; 1865.

Deiters VS, Guillery RW. Otto Friedrich Karl Deiters (1834–1863). J Comp Neurol. 2013;521(9):1929–53.

Del Pino SJ, Krishnan S, Aggison LK, Adams HL, Shrikant MM, Lopez-Giraldez F, Petersen SL. Microarray analysis of neonatal rat anteroventral periventricular transcriptomes identifies the proapoptotic Cugbp2 gene as sex-specific and regulated by estradiol. Neuroscience. 2015;303:312–22. https://doi.org/10.1016/j.neuroscience.2015.07.008

DiCarlo LM, Vied C, Nowakowski RS. The stability of the transcriptome during the estrous cycle in four regions of the mouse brain. J Comp Neurol. 2017;525(15):3360–87.

Dilillo M, Pellegrini D, Ait-Belkacem R, de Graaf EL, Caleo M, McDonnell LA. Mass spectrometry imaging, laser capture microdissection, and LC-MS/MS of the same tissue section. J Proteome Res. 2017;16(8):2993–3001. https://doi.org/10.1021/acs.jproteome.7b00284.

Ding F, Li HH, Li J, Myers RM, Francke U. Neonatal maternal deprivation response and developmental changes in gene expression revealed by hypothalamic gene expression profiling in mice. PLoS One. 2010;5(2):e9402. https://doi.org/10.1371/journal.pone.0009402.

Dong HW. The allen reference atlas: a digital color brain atlas of the C57BL/6J male mouse. New York: Wiley; 2008.

Dong H-W, Swanson LW, Chen L, Fanselow MS, Toga AW. Genomic-anatomic evidence for distinct functional domains in hippocampal field CA1. Proc Natl Acad Sci U S A. 2009;106(28):11794–9. https://doi.org/10.1073/pnas.0812608106.

Dooley GP, Ashley AK, Legare ME, Handa RJ, Hanneman WH. Proteomic analysis of diaminochlorotriazine (DACT) adducts in three brain regions of Wistar rats. Toxicol Lett. 2010;199(1):17–21.

Doubi-Kadmiri S, Benoit C, Benigni X, Beaumont G, Vacher C-M, Taouis M, et al. Substantial and robust changes in microRNA transcriptome support postnatal development of the hypothalamus in rat. Sci Rep. 2016;6:24896. https://doi.org/10.1038/srep24896.

Dougherty JD, Schmidt EF, Nakajima M, Heintz N. Analytical approaches to RNA profiling data for the identification of genes enriched in specific cells. Nucleic Acids Res. 2010;38:4218–30. https://doi.org/10.1101/gad.207654.112.

Doyle JP, Dougherty JD, Heiman M, Schmidt EF, Stevens TR, Ma G, et al. Application of a translational profiling approach for the comparative analysis of CNS cell types. Cell. 2008;135(4):749–62. https://doi.org/10.1016/j.cell.2008.10.029.

Draper S, Kirigiti M, Glavas M, Bernadette G, Angie Chong CN, Betty J, Susan Smith M, Zeltser LM, Grove KL. Differential gene expression between neuropeptide Y expressing neurons of the dorsomedial nucleus of the hypothalamus and the Arcuate nucleus: microarray analysis study. Brain Res. 2010;1350:139–50.

Drnevich J, Replogle KL, Lovell P, Hahn TP, Johnson F, Mast TG, et al. Impact of experience-dependent and –independent factors on gene expression in songbird brain. Proc Natl Acad Sci U S A. 2012;109(Suppl 2):17245–52. https://doi.org/10.1073/pnas.1200655109.

Dunn MJ. Two-dimensional gel electrophoresis of proteins. J Chromatogr B Biomed Sci Appl. 1987;418:145–85.

Eberwine J, Bartfai T. Single cell transcriptomics of hypothalamic warm sensitive neurons that control core body temperature and fever response. Signaling asymmetry and an extension of chemical neuroanatomy. Pharmacol Ther. 2011;129(3):241–59. https://doi.org/10.1016/j.pharmthera.2010.09.010.

Eberwine J, Yeh H, Miyashiro K, Cao Y, Nair S, Finnell R, et al. Analysis of gene expression in single live neurons. Proc Natl Acad Sci U S A. 1992;89:3010–4.

Ekstrand MI, Nectow AR, Knight ZA, Latcha KN, Pomeranz LE, Friedman JM. Molecular profiling of neurons based on connectivity. Cell. 2014;157(5):1230–42. https://doi.org/10.1016/j.cell.2014.03.059.

Elphick MR, Mirabeau O, Larhammar D. Evolution of neuropeptide signaling systems. J Exp Biol. 2018;221(Pt. 3):jeb151092. https://doi.org/10.1242/jeb.151092.

Emmert-Buck MR, Bonner RF, Smith PD, Chuaqui RH, Zhengping Z, Goldstein SR, et al. Laser capture microdissection. Science. 1996;274:998–1001.

Espina V, Wulfkuhle JD, Calvert VS, VanMeter A, Zhou W, Coukos G, et al. Laser-capture microdissection. Nat Protoc. 2006;1:586–603. https://doi.org/10.1038/nprot.2006.85.

Esteve C, Tolner EA, Shyti R, van den Maagdenberg AMJM, McDonnell LA. Mass spectrometry imaging of amino neurotransmitters: a comparison of derivatization methods and application in mouse brain tissue. Metabolomics. 2016;12:30. https://doi.org/10.1007/s11306-015-0926-0.

Fälth M, Sköld K, Norrman M, Svensson M, Fenyö D, Andren PE. SwePep, a database designed for endogenous peptides and mass spectrometry. Mol Cell Proteomics. 2006;5:998–1005. https://doi.org/10.1074/mcp.M500401-MCP200.

Fang X-L, Zhu X-T, Chen S-F, Zhang Z-Q, Zeng Q-J, Deng L, et al. Differential gene expression pattern in hypothalamus of chickens during fasting-induced metabolic reprogramming: functions of glucose and lipid metabolism in the feed intake of chickens. Poult Sci. 2014;93:2841–54. https://doi.org/10.3382/ps.2014-04047.

Farajzadeh L, Hornshøj H, Momeni J, Thomsen B, Larsen K, et al. Pairwise comparisons of ten porcine tissues identify differential transcriptional regulation at the gene, isoform, promoter and transcription start site level. Biochem Biophys Res Commun. 2013;438:346–52. https://doi.org/10.1016/j.bbrc.2013.07.074.

Fassunke J, Majores M, Ullmann C, Elger CE, Schramm J, Wiestler OD, et al. In situ-RT and immunolaser microdissection for mRNA analysis of individual cells isolated from epilepsy-associated glioneuronal tumors. Lab Investig. 2004;84:1520–5. https://doi.org/10.1038/labinvest.3700165.

Fatt P, Katz B. The electrical activity of the motor end-plate. Proc R Soc Lond B Biol Sci. 1952;140:183–6. Available online: http://0-www.jstor.org.lib.utep.edu/stable/82687.

Feldberg W. Central excitation and inhibition from the view point of chemical transmission. Proc R Soc Lond B Biol Sci. 1952;140:199–202. Available online: http://0-www.jstor.org.lib.utep.edu/stable/82690.

Feldberg W, Vogt M. Acetylcholine synthesis in different regions of the central nervous system. J Physiol. 1948;107:372–81.

Feldman SE, Snapir N, Yasuda M, Treuting F, Lepkovsky S. Physiological and nutritional consequences of brain lesions: a functional atlas of the chicken hypothalamus. Hilgardia. 1973;41(19):605–30.

Ferrier D. Experimental researches in cerebral physiology and pathology. West Riding Lunatic Asylum Med Rep. 1873;3:30–96.

Fink L, Kinfe T, Stein MM, Ermert L, Hanze J, Kummer W, et al. Immunostaining and laser-assisted cell picking for mRNA analysis. Lab Investig. 2000;80:327–33.

Firmino M, Weis SN, Souza JMF, Gomes BRB, Mól AR, Mortari MR, Souza GEP, Coca GC, Williams TCR, Fontes W, Ricart CAO, de Sousa MV, Veiga-Souza FH. Label-free quantitative proteomics of rat hypothalamus under fever induced by LPS and PGE 2. J Proteome. 2018;187:182–99.

Flourens P. Recherches expérimentales sur les propriétés et les fonctions du système nerveux, dans les animaux vertébrés. Paris: Chez Crevot; 1824.

Fodor S, Read J, Pirrung M, Stryer L, Lu A, Solas D. Light-directed, spatially addressable parallel chemical synthesis. Science. 1991;251(4995):767–73.

Forssburg A, Larsson S. Studies of isotope distribution and chemical composition in the hypothalamic region of hungry and fed rats. Published as Part II (pp. 41–63) of: Larsson S. 1954. On the hypothalamic organisation of the nervous mechanism regulating food intake. Acta Physiol Scand Suppl. 1954;32(115):7–63.

Fortes MRS, Snelling WM, Reverter A, Nagaraj SH, Lehnert SA, Hawken RJ, et al. Gene network analysis of first service conception in Brangus heifers: use of genome and trait associations, hypothalamic-transcriptome information, and transcription factors. J Anim Sci. 2015;2012(90):2894–906. https://doi.org/10.2527/jas2011-4601

Fortes MRS, Nguyen LT, Weller MMDCA, Cánovas A, Islas-Trejo A, Porto-Neto LR, et al. Transcriptome analyses identify five transcription factors differentially expressed in the hypothalamus of post- versus prepubertal Brahman heifers. J Anim Sci. 2016;94:3693–702. https://doi.org/10.2527/jas2016-0471.

Foster NN, Azam S, Watts AG. Rapid-onset hypoglycemia suppresses Fos expression in discrete parts of the ventromedial nucleus of the hypothalamus. Am J Physiol Regul Integr Comp Physiol. 2016;310:R1177–85. https://doi.org/10.1152/ajpregu.00042.2016.

Fouillen L, Petruzziello F, Veit J, Bhattacharyya A, Kretz R, Rainer G, et al. Neuropeptide alterations in the tree shrew hypothalamus during volatile anesthesia. J Proteome. 2013;80:311–9. https://doi.org/10.1016/j/jprot.2012.11.002.

Frese CK, Boender AJ, Mohammed S, Heck AJR, Adan RAH, Altelaar AFM. Profiling of diet-induced neuropeptide changes in rat brain by quantitative mass spectrometry. Anal Chem. 2013;85:4594–604. https://doi.org/10.1021/ac400232y.

Friede RL. Topographic brain chemistry. New York: Academic Press; 1966.

Gao G, Li Q, Zhao X, Ding N, Han Q, Su J, Wang Q. Transcriptome profiling of the hypothalamus during prelaying and laying periods in Sichuan white geese (Anser cygnoides). Anim Sci J. 2015;86(8):800–5. https://doi.org/10.1111/asj.12356.

Gao Y, Chen S, Xu Q, Yu K, Wang J, Qiao L, et al. Proteomic analysis of differential proteins related to anti-nociceptive effect of electroacupuncture in the hypothalamus following neuropathic pain in rats. Neurochem Res. 2013;38:1467–78. https://doi.org/10.1007/s11064-013-1047-7.

Gao Y-Z, Guo S-Y, Yin Q-Z, Hisamitsu T, Jiang X-H. An individual variation study of electroacupuncture analgesia in rats using microarray. Am J Chin Med. 2007;35(05):767–78.

Gasperini L, Piubelli C, Carboni L. Proteomics of rat hypothalamus, hippocampus and pre-frontal/frontal cortex after central administration of the neuropeptide PACAP. Mol Biol Rep. 2012;30:2921–35. https://doi.org/10.1007/s11033-011-1054-1.

Gauss C, Kalkum M, Löwe M, Lehrach H, Klose J. Analysis of the mouse proteome. (I) Brain proteins: separation by two-dimensional electrophoresis and identification by mass spectrometry and genetic variation. Electrophoresis. 1999;20(3):575–600.

Gautvik KM, de Lecea L, Gautvik VT, Danielson PE, Tranque P, Dopazo A, et al. Overview of the most prevalent hypothalamus-specific mRNAs, as identified by directional tag PCR subtraction. Proc Natl Acad Sci U S A. 1996;93(16):8733–8. https://doi.org/10.1073/pnas.93.16.8733.

Ghorbel MT, Sharman G, Leroux M, Barrett T, Donovan DM, Becker KG, Murphy D. Microarray analysis reveals interleukin-6 as a novel secretory product of the hypothalamo-neurohypophyseal system. J Biol Chem. 2003;278(21):19280–5.

Giacobini E. Histochemical demonstration of AChE activity in isolated nerve cells. Acta Physiol Scand. 1956;36(3):276–90.

Gillespie JW, Best CJ, Bichsel VE, Cole KA, Greenhut SF, Hewitt SM, et al. Evaluation of non-formalin tissue fixation for molecular profiling studies. Am J Pathol. 2002;160:449–57. https://doi.org/10.1016/S0002-9440(10)64864-X.

Ginsberg SD, Che S. Combined histochemical staining, RNA amplification, regional and single-cell cDNA analysis within the hippocampus. Lab Investig. 2004;84:952–62. https://doi.org/10.1038/labinvest.3700110.

Glasgow E, Kusano K, Chin H, Mezey E, Young WS III, Gainer H. Single-cell reverse transcription-polymerase chain reaction analysis of rat supraoptic magnocellular neurons: neuropeptide phenotypes and high voltage-gated calcium channel subtypes. Endocrinology. 1999;140:5391–401. https://doi.org/10.1210/endo.140.11.7136.

Goldsworthy SM, Stockton PS, Trempus CS, Foley JF, Maronpot RR. Effects of fixation on RNA extraction and amplification from laser capture microdissected tissue. Mol Carcinog. 1999;25:86–91.

González CR, Martínez de Morentin PB, Martínez-Sánchez N, Gómez-Díaz C, Lage R, Varela L, et al. Hyperthyroidism differentially regulates neuropeptide S system in the rat brain. Brain Res. 2012;1450:40–8. https://doi.org/10.1016/j.brainres.2012.02.24.

Gouraud SS, Yao ST, Heesom KJ, Paton JFR, Murphy D. 14-3-3 proteins within the hypothalamic-neurohypophyseal system of the osmotically stressed rat: transcriptomic and proteomic studies. J Neuroendocrinol. 2007;19:913–22. https://doi.org/10.1111/j.1365-2826.2007.01604.x.

Govindaraj V, Shridharan RN, Rao AJ. Proteomic changes during adult stage in pre-optic, hypothalamus, hippocampus and pituitary regions of female rat brain following neonatal exposure to estradiol-17Î². Gen Comp Endocrinol. 2018;266:126–34.

Gray PA. Mouse brain organization revealed through direct genome-scale TF expression analysis. Science. 2004;306(5705):2255–7.

Greenberg N. A forebrain atlas and stereotaxic technique for the lizard, Anolis carolinensis. J Morphol. 1982;174(2):217–36.

Greenwood MP, Mecawi AS, Hoe SZ, Mustafa MR, Johnson KR, Al-Mahmoud GA, et al. A comparison of physiological and transcriptome responses to water deprivation and salt loading in the rat supraoptic nucleus. Am J Physiol Regul Integr Comp Physiol. 2015;308:R559–68. https://doi.org/10.1152/ajpregu.00444.2014.

Groseclose MR, Andersson M, Hardesty WM, Caprioli RM. Identification of proteins directly from tissue: in situ tryptic digestions coupled with imaging mass spectrometry. J Mass Spectrom. 2007;42:254–62. https://doi.org/10.1002/jms.1177.

Guest PC, Urday S, Ma D, Stelzhammer V, Harris LW, Amess B, et al. Proteomic analysis of the maternal protein restriction rat model for schizophrenia: identification of translational changes in hormonal signaling pathways and glutamate neurotransmission. Proteomics. 2012;12:3580–9. https://doi.org/10.1002/pmic.201200376.

Habib N, Avraham-Davidi I, Basu A, Burks T, Shekhar K, Hofree M, et al. Massively parallel single-nucleus RNA-seq with DroNc-seq. Nat Methods. 2017;14:955–8. https://doi.org/10.1038/nmeth.4407.

Haendel MA, Balhoff JP, Bastian FB, Blackburn DC, Blake JA, Bradford Y, et al. Unification of multi-species vertebrate anatomy ontologies for comparative biology in Uberon. J Biomed Semantics. 2014;5:21. https://doi.org/10.1186/2041-1480-5-21.

Hahn JD. Comparison of melanin-concentrating hormone and hypocretin/orexin peptide expression patterns in a current parceling scheme of the lateral hypothalamic zone. Neurosci Lett. 2010;468:12–7. https://doi.org/10.1016/j.neulet.2009.10.047.

Hahn JD, Swanson LW. Distinct patterns of neuronal inputs and outputs of the juxtaparaventricular and suprafornical regions of the lateral hypothalamic area in the male rat. Brain Res Rev. 2010;64(1):14–103. https://doi.org/10.1016/j.brainresrev.2010.02.002.

Hahn TM, Breininger JF, Baskin DG, Schwartz MW. Coexpression of Agrp and NPY in fasting-activated hypothalamic neurons. Nat Neurosci. 1998;1:271–2. https://doi.org/10.1038/1082.

Harthoorn LF, Sañé A, Nethe M, Van Heerikhuize JJ. Multi-transcriptional profiling of melanin-concentrating hormone and orexin-containing neurons. Cell Mol Neurobiol. 2005;25:1209–23. https://doi.org/10.1007/s10571-005-8184-8.

Hasin-Brumshtein Y, Khan AH, Hormozdiari F, Pan C, Parks BW, Petyuk VA, Piehowski PD, Brümmer A, Pellegrini M, Xiao X, Eskin E, Smith RD, Lusis AJ, Smith DJ. Hypothalamic transcriptomes of 99 mouse strains reveal trans eQTL hotspots, splicing QTLs and novel non-coding genes. elife. 2016;5:e15614.

Hatcher NG, Atkins N Jr, Annangudi SP, Forbes AJ, Kelleher NL, Gillette MU, et al. Mass spectrometry-based discovery of circadian peptides. Proc Natl Acad Sci U S A. 2008;105:12527–32. https://doi.org/10.1073/pnas.0804340105.

Hatton GI, Johnson JI, Malatesta CZ. Supraoptic nuclei of rodents adapted for mesic and xeric environments: numbers of cells, multiple nucleoli, and their distributions. J Comp Neurol. 1972;145(1):43–59. https://doi.org/10.1002/cne.901450104.

Häusser M. The Hodgkin-Huxley theory of the action potential. Nat Neurosci. 2000;3(Suppl):1165.

Hazell GJG, Hindmarch CC, Pope GR, Roper JA, Lightman SL, Murphy D, et al. G protein-coupled receptors in the hypothalamic paraventricular and supraoptic nuclei—serpentine gateways to neuroendocrine homeostasis. Front Neuroendocrinol. 2012;33:45–66. https://doi.org/10.1016/j.yfrne.2011.07.002.

Heijs B, Carreira RJ, Tolner EA, de Ru AH, van den Maagdenberg AMJM, van Veelen PA, McDonnell LA. Comprehensive analysis of the mouse brain proteome sampled in mass spectrometry imaging. Anal Chem. 2015;87:1867–75. https://doi.org/10.1021/ac503952q.

Heiman M, Schaefer A, Gong S, Peterson JD, Day M, Ramsey KE, et al. A translational profiling approach for the molecular characterization of CNS cell types. Cell. 2008;135(4):738–48. https://doi.org/10.1016/j.cell.2008.10.028.

Heisler LK, Pronchuk N, Nonogaki K, Zhou L, Raber J, Tung L, et al. Serotonin activates the hypothalamic-pituitary-adrenal axis via serotonin 2C receptor stimulation. J Neurosci. 2007;27(26):6956–64. https://doi.org/10.1523/JNEUROSCI.2584-06.2007.

Helmholtz H. Messungen über den zeitlichen Verlauf der Zuckung animalischer Muskeln und die Fortpflanzungsgeschwindigkeit der Reizung in den Nerven. Archiv für Anatomie, Physiologie und Wissenschaftliche Medicin, p. 276–364. Veit et Comp, Berlin; 1850.

Henry FE, Sugino K, Tozer A, Branco T, Sternson SM. Cell type-specific transcriptomics of hypothalamic energy-sensing neuron responses to weight-loss. eLife. 2015;4:e09800. https://doi.org/10.7554/eLife.09800.

Hernandez AE, Khan AM. Migration, spatial alignment, and registration of multi-scale neuroscientific datasets related to the control of motivated behaviors within canonically defined maps of the lateral hypothalamic area. Program No. 453.12. 2016 Society for Neuroscience Meeting Planner. San Diego, CA: Society for Neuroscience; 2016. Online.

Heydorn WE, Creed GJ, Goldman D, Kanter D, Merrill CR, Jacobowitz DM. Mapping and quantitation of proteins from discrete nuclei and other areas of the rat brain by two-dimensional gel electrophoresis. J Neurosci. 1983;3(12):2597–606. https://doi.org/10.1523/JNEUROSCI.03-12-02597.1983.

Heydorn WE, Creed GJ, Patel J, Jacobowitz DM. Distribution of proteins in different subcellular fractions of rat brain studied by two-dimensional gel electrophoresis. Neurochem Int. 1986;9(3):357–70. https://doi.org/10.1016/0197-0186(86)90077-X.

Higgins SE, Ellestad LE, Trakooljul N, McCarthy F, Saliba J, Cogburn LA, Porter TE. Transcriptional and pathway analysis in the hypothalamus of newly hatched chicks during fasting and delayed feeding. BMC Genomics. 2010;11:162. https://doi.org/10.1186/1471-2164-11-162.

Hill AV. Chemical wave transmission in nerve. New York: The Macmillan Company; 1932.

Hindmarch C, Yao S, Beighton G, Paton J, Murphy D. A comprehensive description of the transcriptome of the hypothalamoneurohypophyseal system in euhydrated and dehydrated rats. Proc Natl Acad Sci U S A. 2006;103(5):1609–14. https://doi.org/10.1073/pnas.0507450103.

Hindmarch C, Yao S, Hesketh S, Jessop D, Harbuz M, et al. The transcriptome of the rat hypothalamic-neurohypophyseal system is highly strain-dependent. J Neuroendocrinol. 2007;19:1009–12. https://doi.org/10.1111/j.1365-2826.2007.01612.x.

Hindmarch CCT, Franses P, Goodwin B, Murphy D. Whole transcriptome organisation in the dehydrated supraoptic nucleus. Braz J Med Biol Res. 2013;46:1000–6. https://doi.org/10.1590/1414-431X20133328.
Hindmarch CCT, Ferguson AV. Physiological roles for the subfornical organ: a dynamic transcriptome shaped by autonomic state. J Physiol. 2016;594(6):1581–9.
Hindmarch C, Fry M, Yao ST, Smith PM, Murphy D, Ferguson AV. Microarray analysis of the transcriptome of the subfornical organ in the rat: regulation by fluid and food deprivation. Am J Phys Regul Integr Comp Phys. 2008;295(6):R1914–20.
Hirst BH. Secretin and the exposition of hormonal control. J Physiol (London). 2004;560:339. https://doi.org/10.1113/jphysiol.2004.073056.
Hodgkin AL, Huxley AF. Action potentials recorded from inside a nerve fibre. Nature. 1939;144:710–1. https://doi.org/10.1038/144710a0.
Hodgkin AL, Huxley AF. Resting and action potentials in single nerve fibers. J Physiol Lond. 1945;104:176–95.
Hodgkin AL, Huxley AF. Propagation of electrical signals along giant nerve fibres. Proc R Soc Lond B Biol Sci. 1952a;140:177–83.
Hodgkin AL, Huxley AF. Currents carried by sodium and potassium ions through the membrane of the giant axon of Loligo. J Physiol Lond. 1952b;116:449–72.
Hodgkin AL, Huxley AF. The components of membrane conductance in the giant axon of Loligo. J Physiol. 1952c;116:473–96.
Hodgkin AL, Huxley AF. The dual effect of membrane potential on sodium conductance in the giant axon of Loligo. J Physiol Lond. 1952d;116:497–506.
Hodgkin AL, Huxley AF. A quantitative description of membrane current and its application to conduction and excitation in nerve. J Physiol Lond. 1952e;117:500–44.
Hodgkin AL, Huxley AF. Movement of sodium and potassium ions during nervous activity. Cold Spring Harb Symp Quant Biol. 1952f;17:43–52.
Hodgkin AL, Huxley AF, Katz B. Measurement of current-voltage relations in the membrane of the giant axon of Loligo. J Physiol Lond. 1952;116:424–48.
Hoffmann A. Stereotaxic atlas of the toad's brain. Acta Anat (Basel). 1973;84(3):416–51.
Holland PM, Abramson RD, Watson R, Gelfand DH. Detection of specific polymerase chain reaction product by utilizing the 5′-3′ exonuclease activity of Thermus aquaticus DNA polymerase. Proc Natl Acad Sci U S A. 1991;88:7276–80.
Honda M, Eriksson KS, Zhang S, Tanaka S, Lin L, Salehi A, et al. IGFBP3 colocalizes with and regulates hypocretin (orexin). PLoS One. 2009;4(1):e4254. https://doi.org/10.1371/journal.pone.0004254.
Hu SP, Yang JS, Wu MY, Shen ZY, Zhang KH, Liu JW, et al. Effect of one-step 100% ethanol fixation and modified manual microdissection on high-quality RNA recovery from esophageal carcinoma specimen. Dis Esophagus. 2005;18:190–8. https://doi.org/10.1111/j.1442-2050.2005.00475.x.
Humerick M, Hanson J, Rodriguez-Canales J, Lubelski D, Rashid OM, Salinas YD, et al. Analysis of transcription factor mRNAs in identified oxytocin and vasopressin magnocellular neurons isolated by laser capture microdissection. PLoS One. 2013;8(7):e69407. https://doi.org/10.1371/journal.pone.0069407.
Huxley AF. Hodgkin and the action potential. J Physiol (London). 2002;538:2. https://doi.org/10.1013/jphysiol.2001.014118.
Hydén H. Quantitative assay of compounds in isolated fresh nerve cells and glial cells from control and stimulated animals. Nature. 1959;184(4684):433–5.
Hydén H. Dynamic aspects on the neuron-glia relationship. A study with micro-chemical methods. In: Hydén H, editor. The Neuron. Amsterdam: Elsevier; 1967. p. 179–219.
Ihnatko R, Post C, Blomqvist A. Proteomic profiling of the hypothalamus in a mouse model of cancer-induced anorexia-cachexia. Br J Cancer. 2013;109:1867–75. https://doi.org/10.1038/bjc.2013.525.
Iqbal J, Li W, Ullah K, Hasan M, Linna G, Awan U, et al. Study of rat hypothalamic proteome by HPLC/ESI ion trap and HPLC/ESI-Q-TOF MS. Proteomics. 2013;13:2455–68. https://doi.org/10.1002/pmic.201300073.

Iqbal J, Li W, Hasan M, Li YJ, Ullah K, Yun W, et al. Distortion of homeostatic signaling proteins by simulated microgravity in rat hypothalamus: a $^{16}O/^{18}O$-labeled comparative integrated proteomic approach. Proteomics. 2014a;14:262–73. https://doi.org/10.1002/pmic.201300337.

Iqbal J, Li W, Hasan M, Liu K, Awan U, Saeed Y, et al. Differential expression of specific cellular defense proteins in rat hypothalamus under simulated microgravity induced conditions: comparative proteomics. Proteomics. 2014b;14:1424–33. https://doi.org/10.1002/pmic.201400019.

Ivask M, Pajusalu S, Reimann E, Kõks S. Hippocampus and hypothalamus RNA-sequencing of WFS1-deficient mice. Neuroscience. 2018;374:91–103.

Jacobowitz DM. Removal of discrete fresh regions of the rat brain. Brain Res. 1974;80:111–5. https://doi.org/10.1016/0006-8993(74)90726-4.

Jacobowitz DM. The birth of neurochemical maps. Neurochem Res. 2006;31(2):125–6. https://doi.org/10.1007/s11064-005-9002-x.

Jacobowitz DM, Palkovits M. Topographic atlas of catecholamine and acetylcholinesterase-containing neurons in the rat brain. I. Forebrain (Telencephalon, Diencephalon). J Comp Neurol. 1974;157:13–28. https://doi.org/10.1002/cne.901570103.

Jékely G. Global view of the evolution and diversity of metazoan neuropeptide signaling. Proc Natl Acad Sci U S A. 2013;110(21):8702–7. https://doi.org/10.1073/pnas.1221833110.

Jeong JH, Woo YJ, Chua S Jr, Jo YH. Single-cell gene expression analysis of cholinergic neurons in the arcuate nucleus of the hypothalamus. PLoS One. 2016;11:e0162839. https://doi.org/10.1371/journal.pone.0162839.

Jiang CH, Tsien JZ, Schultz PG, Hu Y. The effects of aging on gene expression in the hypothalamus and cortex of mice. Proc Natl Acad Sci. 2001;98(4):1930–4.

Jin S, Sun D, Xi Q, Dong X, Song D, Fu H, et al. Identification of genes in the hypothalamus-pituitary-gonad axis in the brain of Amur sturgeons (Acipenser schrenckii) by comparative transcriptome analysis in relation to kisspeptin treatment. Gene. 2016;595:53–61. https://doi.org/10.1016/j/.gene.2016.09.026.

Jing X, Ratty AK, Murphy D. Ontogeny of the vasopressin and oxytocin mRNAs in the mouse hypothalamus. Neurosci Res. 1998;30:343–9. https://doi.org/10.1016/S0168-0102(98)00017-0.

Jockusch H, Voigt S, Eberhard D. Localization of GFP I frozen sections from unfixed mouse tissues: immobilization of a highly soluble marker protein by formaldehyde vapor. J Histochem Cytochem. 2003;51:401–4. https://doi.org/10.1177/002215540305100315.

Johnson KR, Hindmarch CCT, Salinas YD, Shi Y, Greenwood M, Hoe SZ, et al. A RNA-Seq analysis of the rat supraoptic nucleus transcriptome: effects of salt loading on gene expression. PLoS One. 2015;10(4):e0124523. https://doi.org/10.1371/journal.pone.0124523.

Johnson SA, Spollen WG, Manshack LK, Bivens NJ, Givan SA, Rosenfeld CS. Hypothalamic transcriptomic alterations in male and female California mice ( ) developmentally exposed to bisphenol A or ethinyl estradiol. Physiol Rep. 2017;5(3):e13133.

Johnson SA, Ellersieck MR, Rosenfeld CS. Hypothalamic gene expression changes in F 1 California mice (Peromyscus californicus) parents developmentally exposed to bisphenol A or ethinyl estradiol. Heliyon. 2018;4(6):e00672.

Johnston PV, Roots BI. The presence of phosphatidylcholine in neurons isolated from the lateral vestibular nucleus of ox brain. Biochem J. 1966;98:157–8.

Johnston PV, Roots BI. Neuronal and glial perikarya preparations: an appraisal of present methods. Int Rev Cytol. 1970;29:265–80.

Jovanovic Z, Tung YCL, Lam BYH, O'Rahilly S, Yeo GSH. Identification of the global transcriptomic response of the hypothalamic arcuate nucleus to fasting and leptin. J Neuroendocrinol. 2010;22(8):915–25. https://doi.org/10.1111/j.1365-2826.2010.02026.x.

Kabra DG, Pfuhlmann K, García-Cáceres C, Schriever SC, García VC, Kebede AF, et al. Hypothalamic leptin action is mediated by histone deacetylase 5. Nat Commun. 2016;7:10782. https://doi.org/10.1038/ncomms10782.

Karas M, Bachmann D, Bahr U, Hillenkamp F. Matrix-assisted ultraviolet laser desorption of non-volatile compounds. Int J Mass Spectrom Ion Proc. 1987;78:53–68.

Kasukawa T, Masumoto K-h, Nikaido I, Nagano M, Uno KD, Tsujino K, Hanashima C, Shigeyoshi Y, Ueda HR, Jothi R. Quantitative expression profile of distinct functional regions in the adult mouse brain. PLoS One. 2011;6(8):e23228. https://doi.org/10.1371/journal.pone.0023228.

Kefaloyianni E, Lyssand JS, Moreno C, Delaroche D, Hong M, Fenyö D, et al. Comparative proteomic analysis of the ATP-sensitive K+ channel complex in different tissue types. Proteomics. 2013;13:368–78. https://doi.org/10.1002/pmic.201200324.
Kerman IA, Buck BJ, Evans SJ, Akil H, Watson SJ. Combining laser capture microdissection with quantitative real-time PCR: effects of tissue manipulation on RNA quality and gene expression. J Neurosci Methods. 2006;153:71–85. https://doi.org/10.1016/j.jneumeth.2005.10.010.
Khan AM. [2 parts] Nerve, muscle, blood, toil, tears, and sweat: England's pioneering biophysicist, soldier, and statesman. J Hist Neurosci. 2009;18:80–1.; 98–105. https://doi.org/10.1080/09647040802105854.
Khan AM. Controlling feeding behavior by chemical or gene-directed targeting in the brain: what's so spatial about our methods? Front Neurosci. 2013;7(Article 182):1–49. https://doi.org/10.3389/fnins.2013.00182.
Khan AM, Hahn JD, Cheng W-C, Watts AG, Burns GAPC. NeuroScholar's electronic laboratory notebook and its application to neuroendocrinology. Neuroinformatics. 2006;4(2):139–61. https://doi.org/10.1385/NI:4:2:139.
Khan AM, Ponzio TA, Sanchez-Watts G, Stanley BG, Hatton GI, Watts AG. Catecholaminergic control of mitogen-activated protein kinase signaling in paraventricular neuroendocrine neurons in vivo and in vitro: a proposed role during glycemic challenges. J Neurosci. 2007;27(27):7344–60. https://doi.org/10.1523/JNEUROSCI.0873-07.2007.
Khan AM, Walker EM, Watts AG. Tracking the coupling of external signals to intracellular programs controlling peptide synthesis and release in hypothalamic neuroendocrine neurons. In: Fink G, editor. Stress: neuroendocrinology and neurobiology. Amsterdam: Elsevier; 2017. p. 67–81. https://doi.org/10.1016/B978-0-12-802175-0.00007-3.
Khan AM, Perez J, Wells CE, Fuentes O. Computer vision evidence supporting craniometric alignment of rat brain atlases to streamline expert-guided, first-order migration of hypothalamic spatial datasets related to behavioral control. Front Syst Neurosci. 2018;12(Article 7):1–29. https://doi.org/10.3389/fnsys.2018.00007.
Khodosevich K, Inta D, Seeburg PH, Monyer H. Gene expression analysis of in vivo fluorescent cells. PLoS One. 2007;2:e1151.
Kim H-J, Park HJ, Hong MS, Song JY, Park H-K, Jo DJ, et al. Effect by acupuncture on hypothalamic expression of maternally separated rats: proteomic approach. Neurol Res. 2010;32(Suppl 1):69–73. https://doi.org/10.1179/016164109X12537002794129.
Kim J-H, Kim J-H, Cho Y-E, Baek M-C, Jung J-Y, Lee M-G, et al. Chronic sleep deprivation-induced proteome changes in astrocytes of the rat hypothalamus. J Proteome Res. 2014;13:4047–61. https://doi.org/10.1021/pr500431j.
Kim KW, Donato J, Berglund ED, Choi Y-H, Kohno D, Elias CF, DePinho RA, Elmquist JK. FOXO1 in the ventromedial hypothalamus regulates energy balance. J Clin Investig. 2012;122(7):2578–89.
Kleinzeller A. Charles Ernest Overton's concept of a cell membrane. In: Deamer DW, Kleinzeller A, Fambrough DM, editors. Membrane permeability: 100 years since ernest overton. San Diego: Academic Press; 1999. p. 1–22.
Klimov LO, Ershov NI, Efimov VM, Markel AL, Redina OE. Genome-wide transcriptome analysis of hypothalamus in rats with inherited stress-induced arterial hypertension. BMC Genet. 2016;17(Suppl 1):13. https://doi.org/10.1186/s12863-015-0307-8.
Knight ZA, Tan K, Birsoy K, Schmidt S, Garrison JL, Wysocki RW, et al. Molecular profiling of activated neurons by phosphorylated ribosome capture. Cell. 2012;151(5):1126–37. https://doi.org/10.1016/j.cell.2012.10.039.
Kobayashi Y. DNA microarray unravels rapid changes in transcriptome of MK-801 treated rat brain. World J Biol Chem. 2015;6(4):389.
Koelle GB, Friedenwald JS. A histochemical method for localizing cholinesterase activity. Proc Soc Exp Biol Med. 1949;70:617–22.
Kohno D, Lee S, Harper MJ, Kim KW, Sone H, Sasaki T, Kitamura T, Fan G, Elmquist JK. Dnmt3a in Sim1 neurons is necessary for normal energy homeostasis. J Neurosci. 2014;34(46):15288–96.
König JFR, Klippel RA. The rat brain: a stereotaxic atlas of the forebrain and lower parts of the Brain Stem. Baltimore: The Williams and Wilkins Co.; 1963.

Korner J, Savontaus E, Chua SC Jr, Leibel RL, Wardlaw SL. Leptin regulation of Agrp and Npy mRNA in the rat hypothalamus. J Neuroendocrinol. 2001;13:959–66. https://doi.org/10.1046/j.1365-2826.2001.00716.x.

Kuenzel WJ, Beck MM, Teruyama R. Neural sites and pathways regulating food intake in birds: a comparative analysis to mammalian systems. J Exp Zool. 1999;283:348–64. https://doi.org/10.1002/(SICI)1097-010X(19990301/01)283:4/5<348::AID-JEZ5>3.0.CO;2-5.

Kuhla B, Kuhla S, Rudolph PE, Albrecht D, Metges CC. Proteomics analysis of hypothalamic response to energy restriction in dairy cows. Proteomics. 2007;7:3602–17. https://doi.org/10.1002/pmic.200700248.

Kuo Y-M, Shiue Y-L, Chen C-F, Tang P-C, Lee Y-P. Proteomic analysis of hypothalamic proteins of high and low egg production strains of chickens. Theriogenology. 2005;64:1490–502. https://doi.org/10.1016/j/theriogenology.2005.03.20.

Kurrasch DM, Cheung CC, Lee FY, Tran PV, Hata K, Ingraham HA. The neonatal ventromedial hypothalamus transcriptome reveals novel markers with spatially distinct patterning. J Neurosci. 2007;27(50):13624–34. https://doi.org/10.1523/JNEUROSCI.2858-07.2007.

Lachuer J, Ouyang L, Legras C, Rio JD, Barlow C. Gene expression profiling reveals an inflammatory process in the anx/anx mutant mice. Mol Brain Res. 2005;139:372–6. https://doi.org/10.1016/j.molbrainres.2005.06.003.

Lake BB, Ai R, Kaeser GE, Salathia NS, Yung YC, Liu R, et al. Neuronal subtypes and diversity revealed by single-nucleus RNA sequencing of the human brain. Science. 2016;352:1586–90. https://doi.org/10.1126/science.aaf1204.

Lam BYH, Cimino I, Polex-Wolf J, Kohnke SN, Rimmington D, et al. Heterogeneity of hypothalamic pro-opiomelanocortin-expressing neurons revealed by single-cell RNA sequencing. Mol Metab. 2017;6:383–92. https://doi.org/10.1016/j.molmet.2017.02.007.

Landmann EM, Schellong K, Melchior K, Rodekamp E, Ziska T, Harder T, Plagemann A. Short-term regulation of the hypothalamic melanocortinergic system under fasting and defined glucose-refeeding conditions in rats: a lasercapture microdissection (LMD)-based study. Neurosci Lett. 2012;515(1):87–91. https://doi.org/10.1016/j.neulet.2012.03.025.

Lee H-C, Chang D-E, Yeom M, Kim G-H, Choi K-D, Shim I, Lee H-J, Hahm D-H. Gene expression profiling in hypothalamus of immobilization-stressed mouse using cDNA microarray. Mol Brain Res. 2005;135(1–2):293–300.

Lee HJ, Schneider RF, Manousaki T, Kang JH, Lein E, Franchini P, Meyer A. Lateralized feeding behavior is associated with asymmetrical neuroanatomy and lateralized gene expressions in the brain in scale-eating cichlid fish. Genome Biol Evol. 2017;9(11):3122–36.

Lee JE, Zamdborg L, Southey BR, Atkins N Jr, Mitchell JW, Li M, et al. Quantitative peptidomics for discovery of circadian-related peptides from the suprachiasmatic nucleus. J Proteome Res. 2013;12:585–93. https://doi.org/10.1021/pr300605p.

Lee JH, Atkins N, Hatcher NG, Zamdborg L, Gillette MU, Sweedler JV, et al. Endogenous peptide discovery of the rat circadian clock: a focused study of the suprachiasmatic nucleus by ultra-high performance tandem mass spectrometry. Mol Cell Proteomics. 2010;9:285–97. https://doi.org/10.1074/mcp.M900362-MCP200.

Lee JH, Daugharthy ER, Scheiman J, Kalhor R, Yang JL, Ferrante TC, et al. Highly multiplexed subcellular RNA sequencing in situ. Science. 2014;343:1360–3. https://doi.org/10.1126/science.1250.21.

Lee JH, Daugharthy ER, Scheiman J, Kalhor R, Ferrante TC, Terry R, et al. Fluorescent in situ sequencing (FISSEQ) of RNA for gene expression profiling in intact cells and tissues. Nat Protoc. 2015;10:442–58. https://doi.org/10.1038/nprot.2014.191.

Lee JH, Kim CH, Kim DG, Ahn YS. Microarray analysis of differentially expressed genes in the brains of tubby mice. Korean J Physiol Pharmacol. 2009;13(2):91.

Lee JY, Lee J-H, Moon YW, Chun B-G, Jahng JW. Proteomic analysis of lithium-induced gene expression in the rat hypothalamus. Int J Neurosci. 2009;119(9):1267–81.

Lee S, Bookout AL, Lee CE, Gautron L, Harper MJ, Elias CF, et al. Laser-capture microdissection and transcriptional profiling of the dorsomedial nucleus of the hypothalamus. J Comp Neurol. 2012;520(16):3617–32. https://doi.org/10.1002/cne.23116.

Lehrer GM, Maker HS. Quantitative histochemical approaches to energy metabolism in nervous tissue. In: Balázs R, Cremer JE, editors. Metabolic compartmentation in the brain. New York: John Wiley & Sons; 1972. p. 235–44.

Lein ES, Hawrylycz MJ, Ao N, Ayres M, Bensinger A, Bernard A, et al. Genome-wide atlas of gene expression in the adult mouse brain. Nature. 2007;445(7124):168–76.

Lenoir T, Giannella E. The emergence and diffusion of DNA microarray technology. J Biomed Discov Collab. 2006;1:11. https://doi.org/10.1186/1747-5333-1-11.

Lerner R, Post JM, Ellis SR, Naomi Vos DR, Heeren RMA, Lutz B, Bindila L. Simultaneous lipidomic and transcriptomic profiling in mouse brain punches of acute epileptic seizure model compared to controls. J Lipid Res. 2018;59(2):283–97.

Levey AI, Wainer BH, Rye DB, Mufson EJ, Mesulam M-M. Choline acetyltransferase-immunoreactive neurons intrinsic to rodent cortex and distinction from acetylcholinesterase-positive neurons. Neuroscience. 1983;13:341–53.

Li HF, Zhu WQ, Chen KW, Xu WJ, Song W. Two maternal origins of Chinese domestic goose. Poult Sci. 2011;90(12):2705–10.

Li J-Y, Lescure PA, Misek DE, Lai Y-M, Chai B-X, Kuick R, et al. Food deprivation-induced expression of minoxidil sulfotransferase in the hypothalamus uncovered by microarray analysis. J Biol Chem. 2002;277(11):9069–76. https://doi.org/10.1074/jbc.M110467200.

Li J-Y, Kuick R, Thompson RC, Misek DE, Lai Y-M, Liu Y-Q, et al. Arcuate nucleus transcriptome profiling identifies ankyrin repeat and suppressor of cytokine signaling box-containing protein 4 as a gene regulated by fasting in central nervous system feeding circuits. J Neuroendocrinol. 2005;17:394–404. https://doi.org/10.1111/j.1365-2826.2005.01317.x.

Li X, Qu F, Xie W, Wang F, Liu H, Song S, et al. Transcriptomic analyses of neurotoxic effects in mouse brain after intermittent neonatal administration of thimerosal. Toxicol Sci. 2014;139(2):452–65. https://doi.org/10.1093/toxsci/kfu049.

Liberini CG, Boyle CN, Cifani C, Venniro M, Hope BT. Lutz TA (2016) Amylin receptor components and the leptin receptor are co-expressed in single rat area postrema neurons. Eur J Neurosci. 2016;43:653–61. https://doi.org/10.1111/ejn.13163.

Lisser H. Hypophysis versus hypothalamus. Calif Western Med. 1927;26(4):490–2.

Liu X, Zeng J, Zhou A, Theodorsson E, Fahrenkrug J, Reinscheid RK. Molecular fingerprint of neuropeptide S-producing neurons in the mouse brain. J Comp Neurol. 2011;519:1847–66. https://doi.org/10.1002/cne.22603.

Loewi O. Über humorale Übertragbarkeit der Herznervenwirkung. I. Mitteilung. Pfluger's Archiv. 1921;189(1):239–42.

Lowry OH. The quantitative histochemistry of the brain. Histological sampling. J Histochem Cytochem. 1953;1(6):420–8. https://doi.org/10.1177/1.6.420.

Luan X, Cao Z, Li R, Liu M, Hu J. Differential expression profiling of hypothalamus genes in laying period and ceased period Huoyan geese. Mol Biol Rep. 2014;41:3401–11. https://doi.org/10.1007/s11033-014-3202-x.

Macosko EZ, Basu A, Satija R, Nemesh J, Shekhar K, Goldman M, et al. Highly parallel genome-wide expression profiling of individual cells using nanoliter droplets. Cell. 2015;161:1202–14. https://doi.org/10.1016/j.cell.2015.05.002.

Magdaleno S, Jensen P, Brumwell CL, Seal A, Lehman K, Asbury A, Cheung T, Cornelius T, Batten DM, Eden C, Norland SM, Rice DS, Dosooye N, Shakya S, Mehta P, Curran T. BGEM: an in situ hybridization database of gene expression in the embryonic and adult mouse nervous system. PLoS Biol. 2006;4(4):e86.

Manciu FS, Lee KH, Durrer WG, Bennet KE. Detection and monitoring of neurotransmitters—a spectroscopic analysis. Neuromodulation. 2013;16:192–9. https://doi.org/10.1111/j.1525-1403.2012.00502.x.

Manousopoulou A, Koutmani Y, Karaliota S, Woelk CH, Manolakos ES, Karalis K, et al. Hypothalamus proteomics from mouse models with obesity and anorexia reveals therapeutic targets of appetite regulation. Nutr Diabetes. 2016;6:e204. https://doi.org/10.1038/nutd.2016.10.

Martinez A, Pinales BE, Khan AM. Connections of the rostral portion of the hypothalamic arcuate nucleus: a combined anterograde and retrograde study in the adult male rat. Program No.

616.10. 2015 Neuroscience Meeting Planner. Chicago, IL: Society for Neuroscience; 2015. Online.
Martinez A, Pinales BE, Khan AM. Further elaboration of arcuate hypothalamic nucleus circuitry based on retrograde studies in the adult male rat. Program No. 453.11. 2016 Neuroscience Meeting Planner. San Diego, CA: Society for Neuroscience; 2016. Online.
Martyniuk CJ, Feswick A, Spade DJ, Kroll KJ, Barber DS, Denslow ND. Effects of acute dieldrin exposure on neurotransmitters and global gene transcription in largemouth bass (Micropterus salmoides) hypothalamus. Neurotoxicology. 2010a;31:356–66. https://doi.org/10.1016/j.neuro.2010.04.008.
Martyniuk CJ, Kroll KJ, Doperalski NJ, Barber DS, Denslow ND. Genomic and proteomic responses to environmentally relevant exposures to dieldrin: indicators of neurodegeneration? Toxicol Sci. 2010b;117(1):190–9. https://doi.org/10.1093/toxsci/kfq192.
Martyniuk CJ, Doperalski NJ, Kroll KJ, Barber DS, Denslow ND. Sexually dimorphic transcriptomic responses in the teleostean hypothalamus: a case study with the organochlorine pesticide dieldrin. Neurotoxicology. 2013;34:105–17. https://doi.org/10.1016/j.neuro.2012.09.012.
Maskos U, Southern EM. Oligonucleotide hybridisations on glass supports: a novel linker for oligonucleotide synthesis and hybridisation properties of oligonucleotides synthesised. Nucleic Acids Res. 1992;20(7):1679–84.
Mathieson WB, Taylor SW, Marshall M, Neumann PE. Strain and sex differences in the morphology of the medial preoptic nucleus of mice. J Comp Neurol. 2000;428(2):254–65.
McIlwain H. Biochemistry and the central nervous system. 2nd ed. London: J & A Churchill, Ltd.; 1959.
Mennigen JA, Martyniuk CJ, Crump K, Xiong H, Zhao E, Popesku J, et al. Effects of fluoxetine on the reproductive axis of female goldfish (Carassius auratus). Physiol Genomics. 2008;35:273–82. https://doi.org/10.1152/physiolgenomics.90263.2008.
Mercader JM, Lozano JJ, Sumoy L, Dierssen M, Visa J, Gratacòs M, et al. Hypothalamus transcriptome profile suggests an anorexia-cachexia syndrome in the anx/anx mouse model. Physiol Genomics. 2008;35:341–50. https://doi.org/10.1152/physiolgenbomics.90255.2008.
Mickelsen LE, Kolling FW IV, Chimileski B, Norris CE, Nelson CE, Jackson AC. Neurochemical heterogeneity among lateral hypothalamic hypocretin/orexin and melanin-concentrating hormone neurons identified through single cell gene expression analysis. eNeuro. 2017;4(5):pii: ENEURO.0013-17.2017. https://doi.org/10.1523/ENEURO.0013-17.2017.
Middleton FA, Ramos EJB, Xu Y, Diab H, Zhao X, Das UN, et al. Application of genomic technologies: DNA microarrays and metabolic profiling of obesity in the hypothalamus and in subcutaneous fat. Nutrition. 2004;20:14–25. https://doi.org/10.1016/j.nut.2003.10.002.
Mihailova A, Karaszewski B, Faergestad EM, Hauser R, Nyka WM, Lundanes E, Greibrokk T. Twodimensional LC-MS/MS in detection of peptides in hypothalamus of the rat subjected to hypoxic stress. J Sep Sci. 2008;31(3):468–79.
Miller JA, Ding S-L, Sunkin SM, Smith KA, Ng L, Szafer A, et al. Transcriptional landscape of the prenatal human brain. Nature. 2014;508:199–206. https://doi.org/10.1038/nature13185.
Mirabeau O, Joly J-S. Molecular evolution of peptidergic signaling systems in bilaterians. Proc Natl Acad Sci U S A. 2013;110(22):E2028–37. https://doi.org/10.1073/pnas.1219956110.
Mishra A, Cheng C-H, Lee W-C, Tsai L-L. Proteomic changes in the hypothalamus and retroperitoneal fat from male F344 rats subjected to repeated light-dark shifts. Proteomics. 2009;9:4017–28. https://doi.org/10.1002/pmc.200800813.
Mitchell JW, Atkins NA Jr, Sweedler JV, Gillette MU. Direct cellular peptidomics of hypothalamic neurons. Front Neuroendocrinol. 2011;32(4):377–86. https://doi.org/10.1016/j.yfrne.2011.02.005.
Mizuno TM, Mobbs CV. Hypothalamic agouti-related protein messenger ribonucleic acid is inhibited by leptin and stimulated by fasting. Endocrinology. 1999;140:814–7. https://doi.org/10.1210/endo.140.2.6491.
Mo B, Callegari TM, Renner KJ. Proteomic analysis of the ventromedial nucleus of the hypothalamus (pars lateralis) in the female rat. Proteomics. 2006;6:6066–74. https://doi.org/10.1002/pmic.200600072.

Mo B, Callegari E, Telefont M, Renner KJ. Estrogen regulation of proteins in the rat ventromedial nucleus of the hypothalamus. J Proteome Res. 2008;7(11):5040–8.

Moore RY, Speh JC, Leak RK. Suprachiasmatic nucleus organization. Cell Tissue Res. 2002;309:89–98. https://doi.org/10.1007/s00441-002-0575-2.

Mortazavi A, Williams BA, McCue K, Schaeffer L, Wold B. Mapping and quantifying mammalian transcriptomes by RNA-Seq. Nat Methods. 2008;5(7):621–8.

Moulédous L, Hunt S, Harcourt R, Harry J, Williams KL, Gutstein HB. Lack of compatibility of histological staining methods with proteomic analysis of laser-captured microdissected brain samples. J Biomol Tech. 2002;13:258–64.

Moulédous L, Hunt S, Harcourt R, Harry J, Williams KL, Gutstein HB. Navigated laser capture microdissection as an alternative to direct histological staining for proteomic analysis of brain samples. Proteomics. 2003;3:610–5. https://doi.org/10.1002/pmic.200300398.

Mountjoy KG, Mortrud MT, Low MJ, Simerly RB, Cone RD. Localization of the melanocortin-4 receptor (MC4-R) in neuroendocrine and autonomic control circuits in the brain. Mol Endocrinol. 1994;8:1298–308.

Mozhui K, Lu L, Armstrong WE, Williams RW. Sex-specific modulation of gene expression networks in murine hypothalamus. Front Neurosci. 2012;6:63.

Mungall CJ, McMury JA, Köhler S, Balhoff JP, Borromeo C, Brush M, et al. The Monarch Initiative: an integrative data and analytics platform connecting phenotypes to genotypes across species. Nucleic Acids Res. 2017;45:D712–22. https://doi.org/10.1093/nar/gkw1128.

Mutsuga N, Gainer H. Molecular analysis of the magnocellular neuroendocrine phenotype: from the micropunch to laser microdissection. Neurochem Res. 2006;31:189–99. https://doi.org/10.1007/s11064-005-9008-4.

Mutsuga N, Shahar T, Verbalis JG, Brownstein MJ, Xiang CC, Bonner RF, et al. Selective gene expression in magnocellular neurons in rat supraoptic nucleus. J Neurosci. 2004;24(32):7174–85. https://doi.org/10.1523/JNEUROSCI.2022-04.2004.

Nadler JJ, Zou F, Huang H, Moy SS, Lauder J, Crawley JN, et al. Large-scale gene expression differences across brain regions and inbred strains correlate with a behavioral phenotype. Genetics. 2006;174:1229–36. https://doi.org/10.1534/genetics.106.061481.

Nakazawa CM, Shikata K, Uesugi M, Katayama H, Aoshima K, Tahara K, Takahashi E, Hida T, Shibata H, Ogura H, Seiki T, Oda Y, Kuromitsu J, Miyamoto N. Prediction of relaxin-3-induced downstream pathway resulting in anxiolytic-like behaviors in rats based on a microarray and peptidome analysis. J Recept Signal Transduct. 2013;33(4):224–33.

Navarro JF, Sjöstrand J, Salmén F, Lundeberg J, Ståhl PL. ST pipeline: an automated pipeline for spatial mapping of unique transcripts. Bioinformatics. 2017;33:2591–3. https://doi.org/10.1093/bioinformatics/btx211.

Nectow AR, Ekstrand MI, Friedman JM. Molecular characterization of neuronal cell types based on patterns of projection with Retro-TRAP. Nat Protoc. 2015;10(9):1319–27. https://doi.org/10.1038/nprot.2015.087.

Nectow AR, Moya MV, Ekstrand MI, Mousa A, McGuire KL, Sferrazza CE, et al. Rapid molecular profiling of defined cell types using viral TRAP. Cell Rep. 2017;19(3):655–67. https://doi.org/10.1016/j.celrep.2017.03.048.

Nedungadi TP, Cunningham JT. Differential regulation of TRPC4 in the vasopressin magnocellular system by water deprivation and hepatic cirrhosis in the rat. Am J Physiol Regul Integr Comp Physiol. 2014;306(5):R304–14. https://doi.org/10.1152/ajpregu.00388.2013.

Nedungadi TP, Carreño FR, Walch JD, Bathina CS, Cunningham JT. Region-specific changes in transient receptor potential vanilloid channel expression in the vasopressin magnocellular system in hepatic cirrhosis-induced hyponatraemia. J Neuroendocrinol. 2012a;24:642–52. https://doi.org/10.1111/j.1365-2826.2011.02273.x.

Nedungadi TP, Dutta M, Bathina CS, Caterina MJ, Cunningham JT. Expression and distribution of TRPV2 in rat brain. Exp Neurol. 2012b;237(1):223–37. https://doi.org/10.1016/j.expneurol.2012.06.017.

Nernst W. Zur Kinetik der in Lösung befindlichen Körper. I. Theorie der Diffusion. Z Phys Chem. 1888;2:613–37.

Nichols N, Perlmutter A, Mejino JLV Jr, Brinkley JF. Representing neural connectivity in the foundational model of anatomy ontology. In: Proceedings of the Annual Symposium. American Medical Informatics Association. Washington DC; 2010.

Nilaweera KN, Archer ZA, Campbell G, Mayer C-D, Balik A, Ross AW, et al. Photoperiod regulates genes encoding melanocortin 3 and serotonin receptors and secretogranins in the dorsomedial posterior arcuate of the Siberian hamster. J Neuroendocrinol. 2009;21:123–31. https://doi.org/10.1111/j.1365-2826.2008.01810.x.

Nilsson A, Stroth N, Zhang X, Qi H, Fälth M, et al. Neuropeptidomics of mouse hypothalamus after imipramine treatment reveal somatostatin as a potential mediator of antidepressant effects. Neuropharmacology. 2012;62:347–57. https://doi.org/10.1016/j.neuropharm.2011.08.004.

Nobis S, Goichon A, Achamrah N, Guérin C, Azhar S, Chan P, Morin A, Bôle-Feysot C, do Rego JC, Vaudry D, Déchelotte P, Belmonte L, Coëffier M. Alterations of proteome, mitochondrial dynamic and autophagy in the hypothalamus during activity-based anorexia. Sci Rep. 2018;8(1):7233.

Novoselova TV, Larder R, Rimmington D, Lelliott C, Wynn EH, Gorrigan RJ, Tate PH, Guasti L, O'Rahilly S, Clark AJL, Logan DW, Coll AP, Chan LF. Loss of is associated with deficiency and increased circulating cholesterol. J Endocrinol. 2016;230(1):13–26.

O'Farrell PH. High resolution two-dimensional electrophoresis of proteins. J Biol Chem. 1975;250:4007–21.

Okaty BW, Sugino K, Nelson SB. A quantitative comparison of cell-type-specific microarray gene expression profiling methods in the mouse brain. PLoS One. 2011;6:e16493. https://doi.org/10.1371/journal.pone.0016493.

Olszewski PK, Cedernaes J, Olsson F, Levine AS, Schiöth H. Analysis of the network of feeding neuroregulators using the Allen Brain Atlas. Neurosci Biobehav Rev. 2008;32:945–56. https://doi.org/10.1016/j.neubiorev.2008.01.007.

Orozco-Solís R, Matos RJB, Guzmán-Quevedo O, Lopes de Souza S, Bihouée A, Houlgatte R, et al. Nutritional programming in the rat is linked to long-lasting changes in nutrient sensing and energy homeostasis in the hypothalamus. PLoS One. 2010;5(10):e13537. https://doi.org/10.1371/journal.pone.0013537.

Osborne NN. Microchemical analysis of nervous tissue. Oxford: Pergamon; 1974.

Osumi-Sutherland D, Reeve S, Mungall CJ, Neuhaus F, Ruttenberg A, Jefferis GSXE, et al. A strategy for building neuroanatomy ontologies. Bioinformatics. 2012;28:1262–9. https://doi.org/10.1093/bioinformatics/bts113.

Overton E. Beiträge zur allgemeinen Muskel- und Nervenphysiologie. I. Ueber die osmotischen Eigenschaften der Muskeln. Pflüger Arch. 1902a;92:115–280.

Overton E. Beiträge zur allgemeinen Muskel- und Nervenphysiologie. II. Mittheilung. Ueber die Unentbehrlichkeit von Natrium- (oder Lithium-)Ionen für den Contractionsact des Muskels. Pflüger Arch. 1902b;92:346–86.

Palay SL, Chan-Palay V. The structural heterogeneity of central nervous tissue. In: Balázs R, Cremer JE, editors. Metabolic compartmentation in the brain. New York: John Wiley & Sons; 1972. p. 187–207.

Palkovits M. Isolated removal of hypothalamic or other brain nuclei of the rat brain. Brain Res. 1973;59:449–50. https://doi.org/10.1016/0006-8993(73)90290-4.

Palkovits M. Isolated removal of hypothalamic nuclei for neuroendocrinological and neurochemical studies. In: Stumpf WE, Grant LD, editors. Anatomical neuroendocrinology. Basel: Karger; 1975. p. 72–80.

Palkovits M. Microdissection of individual brain nuclei and areas. NeuroMethods. 1986;1:1–17. https://doi.org/10.1385/0-89603-075-x:1.

Palkovits M. Microdissection in combination with biochemical microassays as a tool in tract tracing. In: Heimer L, Zaborszky L, editors. Neuroanatomical tract-tracing methods 2. New York: Springer; 1989. p. 299–310.

Palkovits M, Jacobowitz DM. Topographic atlas of catecholamine and acetylcholinesterase-containing neurons in the rat brain. II. Hindbrain (Mesencephalon, Rhombencephalon). J Comp Neurol. 1974;157(1):29–42. https://doi.org/10.1002/cne.901570104.

Pan H, Che F-Y, Peng B, Steiner DF, Pintar JE, Fricker LD. The role of prohormone convertase-2 in hypothalamic neuropeptide processing: a quantitative neuropeptidomic study. J Neurochem. 2006;98:1763–77. https://doi.org/10.1111/j.1471-4159.2006.04067.x.
Panda S, Antoch MP, Miller BH, Su AI, Schook AB, Straume M, et al. Coordinated transcription of key pathways in the mouse by the circadian clock. Cell. 2002;109:307–20. https://doi.org/10.1016/S0092-8674(02)00722-5.
Pandey AK, Williams RW. Genetics of gene expression in CNS. Int Rev Neurobiol. 2014;116:195–231.
Park CC, Petyuk VA, Qian W-J, Smith RD, Smith DJ. Dual spatial maps of transcript and protein abundance in the mouse brain. Expert Rev Proteomics. 2009;6(3):243–9.
Park J, Zhu H, O'Sullivan S, Ogunnaike BA, Weaver DR, Schwaber JS, et al. Single-cell transcriptional analysis reveals novel neuronal phenotypes and interaction networks involved in the central circadian clock. Front Neurosci. 2016;10:481. https://doi.org/10.3389/fnins.2016.00481.
Paternain L, Batlle MA, De La Garza AL, Milagro FI, Martínez JA, Campión J. Transcriptomic and epigenetic changes in the hypothalamus are involved in an increased susceptibility to a high-fat-sucrose diet in prenatally stressed female rats. Neuroendocrinology. 2012;96:249–60. https://doi.org/10.1159/000341684.
Paulsen SJ, Larsen LK, Jelsing J, Janßen U, Gerstmayer B, Vrang N. Gene expression profiling of individual hypothalamic nuclei from single animals using laser capture microdissection and microarrays. J Neurosci Methods. 2009;177(1):87–93. https://doi.org/10.1016/j.jneumeth.2008.09.024.
Pavlidis P, Noble WS. Analysis of strain and regional variation in gene expression in mouse brain. Genome Biol. 2001;2(10):RESEARCH0042.
Paxinos G, Franklin K. The mouse brain in stereotaxic coordinates. 4th ed. San Diego: Academic Press; 2012.
Paxinos G, Watson C. The rat brain in stereotaxic coordinates. 6th ed. Burlington: Academic Press; 2007.
Pederson T. The Nucleolus. Cold Spring Harb Perspect Biol. 2011;3:a000638. https://doi.org/10.1101/cshperspect.a000638.
Pedroso AP, Watanabe RLH, Albuquerque KT, Telles MM, Andrade MCC, Perez JD, et al. Proteomic profiling of the rat hypothalamus. Proteome Sci. 2012;10:26. https://doi.org/10.1186/1477-5956-10-26.
Pedroso AP, Souza AP, Dornellas APS, Oyma LM, Nascimento CMO, Santos GMS, et al. Intrauterine growth restriction programs the hypothalamus of adult male rats: integrated analysis of proteomic and metabolomics data. J Proteome Res. 2017;16:1515–25. https://doi.org/10.1021/acs.jproteome.6b00923.
Pembroke WG, Babbs A, Davies KE, Ponting CP, Oliver PL. Temporal transcriptomics suggest that twin-peaking genes reset the clock. elife. 2015;4:e10518. https://doi.org/10.7554/eLife.10518.
Perez J, Fuentes O, Khan AM. Towards automatic registration of histological data to canonical brain atlases. Program No. 604.05. 2017 Neuroscience Meeting Planner. Washington, DC: Society for Neuroscience; 2017. Online.
Perkel JM. Neuropeptidomics study profiles hypothalamic "nucleus", individual cells. J Proteome Res. 2008;7:4610. https://doi.org/10.1021/pr800672a.
Petyuk VA, Qian W-J, Chin MH, Wang H, Livesay EA, Monroe ME, et al. Spatial mapping of protein abundances in the mouse brain by voxelation integrated with high-throughput liquid chromatography-mass spectrometry. Genome Res. 2007;17:328–36. https://doi.org/10.1101/gr.5799207.
Petyuk VA, Qian W-J, Smith RD, Smith DJ. Mapping protein abundance patterns in the brain using voxelation combined with liquid chromatography and mass spectrometry. Methods. 2010;50(2):77–84.
Phillipps HR, Ladyman SR, Grattan DR. Maintained expression of genes associated with metabolism in the ventromedial hypothalamic nucleus despite development of leptin resistance during pregnancy in the rat. Physiol Rep. 2013;1(6):e00162. https://doi.org/10.1002/phy2.162.

Pilgrim C. Histochemical differentiation of hypothalamic areas. Prog Brain Res. 1974;41:97–110. https://doi.org/10.1016/S0079-6123(08)61901-9.
Pirrung MC, Southern EM. The genesis of microarrays. Biochem Mol Biol Educ. 2014;42(2):106–13.
Pomeranz LE, Ekstrand MI, Latcha KN, Smith GA, Enquist LW, Friedman JM. Gene expression profiling with Cre-conditional pseudorabies virus reveals a subset of midbrain neurons that participate in reward circuitry. J Neurosci. 2017;37(15):4128–44. https://doi.org/10.1523/JNEUROSCI.3193-16.2017.
Popesku JT, Martyniuk CJ, Denslow ND, Trudeau VL. Rapid dopaminergic modulation of the fish hypothalamic transcriptome and proteome. PLoS One. 2010;5(8):e12338. https://doi.org/10.1371/journal.pone.0012338.
Poplawski MM, Mastaitis JW, Yang X-J, Mobbs CV. Hypothalamic responses to fasting indicate metabolic reprogramming away from glycolysis toward lipid oxidation. Endocrinology. 2010;151(11):5206–17. https://doi.org/10.1210/en.2010-0702.
Porterfield VM, Mintz EM. Temporal patterns of light-induced immediate-early gene expression in the suprachiasmatic nucleus. Neurosci Lett. 2009;463:70–3. https://doi.org/10.1016/j.neulet.2009.07.066.
Porterfield VM, Piontkivska H, Mintz EM. Identification of novel light-induced genes in the suprachiasmatic nucleus. BMC Neurosci. 2007;8:98. https://doi.org/10.1186/1471-2202-8-98.
Poulin J-F, Tasic B, Hjerling-Leffler J, Trimarchi JM, Awatramani R. Disentangling neural cell diversity using single-cell transcriptomics. Nat Neurosci. 2016;19(9):1131–41. https://doi.org/10.1038/nn.4366.
Prima V, Tennant M, Gorbatyuk OS, Muzyczka N, Scarpace PJ, Zolotukhin S. Differential modulation of energy balance by leptin, ciliary neurotrophic factor, and leukemia inhibitory factor gene delivery: microarray deoxyribonucleic acid-chip analysis of gene expression. Endocrinology. 2004;145(4):2035–45.
Qiu F, Qu M, Zhang X, Wang H, Ding S. Hypothalamus and pituitary transcriptome profiling of male and female Hong Kong grouper (Epinephelus akaara). Gene. 2018;656:73–9. https://doi.org/10.1016/j.gene.2018.02.057.
Qiu J, Hindmarch CCT, Yao ST, Tasker JG, Murphy D. Transcriptomic analysis of the osmotic and reproductive remodeling of the female rat supraoptic nucleus. Endocrinology. 2011;152:3483–91. https://doi.org/10.1210/en.2011-1044.
Qiu J, Keineidam A, Gouraud S, Yao ST, Greenwood M, Hoe SZ, et al. The use of protein-DNA, chromatin immunoprecipitation, and transcriptome arrays to describe transcriptional circuits in the dehydrated male rat hypothalamus. Endocrinology. 2014;155:4380–90. https://doi.org/10.1210/en.2014-1448.
Rabaglino MB, Richards E, Denslow N, Keller-Wood M, Wood CE. Genomics of estradiol-3-sulfate action in the ovine fetal hypothalamus. Physiol Genomics. 2012;44:669–77. https://doi.org/10.1152/physiolgenomics.00127.2011.
Rabaglino MB, Chang EI, Richards EM, James MO, Keller-Wood M, Wood CE. Genomic effect of triclosan on the fetal hypothalamus: evidence for altered neuropeptide regulation. Endocrinology. 2016;157:2686–97. https://doi.org/10.1210/en.2016-1080.
Rajamani U, Gross AR, Hjelm BE, Sequeira A, Vawter MP, Tang J, Gangalapudi V, Wang Y, Andres AM, Gottlieb RA, Sareen D. Super-obese patient-derived iPSC hypothalamic neurons exhibit obesogenic signatures and hormone responses. Cell Stem Cell. 2018;22(5):698–712.e9.
Replogle K, Arnold AP, Ball GF, Band M, Bensch S, Brenowitz EA, et al. The songbird neurogenomics (SoNG) initiative: community-based tools and strategies for study of brain gene function and evolution. BMC Genomics. 2008;9:131. https://doi.org/10.1186/1471-2164-9-131.
Reyes TM, Walker JR, DeCino C, Hogenesch JB, Sawchenko PE. Categorically distinct acute stressors elicit dissimilar transcriptional profiles in the paraventricular nucleus of the hypothalamus. J Neurosci. 2003;23(13):5607–16.
Richter CA, Martyniuk CJ, Annis ML, Brumbaugh WG, Chasar LC, Denslow ND, et al. Methylmercury-induced changes in gene transcription associated with neuroendocrine disrup-

tion in largemouth bass (Micropterus salmoides). Gen Comp Endocrinol. 2014;201:215–24. https://doi.org/10.1016/j.ygcen.2014.03.029.

Richter D, editor. Metabolism of the nervous system. New York: Pergamon Press; 1957.

Roberts E, Baxter CF. Neurochemistry. Annu Rev Biochem. 1963;32:513–52. https://doi.org/10.1146/annurev.bi.32.070163.002501.

Roberts S, Keller MR. Respiration and glycolysis in the hypophysis and hypothalamus of the rat. Arch Biochem Biophys. 1953;44(1):9–14. https://doi.org/10.1016/0003-9861(53)90003-4.

Roberts S, Keller MR. Influence of epinephrine and cortisone on the metabolism of the hypophysis and hypothalamus of the rat. Endocrinology. 1955;57(1):64–9. https://doi.org/10.1210/endo-57-1-64.

Robinson SM, Fox TO, Sidman RL. A genetic variant in the morphology of the medial preoptic area in mice. J Neurogenet. 1985;2(6):381–8.

Romanov RA, Alpár A, Zhang M-D, Zeisel A, Calas A, Landry M, et al. A secretagogin locus of the mammalian hypothalamus controls stress hormone release. EMBO J. 2014;34(1):36–54. https://doi.org/10.15252/embj.201488977.

Romanov RA, Zeisel A, Bakker J, Girach F, Hellysaz A, Tomer R, et al. Molecular interrogation of hypothalamic organization reveals distinct dopamine neuronal subtypes. Nat Neurosci. 2017;20(2):176–88. https://doi.org/10.1038/nn.4462.

Romanova EV, Sweedler JV. Peptidomics for the discovery and characterization of neuropeptides and hormones. Trends Pharmacol Sci. 2015;36(9):579–86. https://doi.org/10.1016/j.tips.2015.05.009.

Römpp A, Guenther S, Schober Y, Schulz O, Takats Z, Kummer W, et al. Histology by mass spectrometry: label-free tissue characterization obtained from high-accuracy bioanalytical imaging. Angew Chem Int Ed. 2010;49:3834–8. https://doi.org/10.1002/anie.20095559.

Roots BI, Johnston PV. Lipids of isolated neurons. Biochem J. 1965;94:61–3.

Ropp SA, Grunwald WC Jr, Morris M, Cool DR. Pyridostigmine crosses the blood-brain barrier to induce cholinergicger and non-cholinergic changes in mouse hypothalamus. J Med CBR Def. 2008;6:24.

Rose SPR. Holger Hyden and the biochemistry of memory. Brain Res Bull. 1999;50(5/6):443. https://doi.org/10.1016/S0361-9230(99)00125-2.

Rossner MJ, Hirrlinger J, Wichert SP, Boehm C, Newrzella D, Hiemisch H, et al. Global transcriptome analysis of genetically identified neurons in the adult cortex. J Neurosci. 2006;26:9956–66. https://doi.org/10.1523/JNEUROSCI.0468-06.2006.

Roth CL, McCormack AL, Lomniczi A, Mungenast AE, Ojeda SR. Quantitative proteomics identifies a change in glial glutamate metabolism at the time of puberty. Mol Cell Endocrinol. 2006;254–255:51–9. https://doi.org/10.1016/j.mce.2006.04.017.

Roy M, Kim N, Kim K, Chung W-H, Achawanantakun R, Sun Y, et al. Analysis of the canine brain transcriptome with an emphasis on the hypothalamus and cerebral cortex. Mamm Genome. 2013;24:484–99. https://doi.org/10.1007/s00335-013-9480-0.

Russ T, Ramakrishnan C, Hovy E, Bota M, Burns G. Knowledge engineering tools for reasoning with scientific observations and interpretations: a neural connectivity use case. BMC Bioinformatics. 2011;12:351. https://doi.org/10.1186/1471-2105-12-351.

Sakakibara M, Uenoyama Y, Minabe S, Watanabe Y, Deura C, Nakamura S, et al. Microarray analysis of perinatal-estrogen-induced changes in gene expression related to brain sexual differentiation in mice. PLoS One. 2013;8(11):e79437. https://doi.org/10.1371/journal.pone.0079437.

Sandberg R, Yasuda R, Pankratz DG, Carter TA, Del Rio JA, Wodicka L, et al. Regional and strain-specific gene expression mapping in the adult mouse brain. Proc Natl Acad Sci U S A. 2000;97(20):11038–43. https://doi.org/10.1073/pnas.97.20.11038.

Sangiao-Alvarellos S, Pena-Bello L, Manfredi-Lozano M, Tena-Sempere M, Cordido F. Perturbation of hypothalamic MicroRNA expression patterns in male rats after metabolic distress: impact of obesity and conditions of negative energy balance. Endocrinology. 2014;155(5):1838–50.

Sanna PP, King AR, van der Stap LD, Repunte-Canoningo V. Gene profiling of laser-microdissected brain regions and sub-regions. Brain Res Protocol. 2005;15:66–74. https://doi.org/10.1016/j.brainresprot.2005.04.002.

Sarkar P, Sarkar S, Ramesh V, Kim H, Barnes S, Kulkarni A, et al. Proteomic analysis of mouse hypothalamus under simulated microgravity. Neurochem Res. 2008;33:2335–41. https://doi.org/10.1007/s11064-008-9738-1.

Sawyer CH, Everett JW, Green JD. The rabbit diencephalon in stereotaxic coordinates. J Comp Neurol. 1954;101(3):801–24.

Schmidlin T, Boender AJ, Frese CK, Heck AJR, Adan RAH, Altelaar AFM. Diet-induced neuropeptide expression: feasibility of quantifying extended and highly charged endogenous peptide sequences by selected reaction monitoring. Anal Chem. 2015;87:9966–73. https://doi.org/10.1021/acs.analchem.5b03334.

Schneeberger M, Altirriba J, García A, Esteban Y, Castaño C, García-Lavandeira M, Alvarez CV, Gomis R, Claret M. Deletion of miRNA processing enzyme dicer in POMC-expressing cells leads to pituitary dysfunction, neurodegeneration and development of obesity. Mol Metab. 2013;2(2):74–85.

Schrader M, Schulz-Knappe P, Fricker LD. Historical perspective of peptidomics. EuPA Open Proteom. 2014;3:171–82. https://doi.org/10.1016/j.euprot.2014.02.014.

Schwartz S, Reyzer ML, Caprioli RM. Direct tissue analysis using matrix-assisted laser desorption/ionization mass spectrometry: practical aspects of sample preparation. J Mass Spectrom. 2003;38:699–708. https://doi.org/10.1002/jms.505.

Schwiening CJ. A brief historical perspective: Hodgkin and Huxley. J Physiol Lond. 2012;590:2571–5.

Secher A, Kelstrup CD, Conde-Frieboes KW, Pyke C, Raun K, Wulff BS, et al. Analytic framework for peptidomics applied to large-scale neuropeptide identification. Nat Commun. 2016;7:11436. https://doi.org/10.1038/ncomms11436.

Segal JP, Stallings NR, Lee CE, Zhao L, Socci N, Viale A, et al. Use of laser-capture microdissection for the identification of marker genes for the ventromedial hypothalamic nucleus. J Neurosci. 2005;25(16):4181–8. https://doi.org/10.1523/JNEUROSCI.0158-05.2005.

Semmens DC, Mirabeau O, Moghul I, Pancholi MR, Wurm Y, Elphick MR. Transcriptomic identification of starfish neuropeptide precursors yields new insights into neuropeptide evolution. Open Biol. 2016;6(2):150224.

Shariatgorji M, Nilsson A, Goodwin RJA, Källback P, Schintu N, Zhang X, et al. Direct targeted quantitative molecular imaging of neurotransmitters in brain tissue sections. Neuron. 2014a;84:697–707. https://doi.org/10.1016/j.neuron.2014.10.011.

Shariatgorji M, Svenningsson P, Andrén P. Mass spectrometry imaging, an emerging technology in neuropsychopharmacology. Neuropsychopharmacol Rev. 2014b;39:34–49. https://doi.org/10.1038/npp.2013.215.

Sharma A, Singh D, Das S, Kumar V. Hypothalamic and liver transcriptome from two crucial life-history stages in a migratory songbird. Exp Physiol. 2018;103(4):559–69.

Sharma K, Schmitt S, Bergner CG, Tyanova S, Kannaiyan N, Manrique-Hoyos N, et al. Cell type- and brain region-resolved mouse brain proteome. Nat Neurosci. 2015;18:1819–31. https://doi.org/10.1038/nn.4160.

Shimogori T, Lee DA, Miranda-Angulo A, Yang Y, Wang H, Jiang L, Yoshida AC, Kataoka A, Mashiko H, Avetisyan M, Qi L, Qian J, Blackshaw S. A genomic atlas of mouse hypothalamic development. Nat Neurosci. 2010;13(6):767–75.

Shiue Y-L, Chen L-R, Chen C-F, Chen Y-L, Jhy-Phen J, Chao C-H, Lin Y-P, Kuo Y-M, Tang P-C, Lee Y-P. Identification of transcripts related to high egg production in the chicken hypothalamus and pituitary gland. Theriogenology. 2006;66(5):1274–83.

Simmons DM, Swanson LW. Comparing histological data from different brains: sources of error and strategies for minimizing them. Brain Res Rev. 2009;60(2):349–67. https://doi.org/10.1016/j.brainresrev.2009.02.002.

Sköld K, Svensson M, Kaplan A, Björkesten L, Åström J, Andren PE. A neuroproteomic approach to targeting neuropeptides in the brain. Proteomics. 2002;2:447–54. https://doi.org/10.1002/1615-9861(200204)2:4<447::AID-PROT447>3.0.CO;2-A.

Sköld K, Svensson M, Norrman M, Sjögren B, Svenningsson P, Andrén PE. The significance of biochemical and molecular sample integrity in brain proteomics and peptidomics: stathmin 2-2- and peptides as sample quality indicators. Proteomics. 2007;7:4445–56. https://doi.org/10.1002/pmic.200700142.

Skynner HA, Amos DP, Murray F, Salim K, Knowles MR, et al. Proteomic analysis identifies alterations in cellular morphology and cell death pathways in mouse brain after chronic corticosterone treatment. Brain Res. 2006;1102:12–26. https://doi.org/10.1016/j.brainres.2006.04.112.

Smithies O, Poulik MD. Two-dimensional electrophoresis of serum proteins. Nature. 1956;177(4518):1033. https://doi.org/10.1038/1771033a0.

Soga T, Dalpatadu SL, Wong DW, Parhar IS. Neonatal dexamethasone exposure down-regulates GnRH expression through the GnIH pathway in female mice. Neuroscience. 2012;218:56–64. https://doi.org/10.1016/j.neuroscience.2012.05.023.

Sotelo C, Palay SL. The fine structure of the lateral vestibular nucleus in the rat. I. Neurons and neuroglial cells. J Cell Biol. 1968;36:151–79. https://doi.org/10.1083/jcb.36.1.151.

Soukup J, Krskova L, Hilska I, Kodet R. Ethanol fixation of lymphoma samples as an alternative approach for preservation of the nucleic acids. Neoplasma. 2003;50:300–4.

Southey BR, Lee JE, Zamdborg L, Atkins N Jr, Mitchell JW, Li M, et al. Comparing label-free quantitative peptidomics approaches to characterize diurnal variation of peptides in the rat suprachiasmatic nucleus. Anal Chem. 2014;86:443–52. https://doi.org/10.1021/ac4023378l.

Spengler B, Hubert M, Kaufmann R. MALDI ion imaging and biological ion imaging with a new scanning UV-laser microprobe. 42nd Annual Conference on Mass Spectrometry and Allied Topics, ASMS 1994, May 29–Jun 3, Chicago, IL; 1994.

Spraggins JM, Rizzo DG, Moore JL, Noto MJ, Skaar EP, Caprioli RM. Next-generation technologies for spatial proteomics: integrating ultra-high speed MALDI-TOF and high mass resolution MALDI FTICR imaging mass spectrometry for protein analysis. Proteomics. 2016;16:11–2. 1678–1689. DOI https://doi.org/10.1002/pmic.201600003.

Ståhl PL, Salmén F, Vickovic S, Lundmark A, Navarro JF, Magnusson J, et al. Visualization and analysis of gene expression in tissue sections by spatial transcriptomics. Science. 2016;353:78–82. https://doi.org/10.1126/science.aaf2403.

St-Amand J, Yoshioka M, Tanaka K, Nishida Y. Transcriptome-wide identification of preferentially expressed genes in the hypothalamus and pituitary gland. Front Endocrinol. 2012;2:111. https://doi.org/10.3389/fendo.2011.00111.

Stelzhammer V, Amess B, Martins-de-Souza D, Levin Y, Ozanne SE, Martin-Gronert MS, et al. Analysis of the rat hypothalamus proteome by data-independent label-free LC-MS/MS. Proteomics. 2012;12:3386–92. https://doi.org/10.1002/pmic.201100642.

Stewart L, Hindmarch CCT, Qiu J, Tung Y-CL, Yeo GSH, Murphy D. Hypothalamic transcriptome plasticity in two rodent species reveals divergent differential gene expression but conserved pathways. J Neuroendocrinol. 2011;23:177–85. https://doi.org/10.1111/j.1365-2826.2010.02093.x.

Stocker CJ, Wargent ET, Martin-Gronert MS, Cripps RL, O'Dowd JF, Zaibi MS, et al. Leanness in postnatally nutritionally programmed rats is associated with increased sensitivity to leptin and a melanocortin receptor agonist and decreased sensitivity to neuropeptide Y. Int J Obes. 2012;36(8):1040–6. https://doi.org/10.1038/ijo.2011.226.

Su JM, Perlaky L, Li XN, Leung HC, Antalffy B, Armstrong D, et al. Comparison of ethanol versus formalin fixation on preservation of histology and RNA in laser capture microdissected brain tissues. Brain Pathol. 2004;14:175–82. https://doi.org/10.1111/j.1750-3639.2004.tb00050.x.

Su YA, Zhang Q, Su DM, Tang MX. Rat mitochondrion-neuron focused microarray (rMNChip) and bioinformatics tools for rapid identification of differential pathways in brain tissues. Int J Biol Sci. 2011;7(3):308–22.

Sun H, Jiang R, Xu S, Zhang Z, Xu G, et al. Transcriptome responses to heat stress in hypothalamus of a meat-type chicken. J Anim Sci Biotechnol. 2015;6:6. https://doi.org/10.1186/s40104-015-0003-6.

Sun W, Lee S, Zhabotynsky V, Zou F, Wright FA, Crowley JJ, et al. Transcriptome atlases of mouse brain reveals differential expression across brain regions and genetic backgrounds. Genes, Genomes, Genet. 2012;2:203–11. https://doi.org/10.1534/g3.111.001602.

Sung HJ, Kim YS, Kim IS, Jang S-W, Kim YR, Na DS, Han KH, Hwang BG, Park DS, Ko J. Proteomic analysis of differential protein expression in neuropathic pain and electroacupuncture treatment models. Proteomics. 2004;4(9):2805–13.

Sunkin SM, Ng L, Lau C, Dolbeare T, Gilbert TL, Thompson CL, et al. Allen Brain Atlas: an integrated spatio-temporal portal for exploring the central nervous system. Nucleic Acids Res. 2013;41:D996–D1008. https://doi.org/10.1093/nar/gks1042.
Sutcliffe JG. Open-system approaches to gene expression in the CNS. J Neurosci. 2001;21(21):8306–9.
Sutcliffe JG, de Lecea L. The hypocretins: setting the arousal threshold. Nat Rev Neurosci. 2002;3:339–49. https://doi.org/10.1038/nrn808.
Svensson M, Sköld K, Svenningsson P, Andren PE. Peptidomics-based discovery of novel neuropeptides. J Proteome Res. 2003;2:213–9. https://doi.org/10.1021/pr020010u.
Swanson LW. Brain maps: structure of the rat brain. 3rd ed. Amsterdam: Elsevier; 2004.
Swanson LW. Quest for the basic plan of nervous system circuitry. Brain Res Rev. 2007;55(2):356–72. https://doi.org/10.1016/j.brainresrev.2006.12.006.
Swanson LW. Brain Maps 4.0—Structure of the Rat Brain: an open access atlas with global nervous system nomenclature ontology and flatmaps. J Comp Neurol. 2018;526(6):935–43. https://doi.org/10.1002/cne.24381.
Swanson LW, Sanchez-Watts G, Watts AG. Comparison of melanin-concentrating hormone and hypocretin/orexin mRNA expression patterns in a new parceling scheme of the lateral hypothalamic zone. Neurosci Lett. 2005;387(2):80–4. https://doi.org/10.1016/j.neulet.2005.06.066.
Swart I, Jahng JW, Overton JM, Houpt TA. Hypothalamic NPY, AGRP, and POMC mRNA responses to leptin and refeeding in mice. Am J Physiol Regul Integr Comp Physiol. 2002;283:R1020–6. https://doi.org/10.1152/ajpregu.00501.2001.
Tallis M, Thompson R, Russ TA, Burns GAPC. Knowledge synthesis with maps of neural connectivity. Front Neuroinform. 2011;5:24. https://doi.org/10.3389/fninf.2011.00024.
Taouis M. MicroRNAs in the hypothalamus. Best Pract Res Clin Endocrinol Metab. 2016;30:641–51. https://doi.org/10.1016/j.beem.2016.11.006.
Taraslia VK, Kouskoukis A, Anagnostopoulos AK, Stravopodis DJ, Margaritis LH, Tsangaris GT. Proteomic analysis of normal murine brain parts. Cancer Genomics Proteomics. 2013;10:125–54.
Tatemoto K, Mutt V. Isolation of two novel candidate hormones using a chemical method for finding naturally occurring polypeptides. Nature. 1980;285:417–8. https://doi.org/10.1038/285417a0.
Tatemoto K, Carlquist M, Mutt V. Neuropeptide Y—a novel brain peptide with structural similarities to peptide YY and pancreatic polypeptide. Nature. 1982;296:659–60. https://doi.org/10.1038/296659a0.
Tatemoto K, Rökaeus Å, Jörnvall H, McDonald TJ, Mutt V. Galanin—a novel biologically active peptide from porcine intestine. FEBS Lett. 1983;164:124–8. https://doi.org/10.1016/0014-5793(83)80033-7.
Ten Donkelaar HJ, Nicholson C. Appendix—(Stereotaxic) Atlases. A bibliography of (stereotaxic) brain atlases arranged by chapter. In: Nieuwenhuys R, Ten Donkelaar HJ, Nicholson C, editors. The central nervous system of vertebrates, vol. 1. Berlin: Springer; 1998. p. 354–5.
Thudicum JLW. A treatise on the chemical constitution of the brain. London: Baillière, Tindall, and Cox; 1884.
Tindal JS. The forebrain of the guinea pig in stereotaxic coordinates. J Comp Neurol. 1965;124(2):259–66.
Trivedi C, Shan X, Tung YC, Kabra D, Holland J, Amburgy S, et al. Tachykinin-1 in the central nervous system regulates adiposity in rodents. Endocrinology. 2015;156(5):1714–23. https://doi.org/10.1210/en.2014-1781.
Tu W-L, Cheng C-Y, Wang S-H, Tang P-C, Chen C-F, Chen H-H, et al. Profiling of differential gene expression in the hypothalamus of broiler-type Taiwan country chickens in response to acute heat stress. Theriogenology. 2016;85:483–94. https://doi.org/10.1016/j.theriogenology.2015.09.028.
Tung YC, Ma M, Piper S, Coll A, O'Rahilly S, Yeo GS. Novel leptin-regulated genes revealed by transcriptional profiling of the hypothalamic paraventricular nucleus. J Neurosci. 2008;28:12419–26. https://doi.org/10.1523/JNEUROSCI.3412-08.2008.

Udvari EB, Völgyi K, Gulyássy P, Dimén D, Kis V, Barna J, Szabó ÉR, Lubec G, Juhász G, Kékesi KA, Dobolyi Á. Synaptic proteome changes in the hypothalamus of mother rats. J Proteome. 2017;159:54–66.
Uz T, Arslan AD, Kurtuncu M, Imbesi M, Akhisaroglu M, Dwivedi Y, et al. The regional and cellular expression profile of the melatonin receptor MT1 in the central dopaminergic system. Mol Brain Res. 2005;136:45–53. https://doi.org/10.1016/j.molbrainres.2005.01.002.
van de Plas R, Yang J, Spraggins J, Caprioli RM. Image fusion of mass spectrometry and microscopy: a multimodality paradigm for molecular tissue mapping. Nat Methods. 2015;12:366–74. https://doi.org/10.1038/nmeth.3296.
van Tienhoven A, Juhász LP. The chicken telencephalon, diencephalon and mesencephalon in stereotaxic coordinates. J Comp Neurol. 1962;118(2):185–97. https://doi.org/10.1002/cne.901180205.
Vasilache AM, Anderson J, Nilsberth C. Expression of PGE_2 EP_3 receptor subtypes in the mouse preoptic region. Neurosci Lett. 2007;423:179–83. https://doi.org/10.1016/j.neulet.2007.06.048.
Vasilache AM, Kugelberg U, Blomqvist A, Nilsberth C. Minor changes in gene expression in the mouse preoptic hypothalamic region by inflammation-induced prostaglandin E_2. J Neuroendocrinol. 2013;25:635–43. https://doi.org/10.1111/jne.12044.
Volgin DV, Swan J, Kubin L. Single-cell RT-PCR gene expression profiling of acutely dissociated and immunocytochemically identified central neurons. J Neurosci Methods. 2004;136:229–36. https://doi.org/10.1016/j.jneumeth.2004.01.013.
Wada M, Urano A, Gorbman A. A stereotaxic atlas for diencephalic nuclei of the frog, Rana pipiens. Arch Histol Jpn. 1980;43(2):157–73.
Walch JD, Nedungadi TP, Cunningham JT. ANG II receptor subtype 1a gene knockdown in the subfornical organ prevents increased drinking behavior in bile duct-ligated rats. Am J Physiol Regul Integr Comp Physiol. 2014;307:R597–607. https://doi.org/10.1152/ajpregu.00163.2014.
Waller R, Woodroofe MN, Francese S, Heath PR, Wharton SB, Ince PG, et al. Isolation of enriched glial populations from post-mortem human CNS material by immuno-laser capture microdissection. J Neurosci Methods. 2012;208:108–13. https://doi.org/10.1016/j.jneumeth.2012.04.014.
Wang QM, Yang H, Tian DR, Cai Y, Wei ZN, Wang F, Yu AC, Han JS. Proteomic analysis of rat hypothalamus revealed the role of ubiquitin-proteasome system in the genesis of DR or DIO. Neurochem Res. 2011;36:939–46. https://doi.org/10.1007/s11064-011-0423-4.
Wang S-S, Kamphius W, Huitinga I, Zhou J-N, Swaab DF. Gene expression analysis in the human hypothalamus in depression by laser microdissection and real-time PCR: the presence of multiple receptor imbalances. Mol Psychiatry. 2008;13:786–99. https://doi.org/10.1038/mp.2008.38.
Warburg O, Negelein E, Posener K. Versuche an Überlebendem carcinomgewebe. Klin Wochenschr. 1924;3(24):1062–4.
Wells CE, Khan AM. Data transformations between rat brain atlases: mapping central microinjection sites on stereotaxically aligned and anisotropically scaled digital atlas plates in Paxinos & Watson and Swanson reference spaces. Program No. 198.06. 2013 Neuroscience Meeting Planner. San Diego, CA: Society for Neuroscience; 2013. Online.
Williams DL, Schwartz MW, Bastian LS, Blevins JE, Baskin DG. Immunocytochemistry and laser capture microdissection for real-time quantitative PCR identify hindbrain neurons activated by interaction between leptin and cholecystokinin. J Histochem Cytochem. 2008;56(3):285–93. https://doi.org/10.1369/jhc.7A7331.2007.
Williams EA, Verasztó C, Jasek S, Conzelmann M, Shahidi R, Bauknecht P, Mirabeau O, Jékely G. Synaptic and peptidergic connectome of a neurosecretory center in the annelid brain. elife. 2017;6:e26349.
Williams KW, Margatho LO, Lee CE, Choi M, Lee S, Scott MM, et al. Segregation of acute leptin and insulin effects in distinct populations of arcuate proopiomelanocortin neurons. J Neurosci. 2010;30(7):2472–9. https://doi.org/10.1523/JNEUROSCI.3118-09.2010.
Winrow CJ, Tanis KQ, Rigby AM, Taylor RR, Serikawa K, Tokiwa GY, et al. Refined anatomical isolation of functional sleep circuits exhibits distinctive regional and circadian gene transcriptional profiles. Brain Res. 2009;1271:1–17. https://doi.org/10.1016/j.brainres.2009.02.083.

Wolf L, Goldberg C, Manor N, Sharan R, Ruppin E. Gene expression in the rodent brain is associated with its regional connectivity. PLoS Comput Biol. 2011;7(5):e1002040. https://doi.org/10.1371/journal.pcbi.1002040.

Wood CE, Rabaglino MB, Chang EI, Denslow N, Keller-Wood M, Richards E. Genomics of the fetal hypothalamic cellular response to transient hypoxia: endocrine, immune, and metabolic responses. Physiol Genomics. 2013;45:521–7. https://doi.org/10.1152/physiolgenomics.00005.2013.

Xi D, Kusano K, Gainer H. Quantitative analysis of oxytocin and vasopressin messenger ribonucleic acids in single magnocellular neurons isolated from supraoptic nucleus of rat hypothalamus. Endocrinology. 1999;140:4677–82. https://doi.org/10.1210/endo.140.10.7054.

Xiao XQ, Grove KL, Lau SY, Shannon MW, Susan Smith M. Deoxyribonucleic acid microarray analysis of gene expression pattern in the Arcuate nucleus/ventromedial nucleus of hypothalamus during lactation. Endocrinology. 2005;146(10):4391–8.

Xu J, Huang W, Zhong C, Luo D, Li S, Zhu Z, et al. Defining global gene expression changes of the hypothalamic-pituitary-gonadal axis in female sGnRH-antisense transgenic common carp (Cyprinus carpio). PLoS One. 2011;6(6):e21057. https://doi.org/10.1371/journal.pone.0021057.

Xu RY, Wan YP, Tang QY, Wu J, Cai W. The effects of high fat on central appetite genes in Wistar rats: a microarray analysis. Clin Chim Acta. 2008;397:96–100. https://doi.org/10.1016/j.cca.2008.07.027.

Yamashita M, Glasgow E, Zhang BJ, Kusano K, Gainer H. Identification of cell-specific messenger ribonucleic acids in oxytocinergic and vasopressinergic magnocellular neurons in rat supraoptic nucleus by single-cell differential hybridization. Endocrinology. 2002;143:4464–76. https://doi.org/10.1210/en.2002-220516.

Yang N, Anapindi KDB, Romanova EV, Rubakhin SS, Sweedler JV. Improved identification and quantitation of mature endogenous peptides in the rodent hypothalamus using a rapid conductive sample heating system. Analyst. 2017;142(23):4476–85.

Yao I, Sugiura Y, Matsumoto M, Setou M. In situ proteomics with imaging mass spectrometry and principal component analysis in the Scrapper-knockout mouse brain. Proteomics. 2008;8:3692–701. https://doi.org/10.1002/pmic.200701121.

Yao ST, Gouraud S, Paton JF, Murphy D. Water deprivation increases the expression of neuronal nitric oxide synthase (nNOS) but not orexin-A in the lateral hypothalamic area of the rat. J Comp Neurol. 2005;490:180–93. https://doi.org/10.1002/cne.20662.

Yasuda M, Lepkovsky S. The chicken diencephalon in stereotaxic coordinates. Jap J Zootech Sci. 1969;40(10):417–31.

Yelin-Bekerman L, Elbaz I, Diber A, Dahary D, Gibbs-Bar L, Alon S, et al. Hypocretin neuron-specific transcriptome profiling identifies the sleep modulator Kcnh4a. elife. 2015;4:e08638. https://doi.org/10.7554/eLife.08638.

Yonehara K, Suzuki M, Nishihara M. Sex-related differences in gene expression in neonatal rat hypothalamus assessed by cDNA microarray analysis. Endocr J. 2002;49(2):131–7. https://doi.org/10.1507/endocrj.49.131.

Yue C, Mutsuga N, Verbalis J, Gainer H. Microarray analysis of gene expression in the supraoptic nucleus of normoosmotic and hypoosmotic rats. Cell Mol Neurobiol. 2006;26(4–6):959–78. https://doi.org/10.1007/s10571-006-9017-0.

Zapala MA, Hovatta I, Ellison JA, Wodicka L, Del Rio JA, Tennant R, et al. Adult mouse brain gene expression patterns bear an embryologic imprint. Proc Natl Acad Sci U S A. 2005;102(29):10357–62. https://doi.org/10.1073/pnas.0503357102.

Zettergren A, Karlsson S, Studer E, Sarvimäki A, Kettunen P, Thorsell A, et al. Proteomic analyses of limbic regions in neonatal male, female and androgen receptor knockout mice. BMC Neurosci. 2017;18:9. https://doi.org/10.1186/s12868-016-0332-1.

Zhang D, Xiong H, Mennigen JA, Popesku JT, Marlatt VL, Martyniuk CJ, et al. Defining global neuroendocrine gene expression patterns associated with reproductive seasonality in fish. PLoS One. 2009;4(6):e5816. https://doi.org/10.1371/journal.pone.005816.

Zhang L, Cai Z, Wei S, Zhou H, Zhou H, Jiang X, et al. MicroRNA expression profiling of the porcine developing hypothalamus and pituitary tissue. Int J Mol Sci. 2013;14(10):20326–39. https://doi.org/10.3390/ijms141020326.

Zhang X, Petruzziello F, Zani F, Fouillen L, Andren PE, Solinas G, et al. High identification rates of endogenous neuropeptides from mouse brain. J Proteome Res. 2012;11:2819–27. https://doi.org/10.1021/pr3001699.

Zhang XY, Zhu MK, Yuan C, Zou XT. Proteomic analysis of hypothalamus and liver proteins affected by dietary -arginine supplementation in laying hens. J Anim Physiol Anim Nutr. 2018;

Zhong L, Zhou J, Wang D, Zou X, Lou Y, Liu D, et al. Proteomics and bioinformatics analysis of mouse hypothalamic neurogenesis with or without EPHX2 gene deletion. Int J Clin Exp Pathol. 2015;8:12634–45.

Zhu H, Vadigepalli R, Rafferty R, Gonye GE, Weaver DR, Schwaber JS. Integrative gene regulatory network analysis reveals light-induced regional gene expression phase shift programs in the mouse suprachiasmatic nucleus. PLoS One. 2012;7(5):e37833. https://doi.org/10.1371/journal.pone.0037833.

Zmora N, Stubblefield J, Zulperi Z, Biran J, Levavi-Sivan B, Muñoz-Cueto JA, et al. Differential and gonad stage-dependent roles of kisspeptin1 and kisspeptin2 in reproduction in the modern teleosts, Morone species. Biol Reprod. 2012;86(6):177. https://doi.org/10.1095/biolreprod.111.097667.

Zséli G, Vida B, Martinez A, Lechan RM, Khan AM, Fekete C. Elucidation of the anatomy of a satiety network: Focus on connectivity of the parabrachial nucleus in the adult rat. J Comp Neurol. 2016;524(14):2803–27. https://doi.org/10.1002/cne.23992.

Chapter 7
Genome-Scale Brain Metabolic Networks as Scaffolds for the Systems Biology of Neurodegenerative Diseases: Mapping Metabolic Alterations

Emrah Özcan and Tunahan Çakır

7.1 Introduction

Systems neuroscience, an extension of the systems biology field for the analysis of the central nervous system, stands on the shoulders of two driving forces at molecular level: high-throughput genome-scale data and genome-scale cellular networks. Integrative systems-wide analysis of data and networks at genome-scale enables a holistic understanding of the functionality and working principles of the cell. Metabolic networks are among major cellular network types since it is the metabolism that runs the cellular factory. Metabolism covers the uptake of substrates, such as glucose, by the cell and further conversion of these substances, metabolites, to hundreds of other metabolites required for the functioning of the cell. Therefore, the compilation of biochemical pathways of the brain for functional analysis is on the rise in recent years. Similarly, measurement of hundreds of mRNAs, proteins, and metabolites of brain cells is now possible via high-throughput measurement techniques and is also applied for the analysis of neurological disorders among other brain related diseases (Fig. 7.1).

In this chapter, transcriptome, proteome and metabolome studies of major neurodegenerative disorders will be reviewed with a narrower but essential scope: in terms of the induced changes in brain metabolism. The contribution of omics data to the understanding of metabolism of neurodegenerative diseases will be provided. Later, brain-specific genome-scale metabolic networks and related modeling efforts will be reviewed from the aspect of neurodegenerative diseases. Available studies on the integrative analysis of genome-scale data and metabolic networks for these diseases will be covered.

E. Özcan · T. Çakır (✉)
Department of Bioengineering, Gebze Technical University, Gebze, Kocaeli, Turkey
e-mail: tcakir@gtu.edu.tr

A. Cheung-Hoi Yu, L. Li (eds.), *Systems Neuroscience*, Advances in Neurobiology 21, https://doi.org/10.1007/978-3-319-94593-4_7

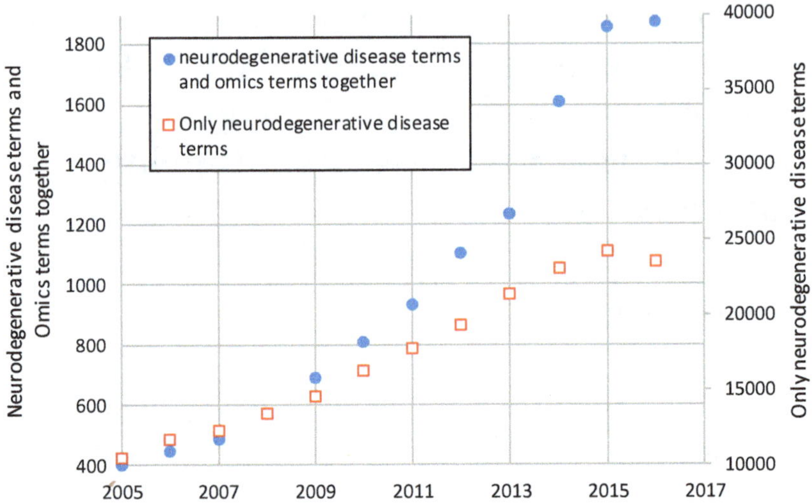

Fig. 7.1 The figure, compiled from PubMed, shows the relative increase over the years in the publications whose abstracts included only neurodegenerative disease related terms (which showed 2.3 fold increase in number between 2005 and 2016) or neurodegenerative disease and omics terms together (which showed 4.5 fold increase in the same period)

Table 7.1 Neurodegeneration related gene expression datasets in Gene Expression Omnibus (GEO), retrieved on July 9, 2017

Neurodegenerative disease	Microarray dataset	RNA-seq dataset
Alzheimer's disease (AD)	128 (human: 61, mouse: 48, rat: 9, others: 10)	26 (human: 9, mouse: 16, others: 1)
Parkinson's disease (PD)	143 (human: 78, mouse: 47, rat: 11, others: 7)	15 (human: 8, mouse: 6, others: 1)
Huntington's disease (HD)	60 (human: 14, mouse: 44, others: 2)	36 (human: 10, mouse: 25, others: 1)
Multiple sclerosis (MS)	130 (human: 84, mouse: 38, rat: 8)	12 (human: 6, mouse: 6)
Amyotrophic lateral sclerosis (ALS)	64 (human: 33, mouse: 23, rat: 5, others: 3)	32 (human: 17, mouse: 17, others: 1)
Total	525	121

7.2 Omics Data Analysis for Neurodegenerative Diseases: Metabolism Perspective

Among the omics approaches, transcriptomics is the frontier in terms of genome coverage and collected number of datasets. In Gene Expression Omnibus (GEO), the public database with transcriptome datasets (Barrett et al. 2011), currently about 86,000 datasets are accessible, among which 36,000 belongs to human. With microarray and RNA-seq technologies, whole-genome coverage of gene expression levels is possible. Table 7.1 lists the number of microarray and RNA-seq datasets stored in

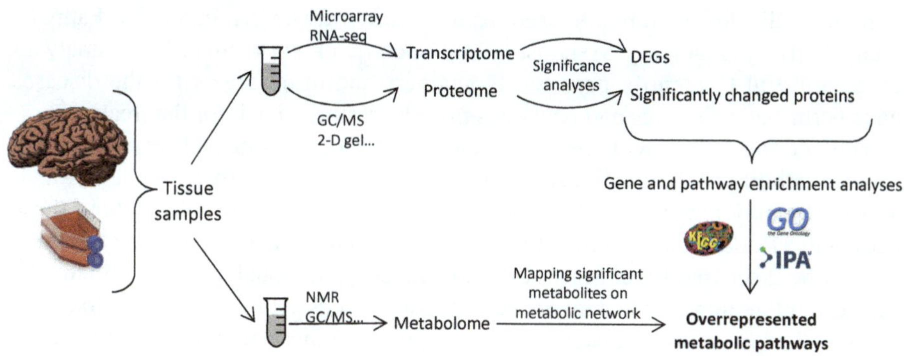

Fig. 7.2 Illustration of the standard omics approach for identifying biologically perturbed metabolic pathways

GEO, specific for major neurodegenerative diseases. On the other hand, mRNA levels may not reflect the real function of proteins because of post-transcriptional modifications. Therefore, proteomics experiments are important, but offer less coverage compared to transcriptomics, and miss the effect of the post-translational modifications. Metabolomics experiments are therefore closest to the functional state of the cell, but, again, have currently lower coverage compared to transcriptomics.

One of the key requirements in omics based data analysis is to collect replicate data such that a significance test can be applied to identify genes, proteins or metabolites which were affected from the disease. Once differentially expressed genes are identified, those who code for enzymes can be selectively analyzed to identify metabolic reactions and pathways perturbed in response to the disease state (Fig. 7.2). The most common and quick way of mapping significantly changed biomolecules to metabolic pathways is by enrichment tests. Enrichment tests use a controlled vocabulary of functional terms associated with the genes to identify terms common for the significantly changed genes. This leads to the identification of cellular functions commonly affected from the disease of interest. In the background, these tests perform an overrepresentation-based significance analysis to assign *p*-values to functional terms associated with the genes (Huang da et al. 2009a). This can be Gene Ontology (GO) Analysis to identify perturbed biological processes or Pathway Enrichment Analysis to identify perturbed KEGG or Reactome pathways. Pathway Commons (Cerami et al. 2011) and WikiPathways (Kelder et al. 2012) are other databases for curated biological pathways, to be used for enrichment analyses. Several softwares and web servers are available for straightforward enrichment-based identification of disease-affected metabolic pathways or processes. AmiGO (Carbon et al. 2009), GOrilla (Eden et al. 2009), g:Profiler (Reimand et al. 2011) and DAVID (Huang da et al. 2009b) are commonly used online free tools for GO enrichment. DAVID and g:Profiler also allow metabolic pathway enrichment. A sound desktop platform to perform such computational analyses is Cytoscape (Smoot et al. 2011), which provides plug-ins for a number of enrichment methods.

A more dedicated commonly used commercial software is Ingenuity Pathway Analysis (IPA) (Ingenuity Systems, USA). Such quick enrichment-based analyses are also useful to identify potential biomarkers and drug targets for the diseases since perturbed processes and pathways must be restored back for the treatment.

This section will provide perturbed metabolic processes or pathways identified by omics studies of the major neurodegenerative diseases mostly by using GO and/ or pathway enrichment tests. The bioinformatic processing of omics data in these studies also leads to the identification of a number of nonmetabolic pathways, which will not be discussed here. More sophisticated computational analyses which take into account genome-scale network structure are discussed in the next section.

As reviewed below, pathways related to lipid metabolism, amino acid metabolism, glycolysis and oxidative phosphorylation are commonly perturbed in neurodegenerative diseases based on the GO and pathway enrichment results. Since even different studies for the same disease can identify different pathways, it is rather challenging to document disease-specific perturbed metabolic pathways based on these results.

7.2.1 Alzheimer's Disease (AD)

Alzheimer's Disease (AD) is characterized by the loss of neurons, leading to dementia and memory loss. It is a complex neurodegenerative disease with active gene deregulation mechanisms. Omics-based investigation of AD is common in recent years. For instance, Ray et al. used microarray analysis of samples from 20 AD patients and identified six co-expressed gene modules using 1663 differentially expressed genes (DEGs) in AD. They preferred to cluster DEGs before enrichment analyses. Six modules obtained by correlating the similarity in expression profiles of DEGs were then used for gene and pathway enrichment analysis (Ray et al. 2008). Phospholipid degradation and nucleotide metabolism pathways (obtained by KEGG pathway enrichment) and protein biosynthesis, macromolecule biosynthesis and membrane lipid metabolism processes (obtained by GO enrichment) were overrepresented as significantly perturbed metabolic pathways in AD.

Sequencing-based transcriptome studies, known as RNA-seq, have recently become more common compared to hybridization-based microarray studies. Issues such as cross-hybridization and biased coverage are reasons for higher preference of the sequencing-based approach (Bradford et al. 2010; Marioni et al. 2008). Satoh et al. investigated a transcriptome dataset of autopsied AD brains derived from two independent cohorts via RNA-seq technology (Satoh et al. 2014). They identified 522 commonly deregulated genes between two cohorts comparing AD and healthy brain using two-tailed Welch t-test. They verified the results by three literature-based AD brain microarray datasets that were derived from different ethnicities, brain regions, and microarray platforms. DEGs identified by the *t*-test were used for gene ontology enrichment by DAVID (Huang da et al. 2009b) and metabolic pathway enrichment by KEGG. The identified GO terms are synaptic transmission and transmission of nerve impulse, showing that production of neurotransmitters and related amino acid pathways are perturbed in AD patients.

Proteome analysis, on the other hand, offers a more functional snapshot of cell metabolism compared to transcriptome analysis since it does not neglect post-transcriptional modifications. Andreev et al. applied proteome analysis for cortical samples of ten AD brains and ten normally aged brains as a control group. They identified 197 significantly changed proteins via the Wilcoxon test, which is a non-parametric significance test (Andreev et al. 2012). The Wilcoxon test was chosen in order to avoid reliance on the assumption of normal distribution of human proteome data. Several standard enrichment analysis approaches were applied to the data to identify perturbed pathway maps and GO processes using MetaCore software (GeneGo, Inc). The metabolism-related changes were mostly related to lipid and amino acid pathways. Stimulation of arachidonic acid production, ubiquinone metabolism, branched chain family amino acid metabolic process, GTP metabolism and oxidative phosphorylation were among the pathways identified.

Begcevic et al. investigated AD via semiquantitative proteomic analysis aiming to identify novel biomarkers that may yield detection of the disease in early stages (Begcevic et al. 2013). Because insufficient number of replicates did not allow a statistical comparison, 204 exclusively detected proteins in AD and 600 exclusively detected proteins in control samples were compared based on fold change. Several gene and pathway enrichment analyses were applied for this study. According to GO enrichment analysis, metabolic network related terms, oxido-reductase activity, adenyl nucleotide binding, cellular ketone metabolic process, oxidative phosphorylation, and positive regulation of ubiquitin activity showed high statistical significance in AD. In addition to the GO enrichment analysis, pathway analysis was also applied by using commercial web-based ProteinCenter tool (Thermo Fisher Scientific, USA). The analysis compares identified proteins, AD proteins in this case, against human proteome database to calculate the enrichment. The predicted over-represented pathways were valine, leucine and isoleucine degradation, oxidative phosphorylation, TCA cycle, glyoxylate and dicarboxylate metabolism, fatty acid metabolism, pyruvate metabolism, beta-alanine metabolism, glycolysis/gluconeogenesis, arginine and proline metabolism, glycine, serine and threonine metabolism, alanine, aspartate and glutamate metabolism, butanoate metabolism, glutathione metabolism, cysteine and methionine metabolism, vitamin B6 metabolism, amino sugar and nucleotide sugar metabolism, histidine metabolism and phenylalanine metabolism.

Manavalan et al. applied significance analysis to proteome data obtained for different brain regions—hippocampus, parietal cortex and cerebellum—to investigate aging-related dementia (Manavalan et al. 2013). Thirty-one aging-related dementia proteins were found as significantly regulated proteins, which are involved in molecular transport, nervous system development, synaptic plasticity and apoptosis. Biocomputational network analysis of the AD brain proteins, performed by IPA, highlighted impairment in glucose utilization as the most important metabolic regulation, which is known as a hallmark of AD. The antioxidant defense enzyme superoxide dismutase was also found to be commonly regulated in the considered brain regions.

Studies mentioned above used control and disease cases to find significantly regulated genes or proteins. Hondius et al. (2016) used tissue samples obtained from different stages (Braak stages) of AD to find significantly regulated proteins. By doing so, the aim was to uncover a chronological pattern in the proteome of hippocampal regions in AD. Of 372 significantly changed proteins found by linear regression analysis, 166 proteins had increased levels and 206 proteins had decreased levels. The most significant overrepresented metabolic pathways determined by IPA with increasing *p*-values were oxidative phosphorylation, glycolysis, serine biosynthesis, glutathione redox reaction and tryptophan degradation.

Kaddurah-Daouk et al. investigated AD via metabolome analysis of cerebrospinal fluid samples of AD, mild cognitive impairment (MCI) and control groups using liquid chromatography electrochemical array (Kaddurah-Daouk et al. 2013). In order to define group differences, the nonparametric Kruskal-Wallis test was employed since quantile-quantile plots of metabolome data showed that most metabolites were not normally distributed. Significantly changed AD metabolites obtained were mapped to the metabolic network to reveal affected biochemical pathways. The perturbed metabolic pathways were methionine, tryptophan, tyrosine and purine pathways. These results provided novel insights for the neurotransmitter and purine alterations in AD which were also reported in the previous studies.

7.2.2 Parkinson's Disease (PD)

Parkinson's disease (PD) is a complex neurodegenerative disease with movement disorders. Dementia is also a common problem in the advanced stages. The motor symptoms of the disease are due to the death of cells in the substantia nigra region in the midbrain. The complexity requires a systems-based approach to characterize the disease at the molecular level.

The first transcriptome analysis of multiple brain regions in PD was carried out by Zhang et al. (2005). Three brain regions—substantia nigra, putamen, and prefrontal cortex—of PD patients and corresponding control groups were analyzed in microarray experiments. Three hundred and twenty nine significantly changed genes were obtained by a simple one-way ANOVA for the three brain areas. They also performed functional group analysis by using groups based on GO terms, KEGG pathways and EC numbers. The enrichment analysis pointed out multiple metabolic energy groups and secretory function groups. Significant terms associated with metabolism with increasing *p*-values were amino acid transporter, branched-chain amino acid aminotransferase, Complex I, NADH dehydrogenase, aerobic respiration, oxidative phosphorylation, glutamate decarboxylation, and citrate metabolism.

Another study (Elstner et al. 2011) is in accordance with Zang et al.'s study, which showed mitochondrial dysfunction in PD characterized by down-regulated genes related to mitochondrial and ubiquitin–proteasome system. Elstner et al. investigated PD using dopaminergic neurons from the substantia nigra region of the brain because dopaminergic neuron degeneration is increased in PD (Elstner et al. 2011). They compared dopaminergic neurons from PD samples with dopaminergic neurons from young and old control samples using microarray analysis. ANOVA was applied between three groups in order to obtain significantly changed genes. Of 1185 significantly changed genes specific for PD, 1045 were also reported before by Simunovic et al.'s study (Simunovic et al. 2009) applying similar design and statistical methods. Overrepresented metabolic pathways in PD obtained by pathway analysis using IPA with increasing *p*-values were oxidative phosphorylation, ubiquinone biosynthesis, citrate cycle, purine metabolism, glycolysis/gluconeogenesis, butanoate metabolism, pentose phosphate pathway and propanoate metabolism.

Glaab and Schneider investigated PD and corresponding control groups via meta-analysis of eight literature-based microarray data (Glaab and Schneider 2015). In order to understand the relation between aging and PD risk, significantly changed genes in PD obtained by meta-analysis were compared at pathway and network level with the significant genes associated with aging adult brain obtained from Human Brain Transcriptome (HBT) project (Kang et al. 2011). Common overrepresented metabolic pathways between PD and the aging adult brain were detected using a network-based pathway analysis approach. In addition to the known overrepresented pathways in PD, such as altered NADH dehydrogenase and oxidoreductase causing mitochondrial dysfunctions, synaptic vesicle endocytosis and phosphatidylinositol metabolism were found as the novel altered pathways in PD.

PD can also be investigated with in-vitro studies using cell lines. 1-methyl-4-phenylpyridinium (MPP^+) is a mitochondrial toxin and is used to trigger biochemical alterations associated with PD in-vitro. Monti et al. performed meta-analysis of literature based proteome data of neuronal alterations due to MPP^+ treatment in addition to the mitochondrial proteome data of SH-SY5Y cell lines treated with MPP^+, and applied several bioinformatic enrichment analyses for these proteome data to obtain overrepresented pathways and molecular functions altered in PD (Monti et al. 2015). Similar to the microarray studies mentioned above, this proteome study also points out mitochondrial and ubiquitin–proteasome system dysfunctionalities in PD. The significantly overrepresented GO terms associated with metabolic pathways were mitochondrial transport (mitochondrial alpha-ketoglutarate/malate transport, mitochondrial aspartate/glutamate transport, mitochondrial sodium/calcium ion exchange), neurotransmitter transport, phosphatase activity, glycolysis/gluconeogenesis and positive regulation of ATPase activity. The overrepresented terms associated with metabolic pathways obtained by several pathway enrichment analysis tools (KEGG, Reactome, Pathway Commons, WikiPathways) were oxidative phosphorylation, ATP production, neurotransmitters metabolism and release, butanoate metabolism, glycolysis/gluconeogenesis, glyoxylate and dicarboxylate metabolism, pyruvate metabolism, alanine, aspartate and glutamate metabolism, pentose phosphate pathway, arginine and proline metabolism, tyrosine metabolism, valine, leucine and isoleucine degradation, citrate cycle (TCA cycle) and respiratory electron transport, GABA synthesis, release, reuptake and degradation and glucose metabolism.

Metabolomics also provides critical insights on metabolic pathways of PD. A recent review (Lei and Powers 2013) states that the electron transport system, choline metabolism, the glutamate-glutamine cycle, energy metabolism and TCA cycle are overrepresented metabolic pathways in PD, based on several NMR metabolomics studies.

7.2.3 Other Disorders

Huntington's disease (HD) is another neurodegenerative disorder associated with mental decline and lack of coordination. A mutation in the Huntingtin protein leads to several alterations at the molecular level since the protein is known to interact with more than a hundred proteins (Goehler et al. 2004). Mastrokolias et al. (2015) applied RNA-seq gene expression analysis to peripheral blood, known as a useful source to identify biomarkers for Huntington's disease. One hundred and sixty seven significantly changed genes were found comparing peripheral blood samples of HD and control groups. Of 167 significantly changed genes, 40 were previously reported as significantly altered in HD. Several bioinformatic enrichment analyses were applied for significantly changed genes. In addition to some common and previously reported processes in HD such as mitochondria-associated metabolic dysfunction and increased glycolytic rate, relatively new overrepresented pathways such as pentose phosphate pathway were also obtained. Overrepresented metabolism-related terms obtained by GO analysis, KEGG pathway enrichment and IPA were NADP binding, positive regulation of interleukin 6, cellular carbohydrate biosynthetic process, glucose catabolic process, pentose phosphate pathway and carbohydrate metabolism. Among these terms, positive regulation of interleukin 6 is associated with cholesterol biosynthesis impairment in HD. Mitochondrial energy dysfunction in HD was also observed in a metabolomics study, which analyzed serum and cerebrospinal fluid (CSF) samples obtained from HD transgenic rats and control groups (Verwaest et al. 2011). In this study, increased levels of glutamine and succinic acid reflect a shutdown in the neuronal-glial glutamate-glutamine cycling and inhibition of the enzyme succinate dehydrogenase. The dual-function enzyme takes part in TCA cycle and electron transport chain. The identified decrease in the level of *N*-acetyl-aspartate reflects impairment of mitochondrial energy production. Furthermore, increased lactate level also indicates the deficiency of oxidative energy metabolism in HD.

The major pathophysiological characteristics of Multiple sclerosis (MS) are demyelination of neurons, inflammation and plaque formation in the central nervous system. These defects lead to symptoms such as muscle weakness and some disabilities since the communication within the brain and between the brain and body is damaged. Reinke et al. analyzed cerebrospinal fluid samples from 15 MS patients and 17 non-MS specimens in terms of metabolome profiles (Reinke et al. 2014). The 1H-NMR spectroscopy produced 15 reproducibly detectable metabolites, providing a more robust profiling approach. Use of hierarchical clustering pointed to increased importance of glycolysis and decreased mitochondrial energy metabolism in MS patients. The analysis showed no difference in amino acids metabolism, while pointing to alterations in biogenic amine and phospholipid metabolisms.

Amyotrophic lateral sclerosis (ALS) causes muscle weakness due to degeneration in motor neurons. This leads to gradual loss of voluntary movements. Aiming to investigate the role of candidate genes and altered pathways in ALS pathology, Lederer et al. performed a comprehensive whole genome microarray analysis of human motor cortex of 11 sporadic ALS and nine control subjects (Lederer et al. 2007). They performed pathway-based gene expression analysis by GenMAPP software package (Doniger et al. 2003), which uses fold changes and p-values of the significance test as inputs and dynamically links GO terms with gene-expression data. Overrepresented metabolic pathways obtained by GenMAPP can be summarized as follows: down regulation in glycolysis, down regulation in mitochondrial energy metabolism (this alteration was represented by both GO terms obtained by GenMAPP and down regulated genes associated with TCA cycle and oxidative phosphorylation), ion hemostasis and solute transport (example GO terms: monovalent inorganic cation transporter activity and hydrogen ion transporter activity), and nucleotide metabolism. Additionally, comparison of the results with previously reported data from spinal cord of sporadic ALS patients showed that the associated changes are highly correlated.

Not only neurodegenerative diseases, but also mental disorders like schizophrenia can be analyzed by gene expression studies to understand transcriptional abnormalities of the disease. Middleton et al. applied gene expression analysis by microarray to postmortem samples of schizophrenia and control group (Middleton et al. 2002). They showed significant decrease in the expression of the gene groups regulating the following metabolic pathways: the ornithine and polyamine metabolism, the mitochondrial malate shuttle, the TCA cycle, aspartate and alanine metabolism and ubiquitin metabolism. In addition to neurodegenerative diseases and mental disorders, metabolic alterations of the traumatic brain injury (Yu et al. 2015) and trauma-related mental disorders (Zhang et al. 2015) can also be investigated by systems biology tools mentioned above.

7.3 Genome-Scale Metabolic Network Reconstructions for Brain

A genome-scale metabolic reconstruction is the list of curated organism-specific metabolic reactions with associated genes, such that all enzyme-coding genes in the genome are covered. Such curated genome-scale metabolic networks can be

Table 7.2 Basic properties of generic and brain-specific genome-scale human metabolic network reconstructions

	Genes	Reactions	Unique metabolites	Pathways	Specificity
Recon1 (Duarte et al. 2007)	1496	3744	1509	70	Generic
EHMN (Ma et al. 2007)	2322	2823	2671	88	Generic
Recon2 (Thiele et al. 2013)	1789	7440	2620	99	Generic
HMR2.0 (Pornputtapong et al. 2015)	3765	8181	2895	130	Generic
iNL403 (Lewis et al. 2010)	403	1070	331	Not specified	Brain-specific
iMS570 (Sertbaş et al. 2014)	570	630	308	45	Brain-specific
Astrocyte network (Martin-Jimenez et al. 2017)	3765	5659	3061	123	Astrocyte-specific

simulated for the quantification of metabolic fluxes to identify impairments in metabolic functions, which is a clear superiority over functional group analyses employed by enrichment methods discussed mainly in the previous sections.

The first reconstructions of a genome-scale human metabolic network appeared in 2007 (Duarte et al. 2007; Ma et al. 2007) (Table 7.2). Duarte et al. demonstrated the use of their reconstructed metabolic network for the assessment of functional metabolic states and identification of alternative drug targets through mathematical analysis of the network structure. They named their reconstruction as human Recon1. Ma et al. focused on the functional connectivity analysis of their reconstruction as well as the distribution of disease related genes in the network. They called their reconstruction EHMN, the Edinburgh human metabolic reconstruction. Several others used these reconstructions later to derive specific models for tissues and cell types (Mardinoglu and Nielsen 2015; Ryu et al. 2015). These two major human reconstructions were later merged and extended, leading to human Recon2 (Thiele et al. 2013). An alternative recent reconstruction with a large genome coverage is termed HMR 2.0 (Human Metabolic Reconstruction) (Pornputtapong et al. 2015). These generic human reconstructions are not tissue specific. However, tissue-specific models are required for the investigation of disease states associated with specific tissues. Currently, iNL403 and iMS570 are the two simulatable genome-scale brain specific reconstructions that are available and both take into account two major brain cell types—neurons and astrocytes (Table 7.2). Recently, an astrocyte-specific comprehensive metabolic reconstruction has been released (Martin-Jimenez et al. 2017). However, it has not yet been used for simulating neurodegeneration related diseases.

Although brain-specific metabolic network models that account for neuron-astrocyte interactions were reconstructed before (Çakır et al. 2007; Occhipinti et al. 2007), the first genome-scale brain-specific metabolic model was reported by Lewis et al. (2010). They started from human Recon1, and, by the use of brain transcriptome data and brain-specific localization information of proteins, they eliminated inactive reactions from the model to reconstruct coupled brain-specific models for glutamatergic neuron-astrocyte, GABAergic neuron-astrocyte and cholinergic neuron-astrocyte cell pairs separately. The final metabolic models, iNL403, included about 1000 reactions controlled by 403 genes. The authors then integrated these brain-specific metabolic networks with AD transcriptome data by using constraint-based modeling framework (see next section).

Later, Sertbaş et al. reconstructed a more comprehensive brain specific metabolic network, which can correctly predict resting state metabolic fluxes based on constraint based modeling (Sertbaş et al. 2014). The reconstruction was based on a previous two-cell 217-reaction model (Çakır et al. 2007), which was used to predict resting state and hypoxic state metabolic fluxes in brain. It included 630 reactions controlled by 570 genes, thus termed iMS570. Although neuron and astrocyte metabolism share many reactions, some reactions are cell-specific in the brain. For example, pyruvate carboxylation is only active in astrocytes, malic enzyme is cytosolic in astrocytes and mitochondrial in neurons, and glutaminase and glutamate decarboxylase are only active in neurons (Çakır 2018). Such cell specific behaviors are covered in iMS570. In addition, the specificity is also reflected in the uptake of metabolites by the two cell types. The authors used flux balance analysis (Lakshmanan et al. 2014; Orth et al. 2010) to apply optimization for the prediction of metabolic fluxes. The optimization first maximizes glutamate-glutamine-GABA exchange fluxes between astrocytes and neurons and subsequently minimizes the Euclidean norm of the fluxes. The first objective ensured a tight coupling between the two cell types in accordance with literature (Gruetter 2002; Shen et al. 1999), and the second objective guaranteed the first objective with minimal investment on enzyme levels (Çakır et al. 2007; Holzhutter 2004; Tarlak et al. 2014). After the reconstruction, they integrated the brain specific metabolic network with transcriptome data of six neurodegenerative diseases (see next section).

With the reconstruction of generic human metabolic networks in 2007, several algorithms appeared to automatize tissue specific network reconstruction from generic networks (Pacheco et al. 2015; Ryu et al. 2015). In one such effort, the authors developed a tool called mCADRE (metabolic Context-specificity Assessed by Deterministic Reaction Evaluation) and used the tool to derive draft genome-scale metabolic models for 126 human tissues and cell types based on Recon1 (Wang et al. 2012). Of those, 30 were for brain tissues. Albeit valuable for being specific for 30 different brain regions from temporal lobe to pituitary gland, the models do not include neuron-astrocyte specificity. In another study based on Recon2, 65 draft cell-specific genome-scale metabolic models were derived based on protein expression data (Thiele et al. 2013). Five of these models were for the following brain tissues: cerebral cortex (separate models for glial and neuronal cells), hippocampus (separate models for glial and neuronal cells), and cerebellum. The brain is represented as a single cell type in all these models (Thiele et al. 2013; Wang et al. 2012). To our knowledge, there hasn't been any use of these draft brain models for the analysis of neurodegenerative diseases in the literature.

7.3.1 *Analysis of Neurodegenerative Diseases via Constraint-Based Modelling of Genome-Scale Reconstructions*

Constraint-based modelling is the most common modeling framework applied to genome-scale metabolic networks to predict cellular fluxes. The approach is crucial to predict relative activity of metabolic pathways, to quantify secretion rates of specific metabolites or to quantify metabolic dysfunctions in a given state. The mathematical formulation and details of the approach are discussed elsewhere (Bordbar et al. 2014; Çakır 2018). Briefly, the approach uses the stoichiometries of the covered reactions as well as their reversibility information as constraints. The stoichiometries are used to represent mass balances around metabolites, assuming that concentration of intracellular metabolites do not change at steady state. The fluxes can be predicted by an optimization framework, which needs the representation of cellular objective as a formula in terms of a subset of covered reactions, or sampling-based approaches can be implemented to sample flux solution space for a high number of flux distributions to reveal common patterns. Such constraint-based modeling approaches gave promising results for a range of human diseases, including neurodegenerative disorders (Sangar et al. 2012).

Reconstruction of genome-scale metabolic networks specific for neurodegenerative diseases is another challenge, which is discussed in detail for Parkinson's disease in a recent review (Mao et al. 2015). The authors provide a framework based on constraint-based modeling of dopaminergic neuronal metabolism to elucidate molecular mechanisms of neuronal degeneration and pathology of the disease. Their proposal includes the processing of generic model, human Recon2, to reconstruct a network specific to dopaminergic neurons and further analysis of the model by constraint-based modelling tools such as COBRA (Schellenberger et al. 2011) and ORCA (Mao and Verwoerd 2014). The incorporation of fluxome and exometabolome data for the refinement of the model is also included in the framework. A previous constraint-based modelling of Parkinson's disease (Buchel et al. 2013) constructs a dopaminergic nerve-cell model of about 100 mostly-abstract reactions and applies FBA to model α-synuclein accumulation. The reactions do not directly represent enzymatic reactions of a metabolic network, but rather lumped reactions representing specific processes in the cell.

In another approach, the author examines the influence of somatic transposition in brain metabolism by using constraint-based modelling (Abrusan 2012). Transposable elements are known to have a role in the central nervous system (Baillie et al. 2011; Reilly et al. 2013). Recon1 was used together with FBA and FVA. The insertion of a transposable element into an exon would lead to a knockout effect for the corresponding gene. The author used the information provided in a high-throughput study on the identification of somatic insertions. That study identified more than 24,000 novel insertions mostly belonging to L1, Alu and SVA families known to be associated with insertional mutagenesis and disease. The author followed an interesting approach and found the intersecting set of genes hit by the insertions reported by (Baillie et al. 2011) and also reported in human Recon1.

Then, FBA was used to test the effect of the lower activity of such metabolic genes on the biosynthesis of metabolites in the network. That is, the reactions which are coded by the genes which contain somatic transposable element insertion were constrained to have lower flux, and the change in the production rate of metabolites in the network was calculated. Their simulation identified 93 genes affecting metabolite production. In total, 256 metabolites were identified to be affected. Among the affected metabolites were key neurotransmitters such as dopamine and glutamate. The author further scanned the Human Metabolome Database for statistically significant associations between the identified metabolites and diseases. Interestingly, Parkinson's Disease was identified to be the most affected disease. Schizophrenia, another disease with neurodegeneration, was also identified. The results led to the development of new hypotheses about the mechanism of neurological diseases (Abrusan 2012).

Several neurodegenerative diseases are associated with the formation and accumulation of intracellular aggregates. These aggregates have a role in the activation of cell death mechanisms. The author used Recon1, and applied an approach known as molecular-crowding FBA to simulate the effect of protein aggregation on neuron metabolism (Vazquez 2013). Macromolecules constitute about 40% of the cell volume (Zimmerman and Trach 1991), and an increase in this ratio causes limitations in the diffusion of metabolites. Including this constraint in the FBA formulation, the author calculates the change in the key fluxes in response to increase in protein aggregates concentration. He identifies three distinct phases during the protein accumulation. The first phase represents normal neuronal behavior where lactate is the carbon source, and the second phase represents mixed oxidative phosphorylation of lactate and glucose. In the last phase, glucose is the source of energy support with lactate secretion. The model also predicts a decrease in the maximal energy production capacity of neurons with the increasing protein aggregate concentration, ultimately leading to the inhibition of neuronal activity and cell death (Vazquez 2013). Another metabolic change observed in the modeling results is the shift in the exchange of ammonia. Initially there is an ammonia uptake whereas it switches to secretion at phase two. Such metabolic changes are reported for Alzheimer's and Huntington's diseases, providing a partial validation of the modelling approach.

A rare childhood neurodegenerative disorder, known as Leigh's syndrome (LS) was also analyzed by constraint-based modelling (Vo et al. 2007). The patients are known to have mutations in pyruvate dehydrogenase complex and in the respiratory chain. The authors combined Recon1 with the syndrome-associated transcriptome data to derive a metabolic network of fibroblast metabolism consisting of 508 reactions between 430 metabolites. They also performed metabolome experiments to identify isotopomer data and the uptake and secretion rates of amino acids. The results were used to constrain the model revealing a slower metabolism for the syndrome and a more restricted flux range for ATP producing reactions. Moreover, succinate cytochrome c reductase was identified as a probable candidate for the mutation causing LS.

AD was analyzed in a study by applying constraint-based modelling to the first brain-specific genome-scale metabolic network, iNL403 (Lewis et al. 2010). By sampling the solution space for metabolic flux distributions, they simulated the deficiencies of three enzymes known to be affected in Alzheimer's disease—pyruvate dehydrogenase, alpha-ketoglutarate dehydrogenase (AKGD), and cytochrome c oxidase. The simulations were repeated separately by using GABAergic, glutamatergic and cholinergic metabolic networks reconstructed. In agreement with the literature, which states that GABAergic neurons are relatively unaffected in AD, sampling-based enzyme deficiency simulations predicted impairments in cholinergic and/or glutamatergic neurons. In AKGD simulations, for example, the deficiency is due to limited oxidative phosphorylation capacity in neurons in the two cell types. However, the capacity is not impaired in GABAergic neurons based on the simulations. Additional simulations identified GAD (glutamate decarboxylase) as the neuroprotective enzyme in GABAergic neurons via contributing to a higher flux through GABA shunt, leading to a bypass for the AKGD deficiency in this cell type. As an additional support for this simulation-based hypothesis, the authors analyzed a compendium of published AD and control microarray datasets from six brain regions. The analysis showed that the brain regions known to be affected severely from AD (hippocampus and entorhinal cortex) had lower GAD expression in control brains whereas the regions associated with less neuronal loss in AD (superior frontal gyrus and visual cortex) showed much higher expression levels of GAD in controls. The results support the neuroprotective role of GAD as hypothesized in the study based on constraint based modeling of genome-scale brain-specific metabolic network.

In a recent study, human Recon1 was integrated with AD transcriptome data to predict biomarkers and drug targets by using constraint-based modelling (Stempler et al. 2014). They obtained the transcriptome data from Gene Expression Omnibus, a public repository for transcriptome data. The data included post-mortem cortical samples from 187 controls and 176 AD patients (Webster et al. 2009). The expression data were discretized based on highly and lowly expressed genes and used as soft-constraints for the constraint-based flux calculation using the genome-scale network. A maximum consistency between the experimental low/high activity of genes and corresponding model-predicted activity of reactions was sought in an optimization framework. Repeating this for both control and AD data separately, two metabolic flux distributions were calculated. Then, the pathways with altered activity were identified by comparing the two flux distributions. Carnitine shuttle was found to have most significantly decreased activity. Folate pathway and neurotransmitter-related pathways were also among pathways with low activity in AD. Later, as an attempt to identify potential biomarkers in the cerebrospinal fluid, the authors compared the changes in the fluxes of extracellular transport of metabolites. Succinate and prostaglandin D2 were metabolites with significantly decreased secretions, as the literature supports. As a final step, the authors used Metabolic Transformation Algorithm (MTA) (Yizhak et al. 2013) to predict candidate drug targets for AD. By performing in silico gene deletions systematically for each gene in the genome-scale network they predicted a single-gene knockout set, which would transform AD metabolic state closer to the healthy state. The algorithm revealed Vitamin D, nucleotides and steroid metabolism as the pathways enriched with the identified gene knockouts.

A group of researchers recently introduced an extension of FBA, called Lsei-FBA (least-squares with equalities and inequalities) and applied it to AD (Gavai et al. 2015). They used R package BiGGR (van Beek et al. 2011) for the simulations. In their analysis, they first obtained an AD transcriptome data from laser captured micro dissected neurons from healthy and AD subjects (Liang et al. 2008). The dataset is freely available in GEO (GSE5281). The data were from six different brain regions, however they focused on data from the hippocampal region since it is more affected during the early stages of AD. They rather used a small-scale metabolic model with 71 reactions between 89 metabolites, derived from Recon1. The model included central carbon metabolism and glutamate and GABA shunts. In their FBA framework, maximal ATP production was used as an objective of the optimization problem to calculate fluxes. They later repeated the analysis by using a flux sampling approach. The related algorithm is available in BiGGR. The use of Lsei-FBA gave metabolic flux patterns in agreement with the measurements reported in literature for the cerebral metabolism of AD patients.

7.3.2 Analysis of Neurodegenerative Diseases via Graph-Based Integration of Genome-Scale Reconstructions and Omics Data

Graph-based analysis of omics data is another branch of bioinformatics approach as opposed to constraint-based modeling discussed above. Here, the cellular network is represented as a graph and used as a scaffold to process omics data (Aittokallio and Schwikowski 2006; Vidal et al. 2011). As an advantage over enrichment methods discussed in Sect. 7.2, metabolic network connectivity is taken into account to identify perturbed cellular pathways.

A genome-scale metabolic network can be represented as a graph where metabolites are represented as nodes and the reactions as edges. Such a graph representation can easily be integrated with omics data to extract information on the perturbed mechanisms for the condition change in question. An approach, called Reporter Metabolite Analysis (Patil and Nielsen 2005), uses graph-represented metabolic network as a scaffold for the *p*-values of genes from transcriptome data and identifies metabolites around which coordinated alterations are observed at transcriptional level (Fig. 7.3). Reporter metabolites are the metabolites which respond to a perturbation by changing the expression levels of the surrounding genes. They can act as a regulatory spot and aim to keep their levels constant by imposing such a change, or the change in the surrounding can bring the level of these metabolites to a new level. The approach was applied to microorganisms first (Patil and Nielsen 2005) and application to diabetes was reported later (Zelezniak et al. 2010). In a systematic study comprehending six neurodegenerative diseases, transcriptome

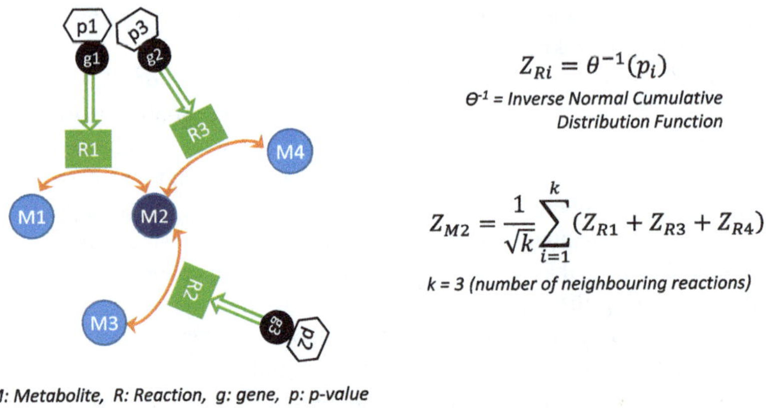

Fig. 7.3 Reporter metabolite analysis. *P*-values of the genes from transcriptome data are assigned to corresponding reactions, and an averaging of the values is performed for the neighbor edges of a metabolite for scoring. Highest scoring metabolites are called reporters

data belonging to the diseases were integrated with a newly reconstructed genome-scale brain-specific metabolic network (iMS570) to identify reporter metabolites (Sertbaş et al. 2014). The concept of reporter metabolite analysis was used as a basis to identify pathways affected from a perturbation, termed reporter pathways (Çakır 2015; Sertbaş et al. 2014).

In the study (Sertbaş et al. 2014), six neurodegenerative diseases were investigated by reporter metabolite analysis and reported pathway analysis approaches: AD, PD, ALS, MS, HD and schizophrenia. The transcriptome data (Durrenberger et al. 2015) included 113 samples from post-mortem brain tissues, with about ten control and ten patient samples for each disease. The authors used the data to identify the effect of these diseases on metabolism. Of the 570 genes covered by their metabolic network, 496 (87%) matched with the transcriptome data and were used in the analysis. They identified reporter metabolites based on a cutoff of *p*-value <0.0.5. This led to the identification of disease-specific reporter metabolites and those shared by disease pairs. Disease-specific reporter metabolites are potential biomarker candidates in the diagnosis of corresponding diseases. Identified metabolites spanned energy, amino-acid and lipid metabolisms, and many of them were supported by literature in terms of their role in the corresponding disease. For example, succinate was identified as an AD-specific reporter metabolite, and there are studies reporting succinate as a potential biomarker. Cholesterol and its precursors lanosterol, desmosterol and lathosterol are identified as reporters only for MS. This is supported by previous experimental studies reporting decrease in the level of these metabolites in MS and suggests their use as potential biomarkers. Identification of pyruvate as a reporter for HD is in agreement with the reported change in the lactate-to-pyruvate ratio in this disease (Sertbaş et al. 2014).

In order to provide a more systematic interpretation of reporter metabolite results, the transcriptome-based *p*-value scores calculated for metabolites in the network were used as input to reporter pathway calculation. The approach, termed metabolite-centric reporter pathway analysis (RPA^m), was first reported by Sertbaş et al. (2014) and applied to the analyzed neurodegenerative diseases. There are pathways which are known to be linked to the diseases but none or few of their associated metabolites were identified as reporter metabolites. Such pathways were able to be captured by RPA^m. This means that metabolites which do not have a significant reporter score can show a significant aggregate score when associated with a pathway. In other words, small nonsignificant individual changes around metabolites can imply a significant overall perturbation of their pathway. For example, the authors identified valine metabolism as reporter pathway for MS, alanine, valine, isoleucine and leucine metabolisms as reporter pathways for PD, and leucine and glutamate metabolisms as reporters for schizophrenia, all of which are supported by experimental studies.

Çakır (2015) took a different approach for the validation of RPA^m as a more promising approach to capture disease-related pathways from transcriptome data (Çakır 2015). They also calculated pathway scores by directly using the averaging of the *p*-values of associated reactions, where a reaction *p*-value is the *p*-value of the controlling gene. The reaction-centric approach, termed RPA^r, does not take into account all reactions which consume or produce the metabolites associated with the pathway of interest. Some reactions are traditionally listed under other pathways although they directly contribute to the level of metabolites in a pathway of interest. Metabolite-centric RPA^m, considers all these reactions associated with a metabolite in the pathway and thus reflects cross-talks between pathways. The author comparatively tested RPA^r and RPA^m on three different datasets, and demonstrated the superiority of the RPA^m approach. One of the datasets analyzed belonged to AD. Although the author used the same AD dataset (Durrenberger et al. 2015), a more comprehensive metabolic network obtained from HumanCyc (Caspi et al. 2014) was used as a scaffold rather than iMS570. HumanCyc includes a higher number of reactions in a brain-nonspecific manner and have much higher pathway specifications compared to iMS570. This human metabolic network from HumanCyc included 2521 genes controlling 2036 reactions associated with 133 pathways. The results were very promising in terms of the power of the metabolite-centric approach over the reaction-centric approach and the identification of AD-perturbed pathways. Among the predicted metabolic pathways were 3-phosphoinositide, myo-inositol, TCA cycle, and melatonin degradation pathways, all with literature support. Among the other predicted pathways, mevalonate pathway is the precursor pathway of AD-affected cholesterol biosynthesis, and retinol biosynthesis pathway is linked to retinoids known to have a role in late-onset AD. The synthesis pathway of *N*-acetylneuraminate, the most common sialic acid in mammalian cells, is also predicted, and it is a structural component of gangliosides known to be linked to AD pathology. These pathways were mostly predicted by the metabolite-centric approach, RPA^m, and could not be captured by the reaction-centric approach (Çakır 2015).

Similar to the reporter pathway analysis, an approach termed PathWave is also based on the integrative processing of graphs and omics data to identify perturbed pathways (Schramm et al. 2010). The approach defines reactions as nodes and metabolites as edges between reactions for each defined pathway separately. Then, two dimensional representation of the pathways is achieved through regular square lattice grid representation. After the optimal arrangement of each pathway grid via an optimization problem, gene expression data were mapped onto the pathways and statistically analyzed. Lewis et al. used Pathwave to analyze AD gene expression data belonging to six different brain regions to further support their constraint-based analysis (Lewis et al. 2010) (see previous section). Pathwave identified a significant suppression of glycolysis, TCA cycle, malate-aspartate shuttle and oxidative phosphorylation in brain regions known to be affected by AD. Additionally, region specific suppression of heme biosynthesis, ethanol metabolism and several amino acid pathways was revealed.

7.4 Final Remarks

Metabolism is one of the key processes in a cell, maintaining several functions via life-sustaining chemical transformations of glucose or other substrates into precursor metabolites, and then building blocks such as amino acids, fatty acids, nucleic acids and carbohydrates. Investigation of the effect of neurodegenerative diseases from a systems-medicine viewpoint by mapping alterations at pathway level is therefore crucial (Ostaszewski et al. 2016). In this chapter, we reviewed a set of studies which process transcriptomic, proteomic and/or metabolomic data to identify perturbed pathways in major neurodegenerative diseases. The first part of the chapter focused on the studies that use enrichment methods such as GO and pathway enrichment. These methods first identify significantly changed sets of genes, proteins or metabolites, and predict the common cellular processes or pathways significantly shared by these biomolecules, without considering the network connectivity information among them. The advantage of the enrichment approaches is their easy and quick implementation to get an initial understanding of the omics data. To get a deeper and more sound understanding, however, consideration of the network structure is important. Therefore, the concept of genome-scale metabolic networks was discussed in the second part of the chapter with a special focus on the application of neurodegenerative diseases. The use of genome-scale networks for the analysis of neurodegenerative diseases is possible by two major approaches: constraint-based modelling and graph-based integration with omics data. Both were reviewed in detail in the related subsections to present a thorough overview of the concept. There are efforts to apply another very useful approach, the kinetic modelling approach, for genome-scale metabolic models (Stanford et al. 2013). Currently, there are only small-scale kinetics-based metabolic models for neurodegenerative diseases (Cloutier and Wellstead 2012; Lloret-Villas et al. 2017; Tiveci et al. 2005). This decade will witness continuously increasing use of genome-scale brain metabolic networks as scaffolds for the systems biology of neurodegenerative diseases to map metabolic alterations via graph-based, constraint-based and kinetics-based approaches with an overall goal of identifying biomarker, drug target and/or drug candidates for better diagnosis and treatment.

Acknowledgements This work was financially supported by TUBITAK, The Scientific and Technological Research Council of Turkey (Project Code: 315S302), and by The Turkish Academy of Sciences–Outstanding Young Scientists Award Program (TUBA-GEBIP).

Conflict of Interest The authors declare no conflict of interest.

References

Abrusan G. Somatic transposition in the brain has the potential to influence the biosynthesis of metabolites involved in Parkinson's disease and schizophrenia. Biol Direct. 2012;7:41.

Aittokallio T, Schwikowski B. Graph-based methods for analysing networks in cell biology. Brief Bioinform. 2006;7:243–55.

Andreev VP, Petyuk VA, Brewer HM, Karpievitch YV, Xie F, Clarke J, Camp D, Smith RD, Lieberman AP, Albin RL, Nawaz Z, El Hokayem J, Myers AJ. Label-free quantitative LC-MS proteomics of Alzheimer's disease and normally aged human brains. J Proteome Res. 2012;11:3053–67.

Baillie JK, Barnett MW, Upton KR, Gerhardt DJ, Richmond TA, De Sapio F, Brennan PM, Rizzu P, Smith S, Fell M, Talbot RT, Gustincich S, Freeman TC, Mattick JS, Hume DA, Heutink P, Carninci P, Jeddeloh JA, Faulkner GJ. Somatic retrotransposition alters the genetic landscape of the human brain. Nature. 2011;479:534–7.

Barrett T, Troup DB, Wilhite SE, Ledoux P, Evangelista C, Kim IF, Tomashevsky M, Marshall KA, Phillippy KH, Sherman PM, Muertter RN, Holko M, Ayanbule O, Yefanov A, Soboleva A. NCBI GEO: archive for functional genomics data sets--10 years on. Nucleic Acids Res. 2011;39:D1005–10.

Begcevic I, Kosanam H, Martinez-Morillo E, Dimitromanolakis A, Diamandis P, Kuzmanov U, Hazrati LN, Diamandis EP. Semiquantitative proteomic analysis of human hippocampal tissues from Alzheimer's disease and age-matched control brains. Clin Proteomics. 2013;10:5.

Bordbar A, Monk JM, King ZA, Palsson BO. Constraint-based models predict metabolic and associated cellular functions. Nat Rev Genet. 2014;15:107–20.

Bradford JR, Hey Y, Yates T, Li Y, Pepper SD, Miller CJ. A comparison of massively parallel nucleotide sequencing with oligonucleotide microarrays for global transcription profiling. BMC Genomics. 2010;11:282.

Buchel F, Saliger S, Drager A, Hoffmann S, Wrzodek C, Zell A, Kahle PJ. Parkinson's disease: dopaminergic nerve cell model is consistent with experimental finding of increased extracellular transport of alpha-synuclein. BMC Neurosci. 2013;14:136.

Çakır T. Reporter pathway analysis from transcriptome data: metabolite-centric versus reaction-centric approach. Sci Rep. 2015;5:14563.

Çakır T. Constraint-based Modeling of metabolic interactions in and between astrocytes and neurons. In: De Pitta M, Berry H, editors. Computational Glioscience. Berlin: Springer; 2018. [accepted].

Çakır T, Alsan S, Saybasili H, Akin A, Ulgen KO. Reconstruction and flux analysis of coupling between metabolic pathways of astrocytes and neurons: application to cerebral hypoxia. Theor Biol Med Model. 2007;4:48.

Carbon S, Ireland A, Mungall CJ, Shu S, Marshall B, Lewis S, Ami GOH, Web Presence Working G. AmiGO: online access to ontology and annotation data. Bioinformatics. 2009;25:288–9.

Caspi R, Altman T, Billington R, Dreher K, Foerster H, Fulcher CA, Holland TA, Keseler IM, Kothari A, Kubo A, Krummenacker M, Latendresse M, Mueller LA, Ong Q, Paley S,

Subhraveti P, Weaver DS, Weerasinghe D, Zhang P, Karp PD. The MetaCyc database of metabolic pathways and enzymes and the BioCyc collection of pathway/genome databases. Nucleic Acids Res. 2014;42:D459–71.

Cerami EG, Gross BE, Demir E, Rodchenkov I, Babur O, Anwar N, Schultz N, Bader GD, Sander C. Pathway commons, a web resource for biological pathway data. Nucleic Acids Res. 2011;39:D685–90.

Cloutier M, Wellstead P. Dynamic modelling of protein and oxidative metabolisms simulates the pathogenesis of Parkinson's disease. IET Syst Biol. 2012;6:65–72.

Doniger SW, Salomonis N, Dahlquist KD, Vranizan K, Lawlor SC, Conklin BR. MAPPFinder: using Gene Ontology and GenMAPP to create a global gene-expression profile from microarray data. Genome Biol. 2003;4:R7.

Duarte NC, Becker SA, Jamshidi N, Thiele I, Mo ML, Vo TD, Srivas R, Palsson BO. Global reconstruction of the human metabolic network based on genomic and bibliomic data. Proc Natl Acad Sci U S A. 2007;104:1777–82.

Durrenberger PF, Fernando FS, Kashefi SN, Bonnert TP, Seilhean D, Nait-Oumesmar B, Schmitt A, Gebicke-Haerter PJ, Falkai P, Grunblatt E, Palkovits M, Arzberger T, Kretzschmar H, Dexter DT, Reynolds R. Common mechanisms in neurodegeneration and neuroinflammation: a BrainNet Europe gene expression microarray study. J Neural Transm. 2015;122:1055–68.

Eden E, Navon R, Steinfeld I, Lipson D, Yakhini Z. GOrilla: a tool for discovery and visualization of enriched GO terms in ranked gene lists. BMC Bioinformatics. 2009;10:48.

Elstner M, Morris CM, Heim K, Bender A, Mehta D, Jaros E, Klopstock T, Meitinger T, Turnbull DM, Prokisch H. Expression analysis of dopaminergic neurons in Parkinson's disease and aging links transcriptional dysregulation of energy metabolism to cell death. Acta Neuropathol. 2011;122:75–86.

Gavai AK, Supandi F, Hettling H, Murrell P, Leunissen JA, van Beek JH. Using bioconductor package BiGGR for metabolic flux estimation based on gene expression changes in brain. PLoS One. 2015;10:e0119016.

Glaab E, Schneider R. Comparative pathway and network analysis of brain transcriptome changes during adult aging and in Parkinson's disease. Neurobiol Dis. 2015;74:1–13.

Goehler H, Lalowski M, Stelzl U, Waelter S, Stroedicke M, Worm U, Droege A, Lindenberg KS, Knoblich M, Haenig C, Herbst M, Suopanki J, Scherzinger E, Abraham C, Bauer B, Hasenbank R, Fritzsche A, Ludewig AH, Bussow K, Coleman SH, Gutekunst CA, Landwehrmeyer BG, Lehrach H, Wanker EE. A protein interaction network links GIT1, an enhancer of huntingtin aggregation, to Huntington's disease. Mol Cell. 2004;15:853–65.

Gruetter R. In vivo 13C NMR studies of compartmentalized cerebral carbohydrate metabolism. Neurochem Int. 2002;41:143–54.

Holzhutter HG. The principle of flux minimization and its application to estimate stationary fluxes in metabolic networks. Eur J Biochem. 2004;271:2905–22.

Hondius DC, van Nierop P, Li KW, Hoozemans JJ, van der Schors RC, van Haastert ES, van der Vies SM, Rozemuller AJ, Smit AB. Profiling the human hippocampal proteome at all pathologic stages of Alzheimer's disease. Alzheimers Dement. 2016;12(6):654–68.

Huang da W, Sherman BT, Lempicki RA. Bioinformatics enrichment tools: paths toward the comprehensive functional analysis of large gene lists. Nucleic Acids Res. 2009a;37:1–13.

Huang da W, Sherman BT, Lempicki RA. Systematic and integrative analysis of large gene lists using DAVID bioinformatics resources. Nat Protoc. 2009b;4:44–57.

Kaddurah-Daouk R, Zhu H, Sharma S, Bogdanov M, Rozen SG, Matson W, Oki NO, Motsinger-Reif AA, Churchill E, Lei Z, Appleby D, Kling MA, Trojanowski JQ, Doraiswamy PM, Arnold SE, Pharmacometabolomics Research N. Alterations in metabolic pathways and networks in Alzheimer's disease. Transl Psychiatry. 2013;3:e244.

Kang HJ, Kawasawa YI, Cheng F, Zhu Y, Xu X, Li M, Sousa AM, Pletikos M, Meyer KA, Sedmak G, Guennel T, Shin Y, Johnson MB, Krsnik Z, Mayer S, Fertuzinhos S, Umlauf S, Lisgo SN, Vortmeyer A, Weinberger DR, Mane S, Hyde TM, Huttner A, Reimers M, Kleinman JE, Sestan N. Spatio-temporal transcriptome of the human brain. Nature. 2011;478:483–9.

Kelder T, van Iersel MP, Hanspers K, Kutmon M, Conklin BR, Evelo CT, Pico AR. WikiPathways: building research communities on biological pathways. Nucleic Acids Res. 2012;40:D1301–7.

Lakshmanan M, Koh G, Chung BK, Lee DY. Software applications for flux balance analysis. Brief Bioinform. 2014;15:108–22.

Lederer CW, Torrisi A, Pantelidou M, Santama N, Cavallaro S. Pathways and genes differentially expressed in the motor cortex of patients with sporadic amyotrophic lateral sclerosis. BMC Genomics. 2007;8:26.

Lei S, Powers R. NMR metabolomics analysis of Parkinson's disease. Curr Metabolomics. 2013;1:191–209.

Lewis NE, Schramm G, Bordbar A, Schellenberger J, Andersen MP, Cheng JK, Patel N, Yee A, Lewis RA, Eils R, Konig R, Palsson BO. Large-scale in silico modeling of metabolic interactions between cell types in the human brain. Nat Biotechnol. 2010;28:1279–85.

Liang WS, Reiman EM, Valla J, Dunckley T, Beach TG, Grover A, Niedzielko TL, Schneider LE, Mastroeni D, Caselli R, Kukull W, Morris JC, Hulette CM, Schmechel D, Rogers J, Stephan DA. Alzheimer's disease is associated with reduced expression of energy metabolism genes in posterior cingulate neurons. Proc Natl Acad Sci U S A. 2008;105:4441–6.

Lloret-Villas A, Varusai TM, Juty N, Laibe C, Le NovEre N, Hermjakob H, Chelliah V. The impact of mathematical modeling in understanding the mechanisms underlying neurodegeneration: evolving dimensions and future directions. CPT Pharmacometrics Syst Pharmacol. 2017;6:73–86.

Ma H, Sorokin A, Mazein A, Selkov A, Selkov E, Demin O, Goryanin I. The Edinburgh human metabolic network reconstruction and its functional analysis. Mol Syst Biol. 2007;3:135.

Manavalan A, Mishra M, Feng L, Sze SK, Akatsu H, Heese K. Brain site-specific proteome changes in aging-related dementia. Exp Mol Med. 2013;45:e39.

Mao L, Verwoerd WS. ORCA: a COBRA toolbox extension for model-driven discovery and analysis. Bioinformatics. 2014;30:584–5.

Mao L, Nicolae A, Oliveira MA, He F, Hachi S, Fleming RM. A constraint-based modelling approach to metabolic dysfunction in Parkinson's disease. Comput Struct Biotechnol J. 2015;13:484–91.

Mardinoglu A, Nielsen J. New paradigms for metabolic modeling of human cells. Curr Opin Biotechnol. 2015;34:91–7.

Marioni JC, Mason CE, Mane SM, Stephens M, Gilad Y. RNA-seq: an assessment of technical reproducibility and comparison with gene expression arrays. Genome Res. 2008;18:1509–17.

Martin-Jimenez CA, Salazar-Barreto D, Barreto GE, Gonzalez J. Genome-scale reconstruction of the human astrocyte metabolic network. Front Aging Neurosci. 2017;9:23.

Mastrokolias A, Ariyurek Y, Goeman JJ, van Duijn E, Roos RA, van der Mast RC, van Ommen GB, den Dunnen JT, t Hoen PA, van Roon-Mom WM. Huntington's disease biomarker progression profile identified by transcriptome sequencing in peripheral blood. Eur J Hum Genet. 2015;23:1349–56.

Middleton FA, Mirnics K, Pierri JN, Lewis DA, Levitt P. Gene expression profiling reveals alterations of specific metabolic pathways in schizophrenia. J Neurosci. 2002;22:2718–29.

Monti C, Bondi H, Urbani A, Fasano M, Alberio T. Systems biology analysis of the proteomic alterations induced by MPP(+), a Parkinson's disease-related mitochondrial toxin. Front Cell Neurosci. 2015;9:14.

Occhipinti R, Puchowicz MA, LaManna JC, Somersalo E, Calvetti D. Statistical analysis of metabolic pathways of brain metabolism at steady state. Ann Biomed Eng. 2007;35:886–902.

Orth JD, Thiele I, Palsson BO. What is flux balance analysis? Nat Biotechnol. 2010;28:245–8.

Ostaszewski M, Skupin A, Balling R. Neurological diseases from a systems medicine point of view. Methods Mol Biol. 2016;1386:221–50.

Pacheco MP, Pfau T, Sauter T. Benchmarking procedures for high-throughput context specific reconstruction algorithms. Front Physiol. 2015;6:410.

Patil KR, Nielsen J. Uncovering transcriptional regulation of metabolism by using metabolic network topology. Proc Natl Acad Sci U S A. 2005;102:2685–9.

Pornputtapong N, Nookaew I, Nielsen J. Human metabolic atlas: an online resource for human metabolism. Database (Oxford). 2015;2015:bav068.

Ray M, Ruan J, Zhang W. Variations in the transcriptome of Alzheimer's disease reveal molecular networks involved in cardiovascular diseases. Genome Biol. 2008;9:R148.

Reilly MT, Faulkner GJ, Dubnau J, Ponomarev I, Gage FH. The role of transposable elements in health and diseases of the central nervous system. J Neurosci. 2013;33:17577–86.

Reimand J, Arak T, Vilo J. g:Profiler--a web server for functional interpretation of gene lists (2011 update). Nucleic Acids Res. 2011;39:W307–15.

Reinke SN, Broadhurst DL, Sykes BD, Baker GB, Catz I, Warren KG, Power C. Metabolomic profiling in multiple sclerosis: insights into biomarkers and pathogenesis. Mult Scler. 2014;20:1396–400.

Ryu JY, Kim HU, Lee SY. Reconstruction of genome-scale human metabolic models using omics data. Integr Biol (Camb). 2015;7:859–68.

Sangar V, Eddy JA, Simeonidis E, Price ND. Mechanistic modeling of aberrant energy metabolism in human disease. Front Physiol. 2012;3:404.

Satoh J, Yamamoto Y, Asahina N, Kitano S, Kino Y. RNA-Seq data mining: downregulation of NeuroD6 serves as a possible biomarker for alzheimer's disease brains. Dis Markers. 2014;2014:123165.

Schellenberger J, Que R, Fleming RM, Thiele I, Orth JD, Feist AM, Zielinski DC, Bordbar A, Lewis NE, Rahmanian S, Kang J, Hyduke DR, Palsson BO. Quantitative prediction of cellular metabolism with constraint-based models: the COBRA Toolbox v2.0. Nat Protoc. 2011;6:1290–307.

Schramm G, Wiesberg S, Diessl N, Kranz AL, Sagulenko V, Oswald M, Reinelt G, Westermann F, Eils R, Konig R. PathWave: discovering patterns of differentially regulated enzymes in metabolic pathways. Bioinformatics. 2010;26:1225–31.

Sertbaş M, Ulgen K, Cakir T. Systematic analysis of transcription-level effects of neurodegenerative diseases on human brain metabolism by a newly reconstructed brain-specific metabolic network. FEBS Open Bio. 2014;4:542–53.

Shen J, Petersen KF, Behar KL, Brown P, Nixon TW, Mason GF, Petroff OA, Shulman GI, Shulman RG, Rothman DL. Determination of the rate of the glutamate/glutamine cycle in the human brain by in vivo 13C NMR. Proc Natl Acad Sci U S A. 1999;96:8235–40.

Simunovic F, Yi M, Wang Y, Macey L, Brown LT, Krichevsky AM, Andersen SL, Stephens RM, Benes FM, Sonntag KC. Gene expression profiling of substantia nigra dopamine neurons: further insights into Parkinson's disease pathology. Brain. 2009;132:1795–809.

Smoot ME, Ono K, Ruscheinski J, Wang PL, Ideker T. Cytoscape 2.8: new features for data integration and network visualization. Bioinformatics. 2011;27:431–2.

Stanford NJ, Lubitz T, Smallbone K, Klipp E, Mendes P, Liebermeister W. Systematic construction of kinetic models from genome-scale metabolic networks. PLoS One. 2013;8:e79195.

Stempler S, Yizhak K, Ruppin E. Integrating transcriptomics with metabolic modeling predicts biomarkers and drug targets for Alzheimer's disease. PLoS One. 2014;9:e105383.

Tarlak F, Sadikoglu H, Cakir T. The role of flexibility and optimality in the prediction of intracellular fluxes of microbial central carbon metabolism. Mol BioSyst. 2014;10:2459–65.

Thiele I, Swainston N, Fleming RM, Hoppe A, Sahoo S, Aurich MK, Haraldsdottir H, Mo ML, Rolfsson O, Stobbe MD, Thorleifsson SG, Agren R, Bolling C, Bordel S, Chavali AK, Dobson P, Dunn WB, Endler L, Hala D, Hucka M, Hull D, Jameson D, Jamshidi N, Jonsson JJ, Juty N, Keating S, Nookaew I, Le Novere N, Malys N, Mazein A, Papin JA, Price ND, Selkov E Sr, Sigurdsson MI, Simeonidis E, Sonnenschein N, Smallbone K, Sorokin A, van Beek JH, Weichart D, Goryanin I, Nielsen J, Westerhoff HV, Kell DB, Mendes P, Palsson BO. A community-driven global reconstruction of human metabolism. Nat Biotechnol. 2013;31:419–25.

Tiveci S, Akin A, Cakir T, Saybasili H, Ulgen K. Modelling of calcium dynamics in brain energy metabolism and Alzheimer's disease. Comput Biol Chem. 2005;29:151–62.

van Beek JH, Supandi F, Gavai AK, de Graaf AA, Binsl TW, Hettling H. Simulating the physiology of athletes during endurance sports events: modelling human energy conversion and metabolism. Philos Transact A Math Phys Eng Sci. 2011;369:4295–315.

Vazquez A. Metabolic states following accumulation of intracellular aggregates: implications for neurodegenerative diseases. PLoS One. 2013;8:e63822.

Verwaest KA, Vu TN, Laukens K, Clemens LE, Nguyen HP, Van Gasse B, Martins JC, Van Der Linden A, Dommisse R. (1)H NMR based metabolomics of CSF and blood serum: a metabolic profile for a transgenic rat model of Huntington disease. Biochim Biophys Acta. 2011;1812:1371–9.

Vidal M, Cusick ME, Barabasi AL. Interactome networks and human disease. Cell. 2011;144:986–98.

Vo TD, Paul Lee WN, Palsson BO. Systems analysis of energy metabolism elucidates the affected respiratory chain complex in Leigh's syndrome. Mol Genet Metab. 2007;91:15–22.

Wang YL, Eddy JA, Price ND. Reconstruction of genome-scale metabolic models for 126 human tissues using mCADRE. BMC Syst Biol. 2012;6:153.

Webster JA, Gibbs JR, Clarke J, Ray M, Zhang W, Holmans P, Rohrer K, Zhao A, Marlowe L, Kaleem M, McCorquodale DS 3rd, Cuello C, Leung D, Bryden L, Nath P, Zismann VL, Joshipura K, Huentelman MJ, Hu-Lince D, Coon KD, Craig DW, Pearson JV, NACC-Neuropathology Group, Heward CB, Reiman EM, Stephan D, Hardy J, Myers AJ. Genetic control of human brain transcript expression in Alzheimer disease. Am J Hum Genet. 2009;84:445–58.

Yizhak K, Gabay O, Cohen H, Ruppin E. Model-based identification of drug targets that revert disrupted metabolism and its application to ageing. Nat Commun. 2013;4:2632.

Yu C, Boutte A, Yu X, Dutta B, Feala JD, Schmid K, Dave J, Tawa GJ, Wallqvist A, Reifman J. A systems biology strategy to identify molecular mechanisms of action and protein indicators of traumatic brain injury. J Neurosci Res. 2015;93:199–214.

Zelezniak A, Pers TH, Soares S, Patti ME, Patil KR. Metabolic network topology reveals transcriptional regulatory signatures of type 2 diabetes. PLoS Comput Biol. 2010;6:e1000729.

Zhang L, Li H, Hu X, Benedek DM, Fullerton CS, Forsten RD, Naifeh JA, Li X, Wu H, Benevides KN, Le T, Smerin S, Russell DW, Ursano RJ. Mitochondria-focused gene expression profile reveals common pathways and CPT1B dysregulation in both rodent stress model and human subjects with PTSD. Transl Psychiatry. 2015;5:e580.

Zhang Y, James M, Middleton FA, Davis RL. Transcriptional analysis of multiple brain regions in Parkinson's disease supports the involvement of specific protein processing, energy metabolism, and signaling pathways, and suggests novel disease mechanisms. Am J Med Genet B Neuropsychiatr Genet. 2005;137B:5–16.

Zimmerman SB, Trach SO. Estimation of macromolecule concentrations and excluded volume effects for the cytoplasm of *Escherichia coli*. J Mol Biol. 1991;222:599–620.

Chapter 8
Synaptic Plasticity and Synchrony in the Anterior Cingulate Cortex Circuitry: A Neural Network Approach to Causality of Chronic Visceral Pain and Associated Cognitive Deficits

Ying Li

Abbreviations

ACC	Anterior cingulate cortex
AP5	Aminophosphonopentanoic acid
AUC	Area under the curve
ANOVA	Analysis of variance
BLA	Basolateral amygdala
CRD	Colorectal distension
DNQX	Cyanonitroquinoxaline dione
EA	Egg albumin
GFP	Green fluorescent protein
IBS	Irritable bowel syndrome
LFP	Local field potential
LTP	Long-term potentiation
MT	Medial thalamus
NMDA	N-methyl-D-aspartate
NVP-AAM077	[(R)-[(S)-1-(4-bromo-phenyl)-ethylamino]-(2,3-dioxo-1,2,3,4-tetrahydroquinoxalin-5-yl)-methyl]-phosphonic acid
pACC	Perigenual anterior cingulate cortex
RNAi	RNA interference
siRNA	Small interfering RNA

Y. Li (✉)
Department of Biomedical Sciences, City University of Hong Kong, Kowloon, Hong Kong

Centre for Biosystems, Neuroscience, and Nanotechnology, City University of Hong Kong, Kowloon, Hong Kong

School of Veterinary Medicine, City University of Hong Kong, Kowloon, Hong Kong
e-mail: yingli@cityu.edu.hk

A. Cheung-Hoi Yu, L. Li (eds.), *Systems Neuroscience*, Advances in Neurobiology 21, https://doi.org/10.1007/978-3-319-94593-4_8

RGT	Rat gambling task
SFC	Spike-field coherence
STA	Spike-triggered average
TBS	Theta burst stimulation
VH	Viscerally hypersensitive
VMR	Visceromotor response

8.1 Brain Targets for Visceral Pain "Memory" Process in the Visceral Hypersensitive State

The International Association for the Study of Pain (IASP) defines pain as "an unpleasant sensory or emotional experience associated with actual or potential tissue damage, or described in terms of such damage". This definition is based on the concept of pain as a perception rather than as a purely sensory modality, and takes into account the fact that for pain to be consciously experienced, cognitive processing is required.

Visceral hypersensitivity is common among patients with irritable bowel syndrome (IBS). Patients with IBS may be presented with persistent severe pain, yet there are no clearly identifiable clinical or radiographic abnormalities. The mechanisms underpinning the transition from acute into chronic pain, such as prolonged "functional visceral pain" remain unclear (Mayer et al. 2000). Abnormalities that up-regulate signal intensity anywhere in the afferent system can induce hypersensitivity and pain. It has been shown that peripheral sensitization results from an increase in the sensitivity and excitability of the afferent nerve itself and/or the dorsal horn of the spinal cord (Gebhart 2000). The brain interprets and influences the perception of pain-sensation signals transmitted from the gut. Recently, functional magnetic resonance imaging (fMRI) studies are beginning to address the possible neural mechanisms of hyperalgesia in patients with IBS. The anterior cingulate cortex (ACC) is a major cortical component of the limbic loop system, and its functional relationship to emotional and motivational responses has been well described (Vogt et al. 2003). Studies of both humans and animals consistently suggest that the ACC and its related areas are important for processing pain perception (Mayer et al. 2000; Vogt and Robert 1993). Patients with IBS show enhanced activation of the dorsal ACC, and a reduction of this pattern is associated with a reduction in IBS symptoms (Vogt and Robert 1993; Fan et al. 2009). These findings suggest that the emotional and sensory components of the brain pain experience system in IBS are dysfunctional. fMRI is blood oxygenation level-dependent (BOLD) imaging, measuring neuronal activity indirectly via its assumed haemodynamic correlate. BOLD fMRI reflects changes in cerebral blood volume, cerebral blood flow and oxygen consumption. Inconsistencies exist in BOLD fMRI studies (Mayer et al. 2000; Mertz et al. 2000; Silverman et al. 1997; Sidhu et al. 2004). Such variation emphasizes the need to learn how to interpret the BOLD fMRI signal in terms of the neuronal synaptic activity and action potentials in the brain (Arthurs and Boniface

2002). However, the literature is conspicuously devoid of a definitive study that demonstrates the neural electrophysiological activity of the ACC during processing of visceral nociceptive stimulation.

8.1.1 Viscerally Hypersensitive Rat Model

Visceral hypersensitivity in rats was induced by colonic anaphylaxis (Gao et al. 2006; Cao et al. 2008). The rats were injected intraperitoneally with 10 μg egg albumin (EA antigen) and 10 mg aluminum hydroxide (adjuvant) in 1 ml saline. From the third day to the fifth day after antigen injection, the rats were given a colonic perfusion with antigen solution followed by 30 mmHg colorectal anaphylaxis for 30 s repeated 5 times with 3-min intervals.

Food allergens may be important in IBS. Previous studies have shown that luminal antigenic challenge in the sensitized rat intestine resulted in induction of contractile activity and diarrhea (Scott et al. 1998; Nanda et al. 1989). Additionally, a clinical trial has shown that patients with IBS improved with dietary exclusions (Nanda et al. 1989). Intestinal anaphylaxis enhances the activity of the intestinal mesenteric nerve (Nozdrachev et al. 1999) and triggers neuronal activations in the nucleus of the solitary tract (NTS) (Scott et al. 1998). In the intestine, the anaphylactic response is characterized by IgE antibody mediated mast cell degranulation. Studies in animals have provided evidence that mast cell activation triggers visceral hypersensitivity and gastrointestinal motor dysfunction. A recent study indicated that colonic mast cell infiltration and mediator release in proximity to mucosal innervation may contribute to abdominal pain perception in IBS patients (Giovanni et al. 2004). Thus, mast cells could be involved in the disturbed sensory-motor function of IBS. Hence the EA rat model of visceral hypersensitivity with chronic presence of mast cells may be a suitable model to study visceral hypersensitivity. However, this VH rat model has certain limitations. For example, studies indicate that psychosocial trauma and psychological distress play important roles in the onset and modulation of IBS symptoms. Nevertheless, it is well recognized that different factors may be involved in the pathogenesis of IBS.

8.1.2 Enhanced ACC Nociceptive Transmission in Viscerally Hypersensitive Rats

Direct electrophysiological evidence of the sensitization of ACC neurons in viscerally hypersensitive rats has been provided (Gao et al. 2006; Wu et al. 2008). Recording single ACC (Cg1, Cg2 and prelimbic cortex) neuronal activities in response to colorectal distension (CRD) showed that viscerally hypersensitive rats have enhanced ACC spontaneous activity, decreased CRD pressure threshold to

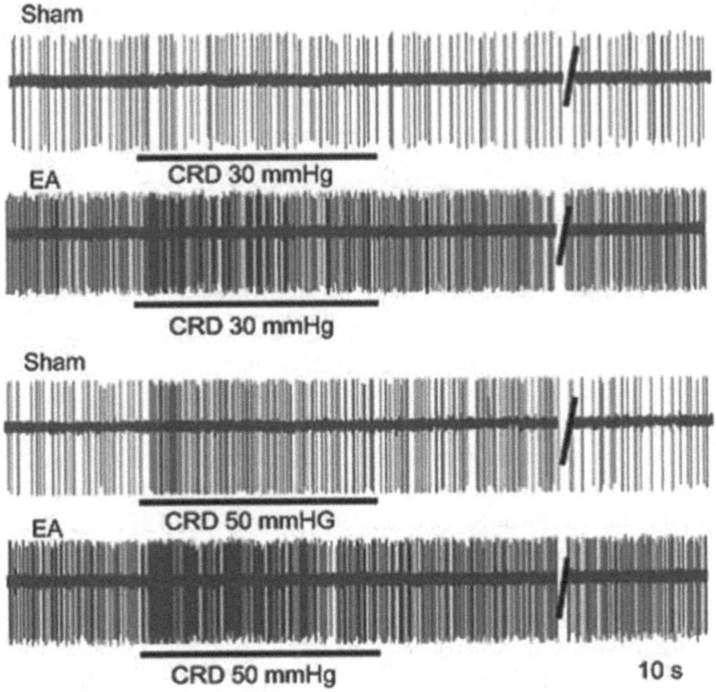

Fig. 8.1 Recordings of colorectal distension (CRD)-excited anterior cingulate cortex (ACC) neurons in response to CRD (30 and 50 mmHg) in sham-treated and viscerally hypersensitive (EA) rats. In sham-treated rats, there was a slight increase in response to 50 mmHg CRD, whereas both low- and high-pressure distension evoked increased responses in the rats sensitized with chicken egg albumin (Adapted from Gao et al., J Physiol (Lond) 570(1):169–184, 2006. With permission.)

stimulate ACC neurons, and increased ACC response magnitude. The ACC neurons of controls failed to respond to 10 or 30 mmHg CRD. In contrast, CRD (10, 30 and 50 mmHg) markedly increased ACC neuronal responses of EA rats (Fig. 8.1). CRD produced greater pressure-dependent increases in ACC spike firing rates in VH rats compared with controls. There are significant increases in the numbers of CRD-excited ACC neurons in the VH rats compared with normal rats (Gao et al. 2006). It appears that colorectal anaphylaxis resulted in sensitization of high-threshold receptors and brought into play previously unresponsive silent nociceptors. Splanchnicectomy combined with pelvic nerve section abolished ACC responses to CRD in VH rats. However, acute nerve section failed to prevent the enhancement of ACC spontaneous firings in the sensitized rats, suggesting that the subspinal peripheral ongoing activity is not required to maintain a higher level of ACC spontaneous activity in visceral hypersensitive rats, a phenomenon that probably originates centrally rather than peripherally (Gao et al. 2006). This electrophysiological evidence suggests that the level of activation in the ACC evoked by noxious CRD in VH state is a determinant in emotional and behavioral reactions to pain.

Visceral stimulation activates a more anterior part of the ACC compared with cutaneous stimulation (Silverman et al. 1997; Lotze et al. 2001). Previous studies revealed that most caudal ACC neurons did not respond to CRD (Gao et al. 2006). fMRI scanning has shown spatially distinct ACC activation during visceral and cutaneous noxious stimulation (Verne et al. 2001). We examined the response of CRD-excited ACC neurons to transcutaneous electrical stimulation (TCES) of the hind paw. One group of CRD-excited ACC neurons was activated exclusively by CRD stimulation. These neurons failed to respond to TCES, which suggests involvement of a population of rostral ACC neurons in visceral nociception and a possible discriminative aspect of visceral nociception. The other group of rostral ACC neurons responded to both CRD and TCES. We showed that a group of ACC neurons in VH rats exhibited enhanced CRD-induced activities. However, neuronal responses evoked by cutaneous noxious heat stimulation did not change significantly (Gao et al. 2006). This is the first demonstration that a population of rostral ACC neurons is capable of discriminative coding for hypersensitivity, specifically, visceral hypersensitivity. It should be noted that the criteria used to define rostral and caudal ACC in current studies do not correspond to the terminology of ACC regions in primates and humans (Vagt 2005).

8.1.3 N-Methyl-D-Aspartate (NMDA) Receptor Mediate ACC Synaptic Responses After the Induction of Visceral Hypersensitivity

Just as in other regions of the central nervous system, fast excitatory synaptic transmission within the ACC is mediated by the excitatory amino acid glutamate (Sah and Nicoll 1991; Wei et al. 2001). Glutamate exerts its signaling role by acting on glutamate receptors, including N-methyl-D-aspartate (NMDA), 3-hydroxy-5-methyl-4-isoxazolepropionate (AMPA)/kainate, and metabotropic glutamate receptors (Dougherty et al. 1992). In the ACC, NMDA receptors are highly expressed, although their function remains unclear. Our early studies have shown that reverse microdialysis of AMPA/kainate receptor antagonist 6-cyano-7-nitroquinoxaline-2,3-dione (CNQX) reduced basal and abolished CRD-induced ACC neuronal firing in normal rats. In contrast, microdialysis of NMDA receptor antagonist AP5 had no effect on ACC neuronal firing in normal rats. However, AP5 produced marked inhibition of ACC neuronal firing evoked by 50 mmHg CRD in VH rats (Wu et al. 2008). It appears that ACC nociceptive transmissions are mediated by glutamate AMPA receptors in the control rats. ACC responses to CRD are enhanced in VH rats. NMDA receptors mediate ACC synaptic responses after the induction of visceral hypersensitivity (Wu et al. 2008). Unlike other ionotropic receptors, NMDA receptors are 5–10 times more permeable to Ca^{2+} by way of NMDA receptors from the extracellular space into the postsynaptic cells, triggering a cascade of signaling molecules, including protein kinases, protein phosphatases, as well as enzymes that produce diffusible retrograde messengers. The function of NMDA receptors as coincidence detectors and their permeability to Ca^{2+} makes these receptors the best candidates for a central mechanism in memory formation (Dougherty et al. 1992; Fan et al. 2009).

8.1.4 The ACC Plays a Critical Role in the Modulation of Behavioral Visceral Pain Responses in VH Rats

Pain contains both sensory and affective dimensions. Teasing apart the mechanisms that control the neural pathways mediating pain effect and sensation is a challenge. It is not clear whether activation of the ACC is only causally involved with the perception of pain-related unpleasantness or if it is also involved with the perceived intensity of pain during noxious visceral stimulation. Rodents do not have the forebrain structures to generate the cognitive feelings of humans. The use of behavioral paradigms to assess spinal pain reflexes that do not include the assessment of cognitive perception in the conscious rat may help to identify the regulatory role of the ACC in visceral pain sensation. Based on brainstem reflexes, which have been described as "pseudoaffective" responses (Ness and Gebhart 1990), the nociceptive response (visceromotor response (VMR) to CRD was recorded. Both control and VH rats showed pressure-dependent increases in the VMR to CRD. A significant VMR to the lowest distention pressure tested in VH rats and an absence of response to the lowest distention pressure in normal rats suggest a reduced pressure threshold (ie, allodynia) in VH rats (Fig. 8.2). These results provide evidence of enhanced visceral pain responses (ie, hyperalgesia) in VH rats (Fan et al. 2009; Cao et al. 2008). In this model, no significant mucosal inflammation in the colon 7 days after the initiation of visceral hypersensitivity were observed. The hypersensitivity to colonic distention, however, can be observed even up to 7 weeks following the initiation of colonic anaphylaxis and appears to be independent of mucosal inflammation (Fan et al. 2009). Hence, this may be a useful model to study post-inflammatory conditions of visceral hyperalgesia such as post-infection IBS, which occurs in up to 20% of patients following an acute bout of gastrointestinal infection.

Electrical stimulation of the rostral ACC in conscious rats enhances the VMR to CRD in a frequency-dependent manner. Furthermore, bilateral ACC lesion does not change the VMR in normal rats but markedly inhibits the VMR to CRD in VH rats. The reduction in the VMR after ablation of the rostral ACC suggests that neural networks in this region mediate allodynia and hyperalgesia in viscerally hypersensitive state (Gao et al. 2006). Further study showed that injection of low-dose glutamate into the ACC has no effect on the pain response in normal rats; however, in VH rats, it has a potent effect on the VMR to CRD, suggesting sensitization of glutamate receptors in ACC neurons in viscerally hypersensitive states (Cao et al. 2008). To determine the role of glutamatergic transmission in the modulation of visceral pain, investigators have shown that microinjection of the NMDA receptor antagonist AP5 into the ACC suppresses the CRD-induced increase in the VMR in VH rats

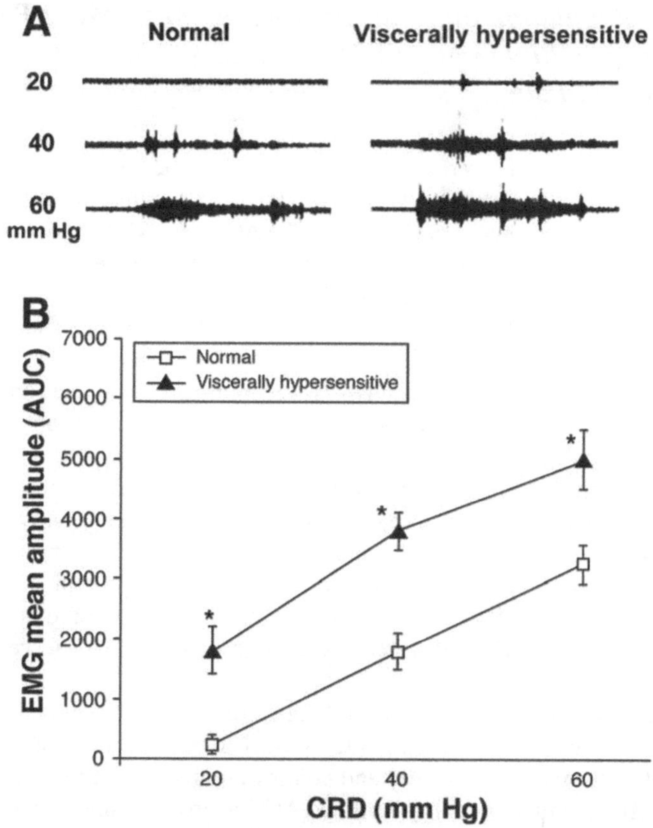

Fig. 8.2 VMR to graded distention pressures in normal control and viscerally hypersensitive rats. Under basal conditions (CRD, 0 mmHg), there was no significant difference between normal control rats and viscerally hypersensitive rats. (**a**) A representative abdominal muscle electromyogram of the VMR to graded-pressure CRD recorded from the external oblique pelvic muscle in normal and viscerally hypersensitive rats. (**b**) Mean amplitude of the abdominal muscle contraction expressed as AUC after baseline subtraction was presented. Data were collected from 7 sham-treated control rats and 8 viscerally hypersensitive rats. Analysis of variance showed a significant effect for distention level, as well as a significant interaction between distention level and group (*$P < 0.05$). Stimulus-response functions were shifted to the left in viscerally hypersensitive rats, indicating group differences in the VMR response. Values are presented as means ± SE (Adapted from Cao et al., Gastroenterology 134:535–543, 2008. With permission.)

(Fan et al. 2009). It appears that visceral hyperalgesia in this rat model involves endogenous activation of descending facilitator pathways mediated by NMDA receptors in the rostral ACC (Gebhart 2004).

What makes glutamatergic synapses unique is that they can sustain synaptic plastic changes that may persist for hours to days (Dudek and Bear 1992). Further studies have to be conducted to explore the cellular and molecular mechanisms underlying these dynamic changes in ACC neurons in the viscerally hypersensitive state.

8.1.5 Up-Regulation and Phosphorylation of CaMKII Post-Synaptic Binding to NR2B Receptors Contributes Visceral Pain

NMDA receptors contain heteromeric combinations of the NR1 subunit plus one or more of the subunits NR2A–2D. Although NR1 is distributed widely in the brain, NR2 subunits show regional specificities. In the ACC, the NMDA receptor containing NR2A or NR2B subunits contributes to most NMDA-receptor currents. In human beings and rodents, the subunits NR2A and NR2B are dominant in forebrain structure (Monyer et al. 1994). In the ACC, the NMDA receptor containing NR2A or NR2B subunits contribute to most NMDA-receptor currents (Zhao et al. 2005). Mice that genetically overexpress the NR2B-receptor subtype in the forebrain show enhanced responsiveness to painful stimuli (Cao et al. 2008) and superior learning ability and memory of different behavioral tasks (Tang et al. 1999).

Considering the distinct roles that NMDA receptors may serve, identification of the receptor subtype in the ACC that mediates visceral hypersensitivity will promote our understanding of the molecular mechanisms underlying nociceptive processes in the VH state. The up-regulation of NR2B-receptor protein was verified by Western blot analysis in VH rats (Fan et al. 2009) (Fig. 8.3). On the other hand, no significant increases in NR1 and NR2A protein expression were observed (Fan et al. 2009).

Electrophysiological studies showed that reverse microdialysis of NVP-AAM077, a specific NR2A-subunit antagonist, had no effect on basal and CRD-induced ACC neuronal firing in VH and control groups. In VH rats, Ro25-6981, a specific NR2B-subunit antagonist, inhibited ACC neuronal firing, evoked by 30 and 50 mm Hg CRD, by 98% and 52% respectively. Behavioral studies showed that neither NVP-AAM077 nor Ro25-6981 changed the VMR to graded-pressure CRD in normal rats (Fan et al. 2009). On the other hand, in VH rats, NVP-AAM077 had no effect on the VMR, whereas Ro25-6981 dose-dependently decreased the VMR to CRD suggesting that NMDA NR2B-receptor activities in the ACC are responsible for allodynia and hyperalgesia in VH rats. To down-regulate NR2B-receptor gene expression, an NR2B-specific small interfering RNA (siRNA) and a plasmid (pEGFP-N1) that expressed the green fluorescent protein were administered into ACC neurons by electroporation. NR2B siRNA-treated VH rats showed a significant reduction in the VMR, compared with controls indicating that the NR2B subunit of NMDA receptor activation of sensitized ACC neurons plays a causative role for the long lasting visceral pain responses, which are independent of inflammation in the colon in the functional viscerally hypersensitive rats (Fan et al. 2009; Cao et al. 2008).

The NMDA receptors may play important roles in plasticity and memory via multiple downstream signaling pathways including calcium/calmodulin-dependent protein kinase II (CaMKII) (Lisman et al. 2002). CaMKII is enriched at the postsynaptic density (PSD). Localization of CaMKII at the PSD has been proposed to play a critical role in long-term potentiation (LTP) (Lisman et al. 2002). CaMKII is

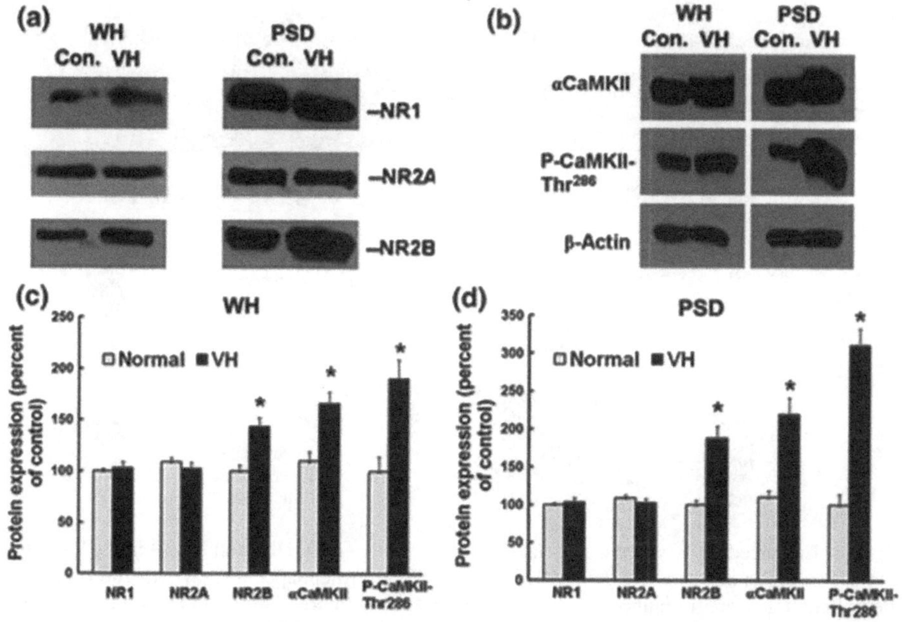

Fig. 8.3 Expression of subtypes of NMDA receptor, aCaMKII and PaCaMKII-Thr286 in the pACC. (**a**) Representative immunoblots from pACC showed the expression of NMDA receptor subunits NR1, NR2A and NR2B in WH and PSD extracts in the normal and VH rats. Western blot showed that the expression level of NR2B, but not NR1 or NR2A was increased significantly in the pACC of VH rats 10 days after the induction of visceral hypersensitivity. (**b**) Representative Western blots showed enhanced expression of aCaMKII and P-α-CaMKII-Thr286 in WH and PSD extract in VH rats compared with control. (**c**, **d**) Relative intensities of NR2A, NR2B, αCaMKII and P-aCaMKII-Thr286 protein were measured by densitometry analysis. Quantification of protein expression in the pACC of VH rats was expressed as the percentage of controls. Each column represented the means ± SEM. Statistical significance was determined by Student's t-test between normal and VH rats for each molecule. $^{*}p < 0.05$, $n = 6$ for each group (Adapted from Li et al. J Neurochemistry, 1111/j.1471-4159. 2012. With permission.)

autophosphorylated at Thr286, and may function as a biochemical 'memory' process. An elegant study showed that NR2B binding targets activated CaMKII in the PSD and maintained persistent activity of this kinase in subcellular compartments (Bayer et al. 2001). A series of studies using a biochemical fractionation approach showed over-expression of NR2B and CaMKI in the ACC, and post-synaptic accumulation of NR2B and CaMKII in the ACC synapses. Further investigation showed the increases in phosphorylated CaMKII (Thr286CaMKII) protein level in the post-synaptic density fraction (PSD) (Triton X-100 insoluble) and extrasynaptic (Triton X-100 soluble) fractions (Li et al. 2012). Western blotting following co-immunoprecipitation showed that phosphorylated-CaMKII-Thr286 bound to NR2B in the PSD, which was increased to 267% of control in VH rats (Fig. 8.4). Administration of CaMKII antagonist Antennapedia-CaMKIINtide suppressed visceromotor response in VH rats, and in parallel, restored the level of NR2B to control

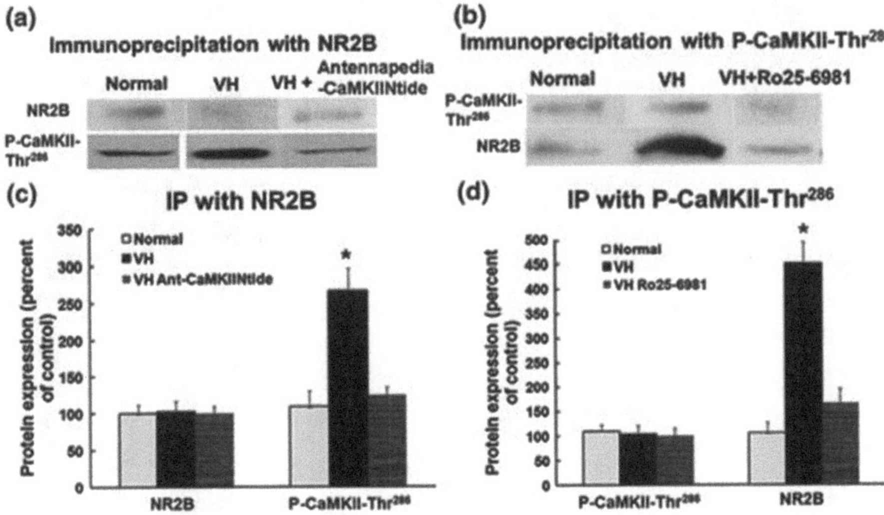

Fig. 8.4 Enhanced post-synaptic association between active αCaMKII and NR2B in the VH rats. (**a**) PSD fractions of control, VH and VH rats treated with Antennapedia-CaMKIINtide were isolated. Samples were immunoprecipitated with NR2B antibody before immunoblotting with P-CaMKII-Thr286 antibody. (**b**) PSD fractions of control, VH and VH rats treated with Ro25-6981 were isolated. Samples were immunoprecipitated with P-CaMKII-Thr286 before immunoblotting with NR2B antibody. (**c**, **d**) Summary graphs showed increased interaction between P-CaMKII-Thr286 and NR2B in the VH rats. Each column represented the means ± SEM. Statistical significance was determined by one-way ANOVA followed Bonferroni post-tests. *$p < 0.05$, $n = 5$ for each group (Adapted from Li et al. J Neurochemistry, 1111/j.1471-4159. 2012. With permission.)

levels and reduced the NR2B-P-CaMKII-Thr286 protein complex in PSD. To down-regulate the gene expression of CaMKII, RNA-specific interfering RNA (siRNA) was administrated into the pACC neurons of VH rats by in vivo electroporation. This treatment completely abolished the increases in CRD-induced VMR (Li et al. 2012). All these data imply the close relationship between CaMKII and NR2B in ACC in chronic pain, namely, CaMKII might act as an amplifier of detrimental cellular calcium signal regulated by NMDA receptors when becoming autophosphorylated and targeting to NR2B; conversely, autophosphorylated CaMKII could modulate post-synaptic NR2B subtype of NMDA receptor localization. Data from these series of experiments supports the hypotheses that CaMKII is a critical signaling molecule in the ACC glutamatergic synaptic transmission, phosphorylation of CaMKII at Thr286, which binds to NR2B subunit at post-synaptic site, modulates visceral pain in viscerally hypersensitive state.

8.1.6 *Perigenual ACC (pACC) Are Necessary for the "Aversiveness" of Visceral Nociceptor Stimulation*

Pain contains both sensory and affective dimensions. Except for human experiments where self report is possible, teasing apart the mechanisms that control the neural pathways mediating pain affect and sensation in an overall animal nociceptive behavioral response is a challenge. It is well documented that the ACC is involved in pain processing and encoding of negative affect in humans, which results in pain-related unpleasantness (Tolle et al. 1999). Research suggests that ACC neuronal activity in rodents is related to stimulus–reward learning (Bussey et al. 1997). Johansen, Fields, and Manning (Johansen et al. 2001), and Johansen and Fields (Johansen and Fields 2004) introduced a formalin-induced conditioned place avoidance (F-CPA) model to distinguish somatic pain emotion from pain sensation in rats. Using a rodent visceral pain assay that combines the colorectal distension (CRD)-induced visceromotor response (VMR) with the conditioning place avoidance (CPA) researchers in the field of visceral pain, Yan et al. measured a learned behavior that directly reflects the affective component of visceral pain (Yan et al. 2012). When CRD was paired with a distinct environment context, the rats spent significantly less time in this compartment on the post-conditioning test days as compared with the pre-conditioning day. Effects lasted for 14 days. Bilateral ACC lesion significantly reduced CPA scores without reducing acute visceral pain behaviors (CRD-induced VMR). Bilateral administration of non-NMDA receptor antagonist CNQX or NMDA receptor antagonist AP5 into the pACC decreased completely abolished the CPA in the day 14 after conditioning. These data suggested that pACC activation is critical for the memory processing involved in long-term negative affective state and prediction of aversive stimuli by contextual cue.

8.2 ACC Synaptic Plasticity Mediates Learning and Long-Lasting Functional Visceral Pain Memory

Using the visceral hypersensitivity rat model, a series of behavioral studies suggested that the facilitation of visceral pain responses following a brief noxious stimulus (colonic anaphylaxis in our model), the hypersensitivity to colorectal distension (CRD) can be observed up to 7 weeks after the initiation of colonic anaphylaxis. The chronic visceral pain was independent of mucosal inflammation, suggesting mediation by a mechanism for the learning and triggering of memory, where information needs to be stored and retrieved. However, the mechanisms underlying how visceral nociceptive input is encoded within the ACC have not been explored. The synaptic substrates in the ACC neuronal circuitry responsible for storing visceral nociceptive information for prolonged periods of time (e.g. by use-dependent change in synaptic strength) have not been identified. A key insight in neuroscience over the past three decades is that synaptic connections between neurons are in a near-continual state of change (Bliss and Collingridge 1993). The responses of neocortical neurons can be persistently modified by alterations in sensory experience. Such modifications reflect changes in synaptic transmission that shape cortical circuits and store information (Martin et al. 2000). At the cellular level, activity-dependent plasticity in synaptic strength, such as long-term potentiation (LTP) and long-term depression (LTD), may serve as key synaptic mechanisms reflecting cortical plasticity (Bliss and Collingridge 1993; Chapman et al. 1998).

8.2.1 Visceral Hypersensitivity Is Associated with Alterations of the Properties of Synaptic Plasticity in the ACC

The medial thalamus (MT) serves as a major relay in the medial pain system and in the conveyance of nociceptive information to the ACC (Vogt et al. 2003; Shyu and Vogt 2009). There is wealth of evidences to support the view that the synapses mechanisms contribute to learning, and memory storage for long lasting functional visceral pain. LFP is a low frequency (40–130 Hz) component of the electrophysiological signal. It reflects the superposition of synchronized dendritic currents, averaged over a large variety of interneurons and intracortical activity. FP recording is one way to study artificially induced synaptic plasticity (Martin et al. 2000; Shyu and Vogt 2009). The ACC FPs elicited by electrical stimulation of the MT were used as a quantitative measure of synaptic strength. I/O curves generated by a gradual increase of the stimulus intensity (50–1000 μA) showed significant increases in LFP in the sensitized rat suggesting enhancement of basal synaptic transmission in the thalamo-ACC synapses after induction of visceral hypersensitivity; this is mediated by both NMDA and AMPA receptor activity (Wang et al. 2013).

Electrophysiological recordings from animals and humans have revealed that ACC neurons are likely to fire action potentials at 4–7 Hz (theta) during various behavioral tests (Nishida et al. 2004). Electrorheological study in the ACC area have shown that theta burst stimulation (TBS) reliably induces LTP-like plasticity in the MT-ACC pathway in normal rats. However, in the VH state, the expression of LTP-like plasticity in MT-ACC synapses was smaller or occluded. An additional study by using low-intensity stimulus, which evoked response in VH rats that is comparable to the response induced by 400 μA (induced 50% of maximum amplitude of the LFP) in control rats. We found that low intensity stimulus in VH rats also failed to elicit increases in LFP amplitude following TBS conditioning suggesting that the induction of LTP-like plasticity in the MT-ACC synapses was blocked in the VH states (Wang et al. 2013). It appears that, in the VH state, transduction signals in the MT-ACC synapses are not available for subsequent electrical recruitment, suggesting that the synaptic strengthening occurring in the VH state engages signal transduction pathways that are in common with those activated by electrical stimulation.

8.2.2 Visceral Hypersensitivity vs ACC Synaptic Plasticity: Chicken or Egg?

To further characterize whether LTP-like synaptic plasticity in the MT-ACC synapses contributes to visceral pain in VH state, we showed that repeated artificial induction of LTP in the MT-ACC synapse by chronic repeated theta-patterned tetanization in the MT in normal rat (Wang et al. 2015) facilitated behavioral visceral pain, which mimic visceral allodynia and hyperalgesia in the VH model (Fan et al. 2009; Cao et al. 2008). These observations lend support to the theory that the enhanced long-lasting transmission at the MT-ACC synapses *causally contributes* to visceral pain. It appears that chronic visceral pain, ACC sensitization and long-lasting enhanced synaptic transmission in the VH state are expressed by the same core mechanisms as TBS-induced canonical LTP in rats (Wang et al. 2013). ACC synaptic strengthening may engage signal transduction pathways that are in common with those activated by TBS, and serves as an attractive cellular model of functional visceral pain. It appears that induction of chronic visceral pain produces a change in the ability to induce subsequent synaptic plasticity at the ACC neural circulatory, we hypothesize that the mechanisms of ACC synaptic metaplasticity not only are involved in the processes of modifying the visceral pain sensitivity (Fan et al. 2009; Cao et al. 2008) and the aversive responses to pain (Yan et al. 2012), but also further affect the processing of learning and memory in chronic pain state (Wang et al. 2013, 2015).

8.3 Visceral Pain and Cognitive Deficits

Pain is a perception rather than as a purely sensory modality, and it is worth mentioning that for pain to be consciously experienced, cognitive processing is required. The conceptualization of pain in humans recognizes the components involved in the encoding and perception of stimulus parameters (e.g., stimulus localization, intensity, and quality), and the affective salience or unpleasantness of the noxious stimulus. Chronic pain (defined as pain persisting for 3–6 months or longer) generally exceeded the duration of the precipitating noxious stimulus or injury, and is associated with the development of affective disorders but the underlying mechanisms are not fully understood. Ample clinical evidence has shown that most of the patients with IBS who seek treatment have psychiatric comorbidity, notably depression and anxiety disorders (Larsson et al. 2012; Longstreth et al. 2006). Symptoms of major depressive disorder (MDD) occur in up to 90% of patients with IBS (Friedrich et al. 2010), clinical evidence has also shown that chronic pain patients usually suffer from memory deficiency, so it is surprising that there has been no experimental animal model to study visceral pain related emotional disorder and cognitive deficits, and little is known about the underlying mechanisms.

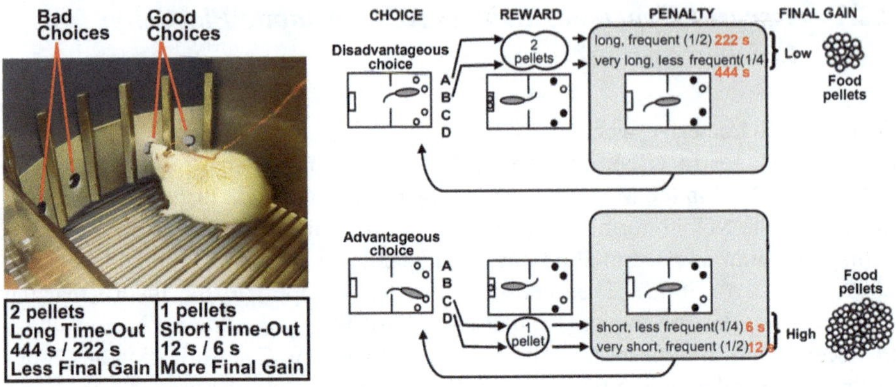

Fig. 8.5 Rat gambling task procedure. During the gambling task, rats can nose-poke either one of the four holes, A, B, C or D to obtain reward. The RGT is used to test the decision-making by choosing between options associated with different amounts of reward in terms of food pellets at different likelihood/probability of penalties, which will be time-outs in this case. With the choice A or B, the rat will have a reward of two food pellets each time but it will either experience a long and frequent or a very long and less frequent timeout, the final gain will be less compare to the choice C or D. Although the rat will only have one food pellet in reward by choosing C or D, it will only experience either a short and less frequent time-out or a very short and frequent time-out. This pattern is advantageous and it will have more food rewarded at the end (Adapted from Cao et al., Experimental Neurology 136 (2016) 74–85. With permission)

Cognition refers to a set of mental processes, including attention, memory, evaluation, decision-making, etc. Making a decision under uncertain conditions is a basic cognitive process for adaption relying on the integration of several executive functions. Impaired decision making has been demonstrated to represent a key symptom in many mental disorders. In humans, decision-making has been accurately modeled using the Iowa gambling task (IGT) (Bechara et al. 2000, 1997). In behavioral tasks for animals, several psychiatric symptoms addressing higher-order cognitive dysfunctions have been reproduced (Nestler 2006). Rodents are individuals which can exhibit human-like cognitive characteristics, such as the ability to learn and reason with causal knowledge (Blaisdell et al. 2006). Rodent models of decision-making, such as rat gambling task (RGT), are particularly valuable as experimental conditions can be controlled (Zeeb and Winstanley 2011; Xu et al. 2015; Mu et al. 2015; Cao et al. 2016).

8.3.1 *Visceral Hypersensitivity Affects Decision-Making Ability in Rats*

The RGT has been developed to test the decision-making capacities in rats via a conflict between immediate and long-term gratification (food reward). Operant chambers (28 × 30 × 34 cm) were used for RGT (Fig. 8.5). During the training stage, rat gradually learned the association between the nose-poke action and the release of food pellet in the food tray. The training phase usually lasted 7–10 days. The 60-min test was performed the following day. Rats were allowed to freely make choices among the four apertures (A–D) as they did in the training phase, however, different choices were associated with different outcomes. Although the immediate reward of choice A and B was two times of C and D, in the long run, the theoretical maximum benefit of C and D will be five times higher than A and B. The rats selected all 4 different options equally during the first 10 min. Over time, good decision makers progressively developed a preference for the advantageous options. The proportion of advantageous choices (%) = numbers of nose-poke (C + D) / nose-poke (A + B + C + D) *100% were used to identify decision-making behavior of rats. Following criterion was used to distinguish the good (>70% preference), delayed-good (>70% preference in the last 20 min), undecided (30–70%), and poor (<30%) decision-makers during the last 20 min (Fig. 8.5). At the end of the testing, good decision-makers earned significantly more food pellets across the session. Studies of brain-lesion or psychiatric patients have discovered the specific prefrontal cortex (PFC) areas mediating decision-making (Bechara et al. 2000; Rushworth et al. 2007). Animal studies have shown that decision-making performances in the RGT depend on the integrated function of several sub-regions of the PFC, especially the prelimbic, cingulate and orbitofrontal cortices, and amygdala (Zeeb and Winstanley 2011). Recently, using chronic visceral pain rat model the difference in the proportions of the subgroups between the control and (VH) groups was reported (Flood et al. 1987) (Fig. 8.6). The significant decreases of the proportion of good decision-makers from 71% in the control to 46% associated marked increases in indecisive decision-makers discovered in the chronic visceral pain rats (Flood et al. 1987). These data provide the first evidence that chronic visceral pain led to decision-making deficits in rats.

8.4 ACC Neuronal Spike Field Phase Locking and Synchrony Cross Areas Associated Involved in the Processing of Chronic Visceral Pain

Neuronal oscillations are likely to be a fundamental mechanism for modulating, filtering, and redirecting information in the nervous system. In the last few years, large scale neural oscillations have been acknowledged to play a primary role in fundamental cognitive functions. Ample evidence suggests that neurons transmit information not only by changing their firing rates but also timing of the spikes corresponding to the ongoing neuronal oscillations (Varela et al. 2001).

Furthermore, the induction of synaptic plasticity is favored by coordinated action potential timing across neuronal networks (Markram et al. 1997), giving rise to oscillations of different frequencies in the local field potential (LFPs). These field potential oscillations have been shown to modulate local spike timing (Jacobs et al. 2007).

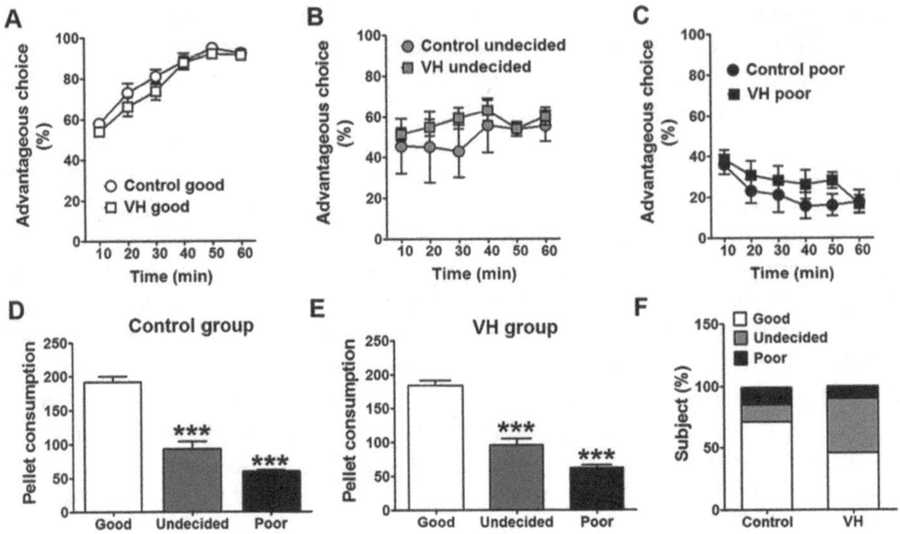

Fig. 8.6 Changes in decision-making behavior induced by visceral hypersensitivity using a rat gambling task. (**a–c**) Mean time-course of advantageous choices of good (**a**), undecided (**b**) and poor (**c**) decision-makers during the RGT. (**d, e**) Total pellet consumption during the 60 min RGT testing for good (white), undecided and poor (black) decision-makers of control rats and VH rats. ***$p < 0.001$ vs. good decision-makers. (**f**) Proportions of good (white bar), undecided and poor (black bar) behavior was represented for control group and VH group. Advantageous choices (%) = numbers of nose-poke for choices (**c** + **d**)/numbers of nose-poke for choices (**a** + **b** + **c** + **d**) × 100%. n = 28 for control group, n = 39 for VH group (Adapted from Cao et al., Experimental Neurology 136 (2016) 74–85. With permission.)

8.4.1 Tight Coordination of Spike Timing with the Local Theta Oscillation Is a Key Index for Predicting Successful Cognitive Function

Rutishauser et al. have shown that memory formation in humans is predicted by close coordination of spikes phase-locking with the theta band local field potentials (Rutishauser et al. 2010). Within individual brain areas, oscillations can synchronize neurons, creating coherent cell assemblies (Harris et al. 2003) and appropriate plasticity depending on the precise timing of pre- and post-synaptic activity (Markram et al. 1997; Fig. 8.7). Evidence points to cortical oscillations as a mechanism for mediating interactions among functionally specialized neurons in distributed brain circuits. A brain function that may use such interactions is declarative memory (Rutishauser et al. 2010)—that is, memory that can be consciously recalled, such as episodes and facts. A growing body of evidence clarified by us and others suggests that cortical oscillation at theta band, in particularly, the synchrony between spike timing and theta oscillation facilitates

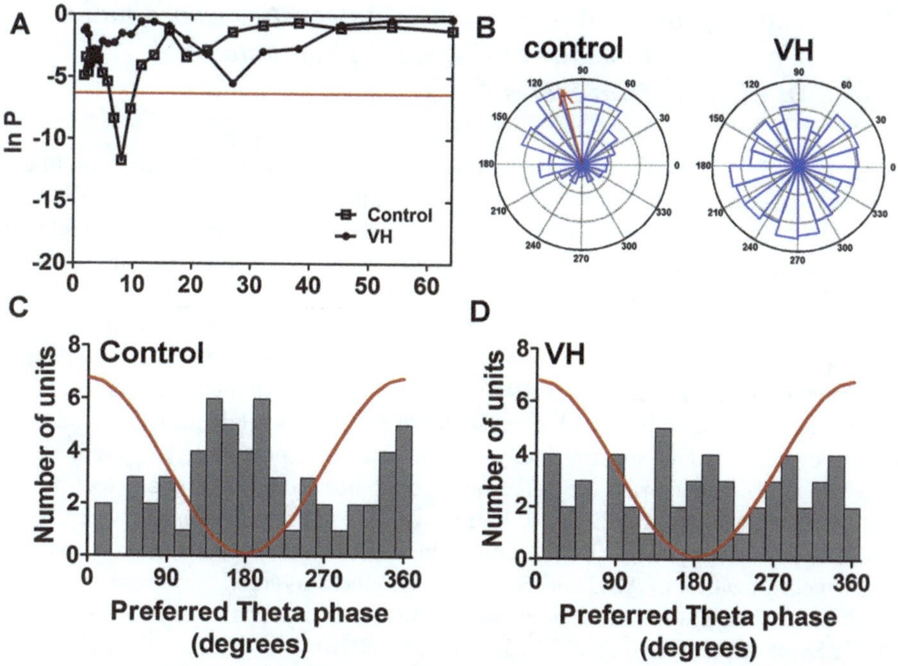

Fig. 8.7 VH disrupted the spikes phase-locking to theta band oscillation in the ACC. (**a**) Test of significance of phase-locking as a function of frequency (1–64 Hz). The threshold (red line) for significant phase-locking was set to p = 0.0023 (0.05/22, Bonferroni corrected). The shown phase-locked neuron in the control rat exhibited maximal phase-locking at 8.0 Hz while the other un-phase-locked neuron in the VH rat showed no significant phase-locking in any frequencies. (**b**) The polar-histogram of the spike-field phase distribution of the phase-locked neuron from the control rat and un-phase-locked neuron from VH rat, which are shown in (**a**). The mean phase showed by red arrows also indicated this neuron preferred firing at 105° of the theta oscillation, while the other un-phase-locked neuron in VH rat showed no significant phase-locking in theta range. (**c**) Histogram of the preferred phase of all phase locked neurons (n = 59 of 114) recorded in the 6 control rats. The figure shows most neurons preferred to fire during the descending phase and at the trough of the oscillations. The red line is a schematic of one theta cycle. (**d**) The phase locked neuron (n = 54 of 217) recorded in 6 VH rats, however, fired action potential at random angles of the theta cycle in the oscillations suggesting disrupted phase-locking relationship between action and field potentials in rats following VH (Adapted from Cao et al., Experimental Neurology 136 (2016) 74–85. With permission.)

neuronal communications, modifies synaptic weights between anatomically distant, but functionally associated brain regions, and related to even behavioral outputs (Xu et al. 2015; Mu et al. 2015; Cao et al. 2016; Cardoso-Cruz et al. 2013).

8.4.2 *Interruption of Amygdala-ACC Integrative Coordination Contribute Causally to Cognitive Dysfunctions in Chronic Pain States*

It has been shown that the basolateral amygdala (BLA) and the ACC form an interconnected neural circuit that may mediate certain types of decision-making processes (Floresco and Ghods-Sharifi 2007). Recently, it was reported that pain related hyperactivity of basolateral amygdala neurons mediates decision making deficits through the amygdala-prefrontal cortex circuit, suggesting that cognitive impairment is caused by amygdala-driven prefrontal cortical deactivation. The reciprocal connections between the BLA and medial PFC including the ACC have been clearly exhibited previously (Bacon et al. 1996). We have performed multiple-channel electrophysiological recordings and adopt standard multi-channel data analyses, such as cross-correlations and spectral analyses for local field potential (LFP) and spike recordings, to characterize the spike-field coherence (SFC), and phase locking of individual neurons to the theta oscillation within each regions, and between ACC and basolateral amygdala (BLA). Our published data showed that phase-locking and synchronization in ACC and between ACC and amygdala play a major role in modulation of cognition function in various preclinical animal models (Wang et al. 2015; Xu et al. 2015; Mu et al. 2015; Cao et al. 2016).

In viscerally hypersensitive rats, recordings of field potential showed facilitation of basal synaptic transmission in the BLA-ACC pathway, suggesting up-regulation of long lasting synaptic transmission in the ACC neural circuitry following induction of visceral hypersensitivity (Zeeb and Winstanley 2011). Previous study showed that BLA efferent exerted a predominantly inhibitory effect (Perez-Jaranay and Vives 1991). In line with this observation, recent study showed that there was a reliable induction of LTP at the BLA-ACC synapses in normal rats. However, the LTP in the BLA-ACC synapses was blocked in VH rats (Zeeb and Winstanley 2011). It appears that induction of visceral hypersensitivity produces a change in the ability to induce subsequent synaptic plasticity at the BLA-ACC pathway. Further, power spectral density analysis showed an increase in accumulative power of the theta band of LFP in both the BLA and ACC in VH rats that was associated with a marked decrease of theta peak frequency (Flood et al. 1987). In fact, the increases in theta power and the shift of the dominant peak of theta to lower frequencies have been proposed as markers of cognitive decline in chronic pain (Cardoso-Cruz et al. 2013; Sarnthein et al. 2006). Cross-correlation analysis revealed visceral hypersensitivity led to suppressed synchronization of theta oscillation between the BLA and ACC (Fig. 8.8) suggesting that they loosely interact for dynamic information transfer, which may in turn disrupt neural network assemblies and affect synaptic plasticity. Finally, we observed suppressed locking of ACC spikes to the phase of the theta oscillations in the BLA in the VH rats. The SFC analysis is independent of the LFP power spectrum and the number of spikes, and is therefore immune to changes in these parameters. These findings are particularly intriguing in view of the recent findings that a tight coordination of spike timing with the local theta oscillation is a key index for predicting successful memory formation in humans (Rutishauser et al. 2010).

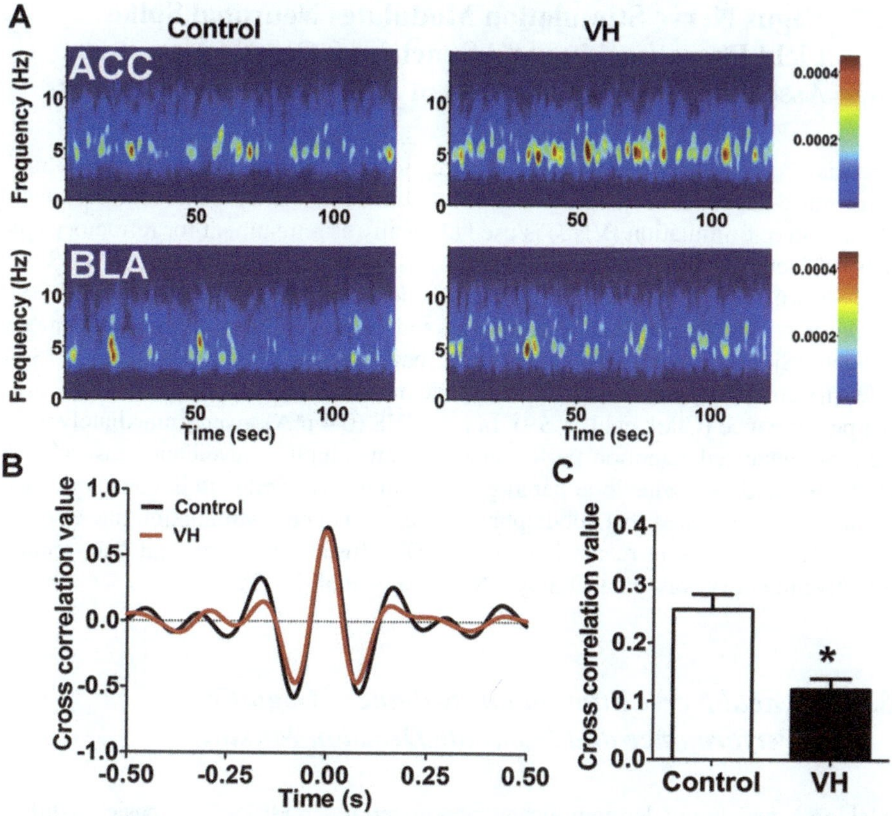

Fig. 8.8 Synchronization between theta oscillations in the basal lateral amygdala (BLA) and anterior cingulate cortex (ACC). (**a**) Typical colored power spectrograms (120 s duration) recorded from ACC (top) and BLA (bottom) in control rats (left) and VH rats (right). Note that theta power decreased in the ACC but increased in the BLA of VH rat compared with control rat. (**b**) The averaged cross-correlograms in control and VH rats at quiet waking state. (**c**) Statistic analysis revealed that the cross-correlation value (the second positive peak in b), which is corresponding to theta activity, decreased in VH rats compared to control rats, *$p < 0.05$

8.5 Vagus Nerve Stimulation Modulates Neuronal Spike Field Phase Locking and Synchrony Cross Areas Associated with Facilitation of Decision-Making in Rats

The viscero-sensation is a faculty of perception that does not depend on any outward sense (Zagon 2001), which acts to influence the elicited behavioral response. Vagus Nerve Stimulation (VNS) is used clinically as a treatment for refractory epilepsy (Elger et al. 2000), and resistant depression (Nemeroff et al. 2006). VNS has also shown several beneficial effects for mood enhancement (Elger et al. 2000; Beekwilder and Beems 2010), and promoted cognitive functions in Alzheimer's patients (Sjögren et al. 2002). Clark et al. have shown in human patients VNS at intensity comparable to that effective in rodents facilitated retention of verbal learning performance (Clark et al. 1999). In rats VNS (0.4 mA) given immediately after training enhanced retention performance on an inhibitory-avoidance task (Clark et al. 1998). Using behavioral paradigm to evaluate visceral pain in conscious rats, we have demonstrated that subdiaphragmatic vagus nerve stimulation has visceral analgesic properties in rats (Chen et al. 2008), furthermore, visceral pain-related affective memory was enhanced by VNS (Zhang et al. 2013).

8.5.1 Vagal Nerve Stimulation Enhances Cognitive Performance and Facilitate Decision Making

Making a decision under complicated and uncertain conditions is a basic cognitive process for adaption relying on the integration of several executive functions. In humans, decision-making has been accurately modeled using the Iowa gambling task (IGT) in the laboratory (Bechara et al. 1999, 2000). Previous report showed that subjects with spinal cord injury (second to sixth cervical vertebra) did not show dysfunctions in decision making (North and O'Carroll 2001) suggesting changes in sympathetic activity are not critical to determining somatic tone, and affect decision making. A preliminary, but impressive experimental study by Martin et al. (Martin et al. 2004) has shown that VNS improved decision making in medical refractory epileptic patients. Together, these lines of evidence provide compelling rational to hypothesize that activation of vagal afferent nerves may play an important role in the process of decision-making.

By employing a conscious rat model equipped with vagus nerve cuff electrode, we assessed ACC the role of chronic VNS on decision-making in rat gambling task (Fell and Axmacher 2011). The average food intake per body weight was not significantly different between the control (sham EVS) and EVS rat groups. Daily VNS, administered immediately following training sessions of RGT, caused an increase in 'good decision-maker' rats. The difference in the proportions of the three types of decision-making behavior (good, bad, undecided) between the two groups was significant. The mean food reward obtained during the RGT by the VNS rats was significantly more than that of the controls (Fell and Axmacher 2011).

8.5.2 Vagal Nerve Stimulation Regulates LFP and Spike Phases, Enhances Spike-Phase Coherence Between Key Brain Areas Involved in Cognitive Performance

Simultaneous multichannel-recordings offer an ideal setup to test the hypothesis that VNS may induce alterations of in both spike-field-coherence and synchronization of theta oscillations across brain areas. Indeed, it has been reported that VNS augmented theta activity in BLA and ACC (Fell and Axmacher 2011). Spike-field coherence (SFC) was used to quantify the alteration in the spike timing-LFP relationship within the BLA and ACC before and after VNS. The SFC value is expressed in percentage and varies as a function of frequency. We found a significant difference in the average SFC after VNS in the theta range. No significant changes were observed in the other three frequency bands. To further examine the functional connectivity between the ACC and BLA, we compared the LFP during 30 s spontaneous periods from before and after VNS in rats. Cross-correlation and time-varying power spectral analysis of the theta oscillations revealed a pattern of dispersion of theta band activity during the basal period, and increases in correlation values immediately following VNS. Moreover, the increased LFP-synchronization between the BLA and the ACC was also associated with greater locking of ACC spikes to the phase of the theta oscillations in the BLA (Fig. 8.9). It is clear that communication between brain areas involves phase synchronization of oscillations (Fries 2005; Palva et al. 2005). The phase of the oscillation regulates exactly when gatherings of neurons spike, thus two brain areas with increased phase synchrony will have improved synaptic interaction and information exchanges (Fell and Axmacher 2011). It appears that the sequential transfer of information via corticipetal BLA/ACC connections may guide response variety when assessing the value of an anticipated outcome (decision making) relative to the costs of a particular action.

These electrophysiological evidences unveil several important roles for VNS in regulating LFP and spike phases, as well as enhancing spike-phase coherence between key brain areas involved in cognitive performance. These data may serve to provide fundamental notions regarding neurophysiological biomarkers for therapeutic VNS in cognitive impairment. Further studies are wanted to clarify what physiological stimuli activate the vagal afferents that modulate decision-making. For instance, Cholecystokinin-octapeptide (CCK-8), which is a gastrointestinal hormone released during feeding (Li et al. 2000; Li and Owyang 1996), acts on vagal afferent fibers. Our previous electrophysiological studies in rats have demonstrated that CCK stimulates vagal afferent fibers (Li et al. 2004, 1999) to modulate various gastrointestinal functions. Flood et al., have shown that administration of CCK-8 acts on vagal afferents to enhance memory retention in the mice after aversive

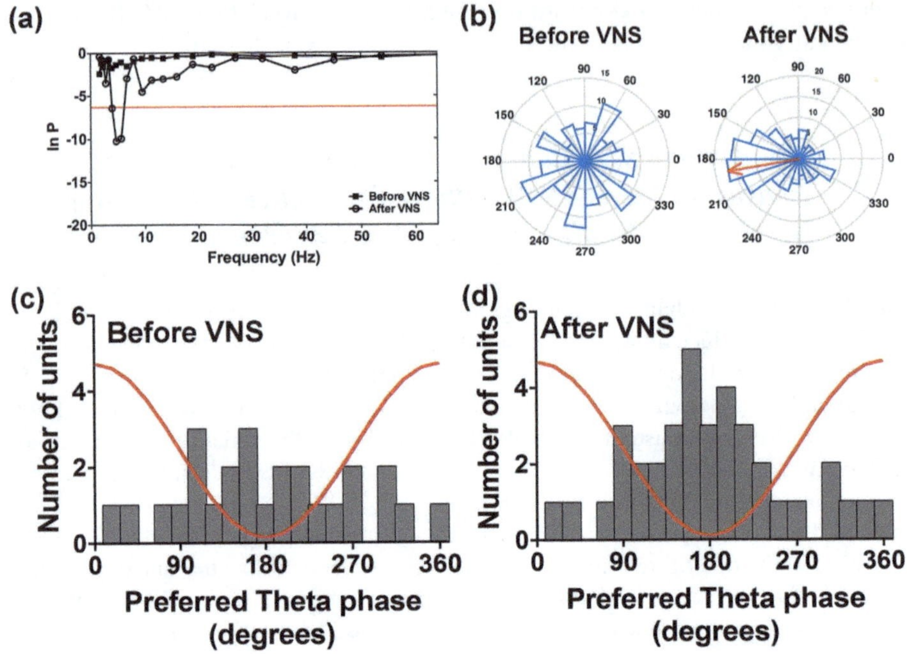

Fig. 8.9 Phase-locking of the spikes of the ACC neurons to the theta oscillations in the BLA. (**a**) Test of significance of phase-locking. The neuron which was un-phase-locked before VNS exhibited maximal phase locking at 5.65 Hz after VNS. (**b**) The polar histograms of the spike-field phase distribution of the same neuron shown in (**a**). The un-phase-locked neuron fired randomly before VNS (left panel) and became phase-locked to the theta cycle at 189° after VNS (right panel, vector length R = 0.53). (**c, d**) Histogram of the distribution of the preferred phases of all phase-locked neurons before and after VNS, n = 26 in (**c**) and n = 37 in (**d**). More phase locked neurons preferred to fire close to the trough of the oscillation (Adapted from Cao et al., Scientific Report)

training (Flood et al. 1987). Further studies are needed to determine if CCK enables or modulates cognitive function, such as decision-making, by acting on vagal afferent fibers.

8.5.3 Final Remark

The pain is likely to be reflected in a matrix of neuronal structures rather than in a fixed pain center. A "neuromatrix" incorporating the ACC, prefrontal cortex and the amygdala may be involved in the processing of pain without any single region unto itself being necessary and sufficient for the pain experience. It is conceivable that the course of the processes of neuron specialization during induction of visceral hypersensitivity are associated with changes in synaptic efficiency in the same cells during the induction of canonical LTP. This suggests the possibility that the same synaptic mechanisms are involved in the processes of modifying ACC cells, reducing the pain threshold, amplifying affective responses to pain, and processing learning and memory in patients with IBS.

The visceral pain experience may be better explained as a biopsychosocial model of pain, although most clinical specialists continue to treat visceral pain as just a symptom and not as a distinct neurological entity. In the chronic visceral pain, state hyperactivity of amygdala effectively blocks the expression of canonical long term potentiation at BLA-ACC synapses. The impaired ACC LTP and the dysfunction of ACC intra-, and between areas spike field coherence play an important role in emotional disorders, and cognitive deficits, such as decision making, in the visceral hypersensitive state.

More recent findings reveal that vagus nerve stimulation induces between-area phase synchronization in theta frequencies and elevated phase locking of neuronal spike firings to theta oscillations across regions, and are perhaps candidates for explaining the neural mechanisms underlying VNS-facilitation of decision-making. The data will serve as a basis for fundamental notions regarding neurophysiological biomarkers for the development of novel therapeutics of cognitive deficits in the chronic visceral pain.

Acknowledgements This work was supported by the National Institute of Neurological Disorders and Stroke grant RO1 NS051466-01 (to Y.L.). National Institutes of Health Grant: RO1 DK51717 (to Y.L.).

Research Grants Council of Hong Kong (grant number 11166116, 11101315, 11100914 and CityU number 160811, 160812, and 160713 to Y. Li), the Innovation and Technology Support Programme (ITS/300/15 to Y. Li), Health and Medical Research Fund of Hong Kong (01122006 to Y. Li0 and City University of Hong Kong Neuroscience Research Infrastructure Grant (9610211 to Y. Li). This work was also supported by City University of Hong Kong Centre for Biosystems, Neuroscience, and Nanotechnology Grant (9360148 to S. Pang and Y. Li). I thanks Prof Chung Owyang Gastroenterology Research Unit, Department of Internal Medicine, University of Michigan, for his long term support.

References

Arthurs OJ, Boniface S. How well do we understand the neural origins of the fMRI BOLD signal? Trends Neurosci. 2002;25:27–31.

Bacon SJ, Headlam AJ, Gabbott PL, Smith AD. Amygdala input to medial prefrontal cortex (mPFC) in the rat: a light and electron microscope study. Brain Res. 1996;720:211–9.

Bayer KU, De Koninck P, Leonard AS, Hell JW, Schulman H. Interaction with the NMDA receptor locks CaMKII in an active conformation. Nature. 2001;411:801–5.

Bechara A, Damasio H, Damasio AR, Lee GP. Different contributions of the human amygdala and ventromedial prefrontal cortex to decision-making. J Neurosci. 1999;19:5473–81.

Bechara A, Damasio H, Tranel D, Damasio AR. Deciding advantageously before knowing the advantageous strategy. Science. 1997;275:1293–5.

Bechara A, Tranel D, Damasio H. Characterization of the decision-making deficit of patients with ventromedial prefrontal cortex lesions. Brain. 2000;123(Pt 11):2189–202.

Beekwilder JP, Beems T. Overview of the clinical applications of vagus nerve stimulation. J Clin Neurophysiol. 2010;27:130–8.

Blaisdell AP, Sawa K, Leising KJ, Waldmann MR. Causal reasoning in rats. Science. 2006;311:1020–2.

Bliss TV, Collingridge GL. A synaptic model of memory: long term potentiation in the hippocampus. Nature. 1993;361:31–9.

Bussey TJ, Everitt BJ, Robbins TW. Dissociable effects of cingulate and medial frontal cortex lesions on stimulus-reward learning using a novel Pavlovian autoshaping procedure for the rat: implications for the neurobiology of emotion. Behav Neurosci. 1997;111:908–19.

Cao B, Wang J, Zhang X, Yang X, Poon D, Jelfs B, Li Y. Impairment of decision making and disruption of synchrony between basolateral amygdala and anterior cingulate cortex in the maternally separated rat. Neurobiol Learn Mem. 2016;136:74–85.

Cao ZJ, Wu XY, Chen SL, Owyang C, Li Y. Anterior cingulate cortex modulates visceral pain as measured by visceromotor responses in viscerally hypersensitive rats. Gastroenterology. 2008;134:535–43.

Cardoso-Cruz H, Sousa M, Vieira JB, Lima D, Galhardo V. Prefrontal cortex and mediodorsal thalamus reduced connectivity is associated with spatial working memory impairment in rats with inflammatory pain. Pain. 2013;154:2397–406.

Chapman CA, Trepel C, Ivanco TL, Froc DJ, Wilson K, Racine RJ. Changes in field potentials and membrane currents in rat sensorimotor cortex following repeated tetanization of the corpus callosum in vivo. Cereb Cortex. 1998;8:730–42.

Chen SL, Wu XY, Cao ZJ, Fan J, Wang M, Owyang C, Li Y. Subdiaphragmatic vagal afferent nerves modulate visceral pain. Am J Phys. 2008;294:G1441–9.

Clark KB, Naritoku DK, Smith DC, Browning RA, Jensen RA. Enhanced recognition memory following vagus nerve stimulation in human subjects. Nat Neurosci. 1999;2:94–8.

Clark KB, Smith DC, Hassert DL, Browning RA, Naritoku DK, Jensen RA. Posttraining electrical stimulation of vagal afferents with concomitant vagal efferent inactivation enhances memory storage processes in the rat. Neurobiol Learn Mem. 1998;70:364–73.

Dougherty PM, Palecek J, Paleckova V, Sorkin LS, Willis WD. The role of NMDA and non-NMDA excitatory amino acid receptors in the excitation of primate spinothalamic tract neurons by mechanical, chemical, thermal, and electrical stimuli. J Neurosci. 1992;12:3025–41.

Dudek SM, Bear MF. Homosynaptic long-term depression in area CA1 of hippocampus and effects of N-methyl-D-aspartate receptor blockade. Proc Natl Acad Sci U S A. 1992;89:4363–7.

Elger G, Hoppe C, Falkai P, Rush AJ, Elger CE. Vagus nerve stimulation is associated with mood improvements in epilepsy patients. Epilepsy Res. 2000;42:203–10.

Fan J, Wu XY, Cao ZJ, Chen SL, Owyang C, Li Y. Upregulation of anterior cingulate cortex NR2B receptors contributes to visceral pain as measured by visceromotor responses in rats. Gastroenterology. 2009;136:1732–40.

Fell J, Axmacher N. The role of phase synchronization in memory processes. Nat Rev Neurosci. 2011;12:105–18.

Flood JF, Smith GE, Morley JE. Modulation of memory processing by cholecystokinin—dependence on the vagus nerve. Science. 1987;236:832–4.

Floresco SB, Ghods-Sharifi S. Amygdala-prefrontal cortical circuitry regulates effort-based decision making. Cereb Cortex. 2007;17:251–60.

Friedrich M, Grady SE, Wall GC. Effects of antidepressants in patients with irritable bowel syndrome and comorbid depression. Clin Ther. 2010;32:1221–33.

Fries P. A mechanism for cognitive dynamics: neuronal communication through neuronal coherence. Trends Cogn Sci. 2005;9:474–80.

Gao J, Wu XY, Owyang C, Li Y. Enhanced responses of the anterior cingulate cortex neurons to colonic distension in viscerally hypersensitive rats. J Physiol. 2006;570:169–84.

Gebhart GF. Pathobiology of visceral pain: molecular mechanisms and therapeutic implications IV. Visceral afferent contributions to the pathobiology of visceral pain. Am J Physiol Gastrointest Liver Physiol. 2000;278:G834–8.

Gebhart GF. Descending modulation of pain. Neurosci Biobehav Rev. 2004;27:729–37.

Giovanni B, Stanghellini V, De Giorgio R, Cremon C, Cottrell GS, Santini D, et al. Activated mast cells in proximity to colonic nerves correlate with abdominal pain in irritable bowel syndrome. Gastroenterology. 2004;126:693–702.

Harris KD, Csicsvari J, Hirase H, Dragoi G, Buzsáki G. Organization of cell assemblies in the hippocampus. Nature. 2003;424:552–6.

Jacobs J, Kahana MJ, Ekstrom AD, Fried I. Brain oscillations control timing of single-neuron activity in humans. J Neurosci. 2007;27:3839–44.

Johansen JP, Fields HL. Glutamatergic activation of anterior cingulate cortex produces an aversive teaching signal. Nat Neurosci. 2004;7:398–403.

Johansen JP, Fields HL, Manning BH. The affective component of pain in rodents: direct evidence for a contribution of the anterior cingulate cortex. Proc Natl Acad Sci U S A. 2001;98:8077–82.

Larsson MB, Tillisch K, Craig AD, Engstrom M, Labus J, Naliboff B, et al. Brain responses to visceral stimuli reflect visceral sensitivity thresholds in patients with irritable bowel syndrome. Gastroenterology. 2012;142:463–72.

Li Y, Hao Y, Owyang C. Diazepam-binding inhibitor mediates feedback regulation of pancreatic secretion and postprandial release of cholecystokinin. J Clin Invest. 2000;105:351–9.

Li Y, Owyang C. Peptone stimulates CCK-releasing peptide secretion by activating intestinal submucosal cholinergic neurons. J Clin Invest. 1996;97:1463–70.

Li Y, Wu XY, Owyang C. Serotonin and cholecystokinin synergistically stimulate rat vagal primary afferent neurones. J Physiol. 2004;559:651–62.

Li Y, Zhang X, Liu H, Cao Z, Chen S, Cao B, et al. Phosphorylated CaMKII post-synaptic binding to NR2B subunits in the anterior cingulate cortex mediates visceral pain in visceral hypersensitive rats. J Neurochem. 2012;121:662–71.

Li Y, Zhu JX, Owyang C. Electrical physiological evidence for high- and low-affinity vagal CCK-A receptors. Am J Physiol Gastrointest Liver Physiol. 1999;277:G469–77.

Lisman J, Schulman H, Cline H. The molecular basis of CaMKII function in synaptic and behavioural memory. Nat Rev Neurosci. 2002;3:175–90.

Longstreth GF, Thompson WG, Chey WD, Houghton LA, Mearin F, Spiller RC. Functional bowel disorders. Gastroenterology. 2006;130:1480–91.

Lotze M, Wietek B, Birbaumer N, Ehrhardt J, Grodd W, Enck P. Cerebral activation during anal and rectal stimulation. NeuroImage. 2001;14:1027–34.

Markram H, Lubke J, Frotscher M, Sakmann B. Regulation of synaptic efficacy by coincidence of postsynaptic APs and EPSPs. Science. 1997;275:213–5.

Martin CO, Denburg NL, Tranel D, Granner MA, Bechara A. The effects of vagus nerve stimulation on decision-making. Cortex. 2004;40:605–12.

Martin SJ, Grimwood PD, Morris RG. Synaptic plasticity and memory: an evaluation of the hypothesis. Ann Rev Neurosci. 2000;23:649–711.

Mayer EA, Derbyshire S, Naliboff BD. Cerebral activation in irritable bowel syndrome. Gastroenterology. 2000;119:1418–9.

Mertz H, Morgan V, Tanner G, Pickens D, Price R, Shyr Y, Kessler R. Regional cerebral activation in irritable bowel syndrome and control subjects with painful and nonpainful rectal distension. Gastroenterology. 2000;118:842–8.

Monyer H, Burnashev N, Laurie DJ, Sakmann B, Seeburg PH. Developmental and regional expression in the rat brain and functional properties of four NMDA receptors. Neuron. 1994;12:529–40.

Mu L, Wang J, Cao B, Jelfs B, Chan RHM, Xu X, Li Y. Impairment of cognitive function by chemotherapy: association with the disruption of phase-locking and synchronization in anterior cingulate cortex. Mol Brain. 2015;8:32. https://doi.org/10.1186/s13041-015-0125-y.

Nanda R, James R, Smith H, Dudley CRK, Jewell DP. Food intolerance and the irritable bowel syndromes. Gut. 1989;30:1098–104.

Nemeroff CB, Mayberg HS, K McNamara J, Frazer A, Henry TR, George MS, et al. VNS therapy in treatment-resistant depression: clinical evidence and putative neurobiological mechanisms. Neuropsychopharmacology. 2006;31:1345–55.

Ness TJ, Gebhart GF. Visceral pain: a review of experimental studies. Pain. 1990;41:167–234.
Nestler EJ. Special issue: animal models of mood and psychotic disorders. Biol Psychiatry. 2006;59:1103.
Nishida M, Hirai N, Miwakeichi F, Maehara T, Kawai K, Shimizu H, et al. Theta oscillation in the human anterior cingulate cortex during all-night sleep: an electrocorticographic study. Neurosci Res. 2004;50:331–41.
North NT, O'Carroll RE. Decision making in patients with spinal cord damage: afferent feedback and the somatic marker hypothesis. Neuropsychologia. 2001;39:521–4.
Nozdrachev AD, Akoev GN, Filippova LV, Sherman NO, Lioudyno MI, Makarov FN. Changes in afferent impulse activity of small intestine mesenteric nerves in response to antigen challenge. J Neurosci. 1999;94:1339–42.
Palva JM, Palva S, Kaila K. Phase synchrony among neuronal oscillations in the human cortex. J Neurosci. 2005;25:3962–72.
Perez-Jaranay JM, Vives F. Electrophysiological study of the response of medial prefrontal cortex neurons to stimulation of the basolateral nucleus of the amygdala in the rat. Brain Res. 1991;564:97–101.
Rushworth MF, Behrens TE, Rudebeck PH, Walton ME. Contrasting roles for cingulate and orbitofrontal cortex in decisions and social behaviour. Trends Cogn Sci. 2007;11:168–76.
Rutishauser U, Ross IB, Mamelak AN, Schuman EM. Human memory strength is predicted by theta-frequency phase locking of single neurons. Nature. 2010;464:903–7.
Sah P, Nicoll RA. Mechanisms underlying potentiation of synaptic transmission in rat anterior cingulate cortex in vitro. J Physiol. 1991;433:615–30.
Sarnthein J, Stern J, Aufenberg C, Rousson V, Jeanmonod D. Increased EEG power and slowed dominant frequency in patients with neurogenic pain. Brain. 2006;129:55–64.
Scott RB, Tan DT, Miampamba M, Sharkey KA. Anaphylaxis-induced alterations in intestinal motility: role of extrinsic neural pathways. Am J Phys. 1998;275:G812–21.
Shyu BC, Vogt BA. Short-term synaptic plasticity in the nociceptive thalamic-anterior cingulate pathway. Mol Pain. 2009;5:51. https://doi.org/10.1186/1744-8069-5-51.
Sidhu H, Kern M, Shaker R. Absence of increasing cortical fMRI activity volume in response to increasing visceral stimulation in IBS patients. Am J Physiol Gastrointest Liver Physiol. 2004;287:G425–35.
Silverman DH, Munakata JA, Ennes H, Mandelkern MA, Hoh CK, Mayer EA. Regional cerebral activity in normal and pathological perception of visceral pain. Gastroenterology. 1997;112:64–72.
Sjögren MJ, Hellstrom PT, Jonsson MA, Runnerstam M, Silander HC, Ben-Menachem E. Cognition-enhancing effect of vagus nerve stimulation in patients with Alzheimer's disease: a pilot study. J Clin Psychiatry. 2002;63:972–80.
Tang YP, Shimizu E, Dube GR, Rampon C, Kerchner GA, Zhuo M, et al. Geneti enhancement of learning and memory in mice. Nature. 1999;401:63–9.
Tolle TR, Kaufmann T, Siessmeier T, Lautenbacher S, Berthele A, Munz F, et al. Region-specific encoding of sensory and affective components of pain in the human brain: a positron emission tomography correlation analysis. Ann Neurol. 1999;45:40–7.
Vagt BA. Pain and emotion interactions in subregions of the cingulate gyrus. Nat Rev Neurosci. 2005;6:533–44.
Varela F, Lachaux JP, Rodriguez E, Martinerie J. The brain web: phase synchronization and large-scale integration. Nat Rev Neurosci. 2001;2:229–39.
Verne GN, Robinson ME, Price DD. Hypersensitivity to visceral and cutaneous pain in the irritable bowel syndrome. Pain. 2001;93:7–14.
Vogt BA, Robert WS. Anterior cingulate cortex and the medial pain system. In: Vogt BA, Gabriel M, editors. Neurobiology of cingulate cortex and limbic thalamus: a comprehensive handbook. Boston: Birkhauser; 1993. p. 313–44.
Vogt BA, Vogt LJ, Farber NB. Cingulate cortex and disease models. In: Paxinos G, editor. The rat nervous system. 3rd ed. San Diego: Elsevier; 2003. p. 705–27.

Wang J, Cao B, Yu TR, Yan J, Li Y. Theta-frequency phase-locking of single anterior cingulate cortex neurons and the synchronization with the medial thalamus are modulated by visceral noxious stimulation in rat. Neuroscience. 2015;298:200–10.

Wang J, Zhang X, Cao B, Liu J, Li Y. Facilitation of synaptic transmission in the anterior cingulate cortex in the viscerally hypersensitive rats. Cereb Cortex. 2013. https://doi.org/10.1093/cercor/bht273.

Wei F, Wang GD, Kerchner GA, Kim SJ, Xu HM, Chen ZF, et al. Genetic enhancement of inflammatory pain by forebrain NR2B overexpression. Nat Neurosci. 2001;4:164–9.

Wu XY, Gao J, Yan J, Fan J, Owyang C, Li Y. Role for NMDA receptors in viscera nociceptive transmission in the anterior cingulate cortex of viscerally hypersensitive rats. Am J Phys. 2008;294:918–27.

Xu X, Cao B, Wang J, Yu T, Li Y. Decision-making deficits associated with disrupted synchronization between basolateral amygdala and anterior cingulate cortex in rats after tooth loss. Prog Neuro-Psychopharmacol Biol Psychiatry. 2015;60:26–35.

Yan N, Cao B, Xu J, Hao C, Zhang X, Li Y. Glutamatergic activation of anterior cingulate cortex mediates the affective component of visceral pain memory in rats. Neurobiol Learn Mem. 2012;97:156–64.

Zagon A. Does the vagus nerve mediate the sixth sense? Trends Neurosci. 2001;24:671–3.

Zeeb FD, Winstanley CA. Lesions of the basolateral amygdala and orbitofrontal cortex differentially affect acquisition and performance of a rodent gambling task. J Neurosci. 2011;31:2197–204.

Zhang X, Cao B, Yan N, Liu J, Wang J, Tung VO, Li Y. Vagus nerve stimulation modulates visceral pain-related affective memory. Behav Brain Res. 2013;236:8–15.

Zhao MG, Toyoda H, Lee YS, Wu LJ, Ko SW, Zhang X–H, et al. Roles of NMDA NR2B subtype receptor in prefrontal long-term potentiation and contextual fear memory. Neuron. 2005;47:859–72.

Chapter 9
Large De Novo Microdeletion in Epilepsy with Intellectual and Developmental Disabilities, with a Systems Biology Analysis

Kai Gao, Yujia Zhang, Ling Zhang, Weijing Kong, Han Xie, Jingmin Wang, Ye Wu, Xiru Wu, Xiaoyan Liu, Yuehua Zhang, Feng Zhang, Albert Cheung-Hoi Yu, and Yuwu Jiang

9.1 Introduction

Epilepsy is a disease which arises from largely unknown cellular and genetic mechanisms. It is a common neurological disease that reflects neuronal hyperexcitability induced by many different factors such as trauma, neurotoxicity and genetic variation (Lee and Heo 2014). ID/DD is one of the most common pediatric neurological diseases and is also one of the most important unsolved problems in health care. Studies have shown that the prevalence rate of ID/DD is 1–3% (Chelly et al. 2006). It is estimated that approximately 30% of patients with ID/DD have seizures (Tuchman et al. 2009). These associations indicate that epilepsy shares a similar pathogenic mechanism with those diseases in some situations (Williams et al. 2009; Cooper et al. 2011; Grayton et al. 2012).

Electronic supplementary material: The online version of this chapter (https://doi.org/10.1007/978-3-319-94593-4_9) contains supplementary material, which is available to authorized users.

K. Gao · Y. Zhang · W. Kong · H. Xie · J. Wang · Y. Wu · X. Wu · X. Liu · Y. Zhang · Y. Jiang (✉)
Department of Pediatrics, Peking University First Hospital, Peking University Health Science Center, Beijing, China
e-mail: jiangyuwu@bjmu.edu.cn

L. Zhang · F. Zhang
Institute of Reproduction and Development, Obstetrics and Gynecology Hospital of Fudan University, Shanghai, China

A. C. H. Yu
Laboratory for Functional Study of Astrocytes, Department of Neurobiology, School of Basic Medical Sciences, Key Laboratory for Neuroscience, Ministry of Education, National Health and Family Planning Commission, Neuroscience Research Institute, Peking University, Beijing, China

Hai Kang Life Corporation Ltd., China

A. Cheung-Hoi Yu, L. Li (eds.), *Systems Neuroscience*, Advances in Neurobiology 21, https://doi.org/10.1007/978-3-319-94593-4_9

Copy number variation (CNV) has been reported to be associated with a group of neuropsychiatric disorders, including epilepsy and ID/DD. Recent studies of CNV in patients with epilepsy have revealed a series of CNV hotspots, such as 1q21.1 (de Kovel et al. 2010; Mefford et al. 2010), 15q11.2 (Zhang et al. 2009; de Kovel et al. 2010; Mefford et al. 2010; Vanlerberghe et al. 2015), 15q13.3 (Mulley et al. 2011; Sisodiya and Mefford 2011; Kogan et al. 2015), 16p11.2 (Mefford et al. 2011; Bassuk et al. 2013; Tiwari et al. 2013), 16p13.11 (Sisodiya and Mefford 2011) and 22q11.2 (Helbig et al. 2013; Kim et al. 2016). At the same time, a group of ID/DD-related CNV hotspots was been found, as 1q21.1 (Harvard et al. 2011), 2q13 (Yu et al. 2012; Riley et al. 2015), 15q11.2 (Derks et al. 2013; Caciotti et al. 2015), 16p11.2 (Bassuk et al. 2013; Derks et al. 2013), 22q11.2 (Mertz et al. 2013; Olszewski et al. 2014). The overlap of those CNV hotpots between epilepsy and ID/DD, indicate these two disease share a similar pathogenic genetic mechanism.

To elucidate whether CNV is a causal factor in epilepsy with ID/DD in Chinese children, we utilized a custom high-density oligonucleotide-based comparative genomic hybridization (CGH) microarrays to detect the CNVs in 96 epilepsy patients with ID/DD.

9.2 Large De Novo Rare Microdeletion Is an Important Pathological Cause of Epilepsy with ID/DD

9.2.1 Ethics and Patients

The study protocol was approved by Medical Ethics Committee of Peking University First Hospital. Informed consent was obtained from the parents. All data of this study were analyzed anonymously. DNA samples were collected from 96 epileptic patients with ID/DD and from their parents. All of the patients were recruited from the Department of Pediatrics, Peking University First Hospital from 2006 to 2014. These samples were prepared from a collection of whole blood samples by DNeasy Blood & Tissue Kit (QIAGEN).

Patient with both epilepsy and ID/DD who fulfilled the following inclusion criteria were assumed to be cryptogenic: (1). no perinatal brain injury (2). no hypoxia, ischemia, trauma or infection of the central nervous system (CNS); (3). no evidence of typical inherited metabolic disorder or specific neurodegenerative disorders, as found by physical examination, cranial neuroimaging and blood/urinary metabolic diseases screening; (4). negative from a gene screen by 300 epilepsy gene panel (Zhang et al. 2015). Finally, according to the inclusion criteria, 96 participating Han ethnicity patients were recruited from Peking University First Hospital.

Table 9.1 Large CNVs identified in 8 of 96 individuals affected by epilepsy and ID/DD

Sample	CNV locus	Start	Stop	Size (bp)	Gene number	de novo or inherited
3940	2q24.1	157,183,677	159,479,627	2,295,951	10	de novo
2332	2q33.1-q34	199,750,753	210,748,712	10,997,960	79	de novo
1549	5q13.2	68,830,699	70,600,323	1,769,625	15	de novo
	Xp22.31	6,705,268	7,942,835	1,237,568	4	de novo
5332	5q13.2	68,828,322	69,732,251	903,930	12	de novo
5319	5q33.1-q34	151,040,072	160,070,141	9,030,070	51	de novo
1277	17p13.3	2,165,369	3,058,821	893,453	17	de novo
1583	22q11.21-q11.22	19,058,829	21,360,978	2,302,150	48	de novo
3568	22q11.21-q11.22	19,058,829	21,360,978	2,302,150	48	de novo

9.2.2 CNV Detection by Array CGH

To detect the changes of CNVs in the genomic DNA, we applied high-density oligonucleotide-based CGH microarrays, a custom-designed Agilent SurePrint G3 Microarray (4 × 180K) was used to verify CNVs. The high-density areas covered the known epilepsy associated genes or related chromosome loci (including genes and CNVs in epilepsy including early infantile epileptic encephalopathy and idiopathic generalized epilepsy, listed in Supplementary Table 9.1). DNA digestion, Cy5-dUTP or Cy3-dUTP labeling, purification, array hybridization, washing, scanning, and data analysis were conducted by following the Agilent oligonucleotide aCGH protocol (version 6.3).

We performed whole-genome array CGH in a series of 96 patients. All of them had a presenting diagnosis of epilepsy with ID/DD. Our goal was to discover novel CNVs associated with epilepsy and ID/DD. In this study, we gathered data from whole-genome analysis and extended our analysis to other idiopathic epilepsy syndromes, such as infantile spasms and early onset epileptic encephalopathy (EOEE). In total 96 patients, we identified 8 individuals (8.3%) with 9 long rare microdeletions. If the CNV is larger than 500 kb, it will be identified as a long/large one.

9.2.3 The Loci of Microdeletions

In this study, we identified 8 of 96 (8.3%) patients with 1 or 2 large microdeletions (more than 500 kb). The biggest deletion was about 11 Mb, and the smallest was 893 kb. The mean CNV size was 3.5 Mb and the median size was 2.9 Mb. The number of deleted genes in each patient was from 12 to 79 (Table 9.1). Figure 9.1 shows the loci of the CNVs in the genome. There were two identical microdeletions at 22q11.21-q11.22 (Fig. 9.2) and two similar microdeletions at 5q13.2 (Fig. 9.3). The other CNVs were located on 2q33.1-q34, 2q24.1, 5q33.1-q34, 17p13.2, and Xp22.31 (Fig. 9.4). All the CNVs were de novo and heterozygous. Microdeletion of 5q13.2

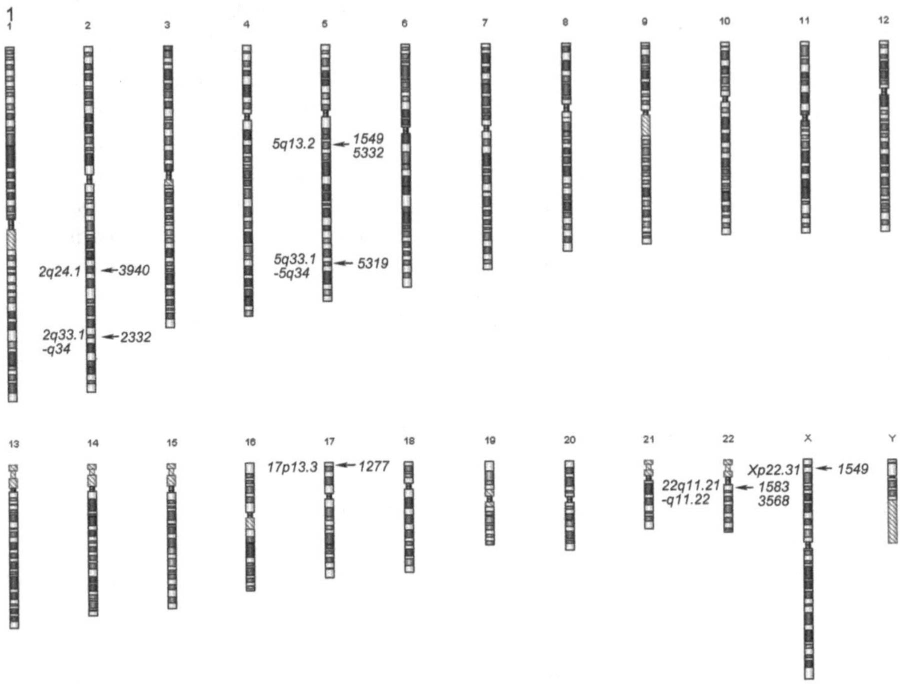

Fig. 9.1 The loci of the 9 CNVs of our cohort in the genome. 2 CNVs (2q24.1 of patient 3940 and 2q33.1-q34 of patient 2332) in chromosome 2, 3 CNVs (5q13.2 of patient 1549 and 5332, 5q33.1-q34 of patient 5319), 1 CNV (17p13.3 of 1277) in chromosome 17, 2 CNVs (22q11.21-q11.22 of patient 1583 and 3568) in chromosome 22, 1 CNV (Xp22.31 of patient 1539)

and 17p13.2 was not found in patients of epilepsy with ID/DD before. The clinical features of the patients with large microdeletion were summarized in Table 9.2. Of the 8 patients with large microdeletions, 5 were male and 3 were female. Besides epilepsy and ID/DD, the phenotypes of the patients were diverse. Six out of 7 patients who had a MRI scan had encephalodysplasia. Of 5 patients who had a psychiatric test, 2 patients suffered from autism. For craniofacial characteristics, 3 patients had facial dysmorphism, 1 patient had cleft lip/palate, and 2 patients had strabismus (Tables 9.2 and 9.3).

From these results, most of patients have only one large microdeletions. Those CNVs should course epilepsy and ID/DD by two different situations, one is the CNVs have both epilepsy-related genes and ID/DD related genes, the other is the CNVs have one or more genes associating both epilepsy and ID/DD. We also found that most of the long rare CNVs in this study were not in the most well-known idiopathic epilepsy CNV hotspots, such as 1q21.1, 15q11.2, 15q13.3, 16p11.2, 16p13.11, 22q11.2 (Table 9.4). The reason for this difference might be that the known CNV hotpots come from studies of idiopathic epilepsy, while our patients suffered from both epilepsy and ID/DD. Most of them are epileptic encephalopathy. The reported CNV regions of epileptic encephalopathy, such as 1q36, 2q32.3,

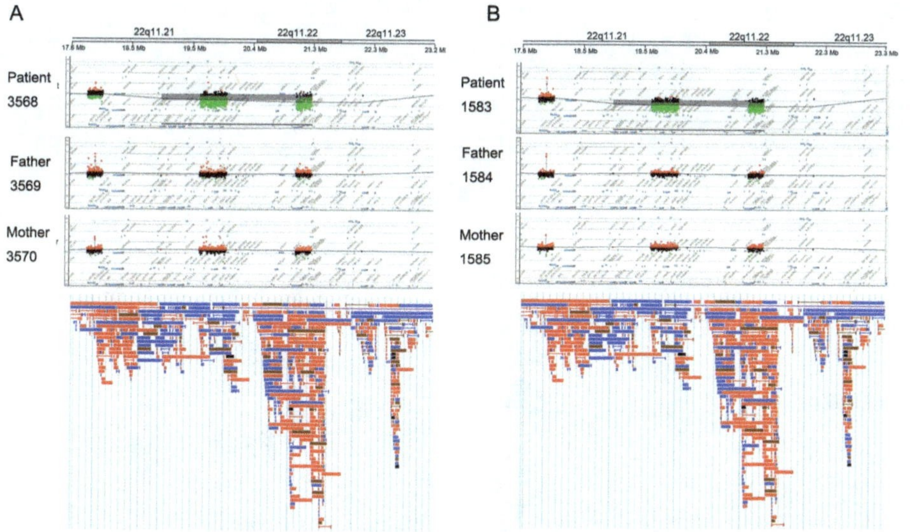

Fig. 9.2 The 2 same CNV in 22q11.21-q11.22. (**a**) The microdeletion in 22q11.21-q11.22 of patient 3568. Compared with the genome DNA of his father and mother, the CNVs is de novo. From the data of SV database in this area, the CNV is rare. (**b**) The microdeletion in 22q11.21-q11.22 of patient 1583. Compared with the genome DNA of his father and mother, the CNVs is de novo. From the data of SV database in this area, the CNV is rare

2q24.3, 3q11, 4q3.1-q3.2, 7q11.23, 14q12, 15q11-13, 16p11.2, Xp22 (Table 9.5) are not in accordance with idiopathic epilepsy.

9.3 Pathogenic Mechanism Analysis

We analyzed the genes in the CNVs and discovered some genes which were candidate pathogenic genes. We screened for candidates by determining whether genes were epilepsy/seizure related, ID/DD related, synapse related, ion channel/receptor related, transmitter related, and neurodevelopment related, or having high expression in the CNS (Table 9.3). We found that 4 out of the 9 CNVs included epilepsy related genes, while 6 out of the 9 CNVs included reported ID/DD related genes. Besides the known epilepsy- or ID/DD-related genes, some novel candidate genes might be involved in epilepsy with ID/DD: *NR4A*, *KCTD18*, *TRAK2*, *UNC80*, *CASP8*, *NRP2*, *KLF7*, *OCLN*, *SMN1*, *SMN2*, *NAIP*, *ADRA1B*, *HAND1*, *SRR*, *PAFAH1B1*, *SEPT5*, *RTN4R*, *TBX1*, *ARVCF*, *RTN4R* and *CRKL* (Table 9.6).

Patient 3940 with a 2q24.1 deletion was suffering from autism with epilepsy and ID/DD. 2q24.1 was reported to be involved in juvenile myoclonic epilepsy (Layouni et al. 2010), ID/DD (Daoud et al. 2009) and schizophrenia (Yamada et al. 2012). Therefore, 2q24.1 should be a potential CNV locus for neuropsychiatric disorders. The pathogenic genes in this patient could be *GPD2* and *NR4A2*. *GPD2*, which is

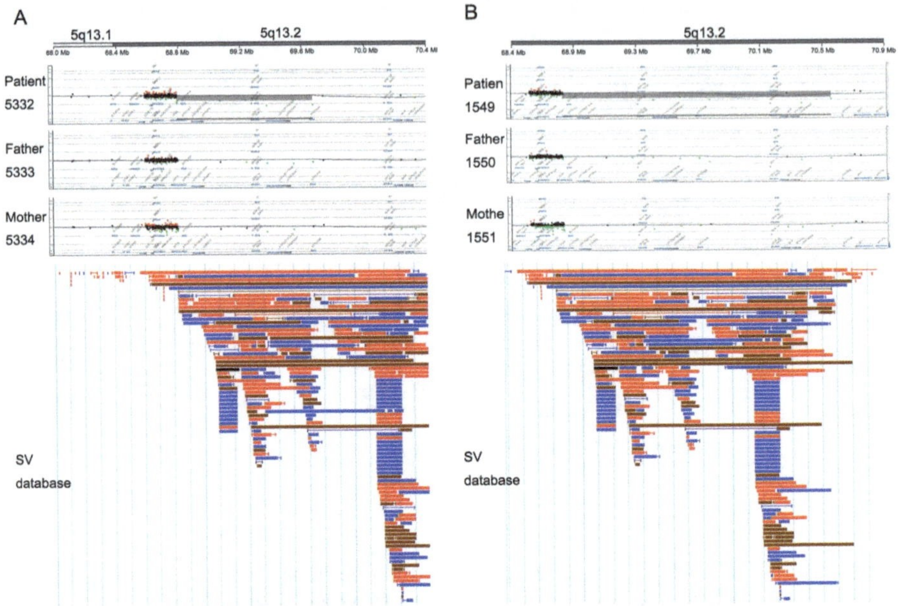

Fig. 9.3 The 2 similar CNV in 5q13.2. (**a**) The microdeletion in 5q13.2 of patient 5332. Compared with the genome DNA of his father and mother, the CNVs is de novo. From the data of SV database in this area, the CNV is not rare. (**b**) The microdeletion in 5q13.2 of patient 1549. Compared with the genome DNA of his father and mother, the CNVs is de novo. From the data of SV database in this area, the CNV is not rare

highly expressed in brain, was reported to be a candidate gene for ID/DD in a female with nonsyndromic ID/DD (Daoud et al. 2009; Barge-Schaapveld et al. 2013). *NR4A2* is the gene for Nuclear Receptor Subfamily 4, Group A, Member 2, which is crucial for expression of a set of genes such as *SLC6A3*, *SLC18A2*, *TH* and *DRD2* for the development of neurons (Messmer et al. 2007).

The clinical features of patient 2332 with a 2q33.1-q34 deletion were similar to 2q32-q33 deletion syndrome. 2q32-q33 deletion syndrome (OMIM # 612313), first reported at 1989, (Glass et al. 1989) is characterized by severe ID/DD, microcephaly and craniofacial dysmorphism. A *STAB2* gene deletion might be the most likely pathogenic gene in the 2q33.1-q34 region. The *SATB2* is a candidate brain developmental gene which should be responsible for the 2q32-q33 deletion syndrome (Van Buggenhout et al. 2005; Rosenfeld et al. 2009; Usui et al. 2013). The *STAB2* gene encodes a transmembrane receptor which has always been a marker of the upper layer of the normal fetal neocortex (Arai et al. 2012). In the 2q33.1-q34 CNV of our patient, there are both epilepsy-related genes and ID/DD related genes. The reported epilepsy-related genes are *ADAM23* (Owuor et al. 2009; Fukata et al. 2010), *MAP2* (Chulanova et al. 2001; Jalava et al. 2007). *SATB2* (Leoyklang et al. 2007) and *CREB1* (Barco et al. 2003) were reported to be ID/DD related gene. Besides those genes, *KCTD18* (Pichler et al. 2013), *TRAK2* (Grishin et al. 2006), and *UNC80*

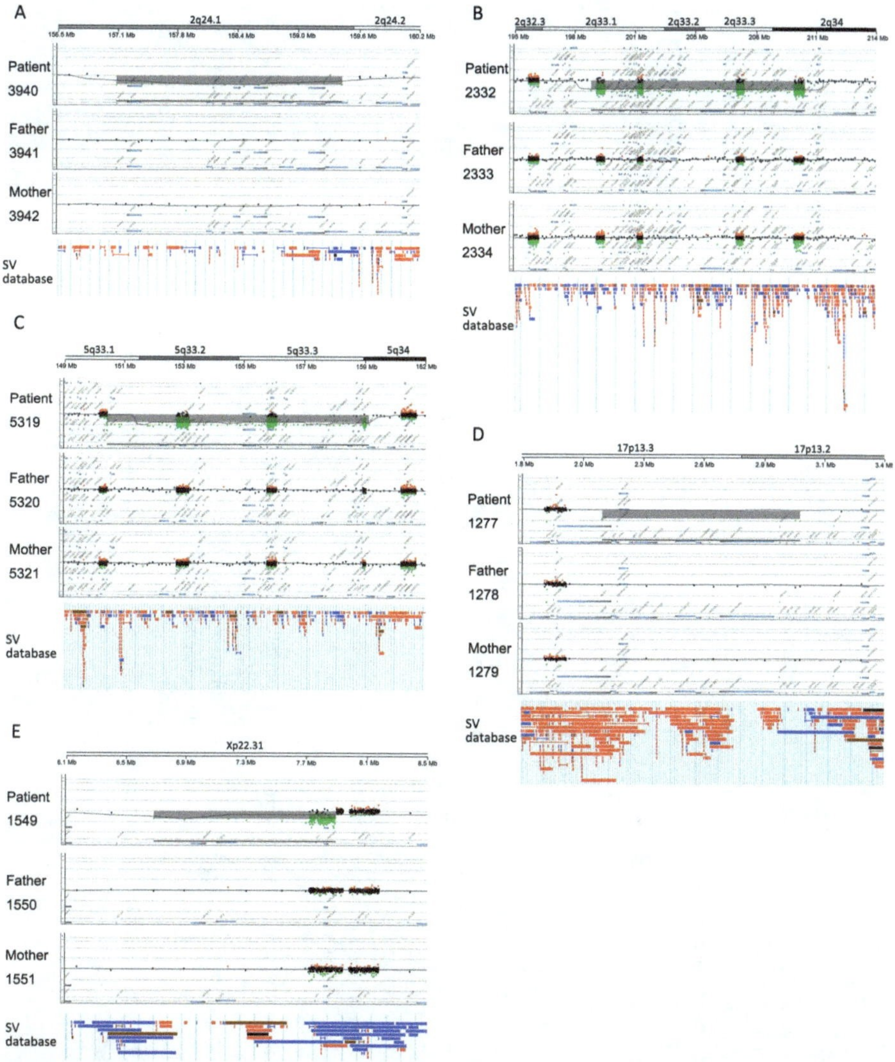

Fig. 9.4 The other CNVs in our cohort. (**a**) The microdeletion in 2q24.1 of patient 3940. Compared with the genome DNA of his father and mother, the CNVs is de novo. From the data of SV database in this area, the CNV is rare. (**b**) The microdeletion in 2q33.1-q34 of patient 2332. Compared with the genome DNA of his father and mother, the CNVs is de novo. From the data of SV database in this area, the CNV is rare. (**c**) The microdeletion in 5q33.1-q34 of patient 5319. Compared with the genome DNA of his father and mother, the CNVs is de novo. From the data of SV database in this area, the CNV is rare. (**d**) The microdeletion in 17p13.2 of patient 1277. Compared with the genome DNA of his father and mother, the CNVs is de novo. From the data of SV database in this area, the CNV is rare. (**e**) The microdeletion in Xp22.31 of patient 1549. Compared with the genome DNA of his father and mother, the CNVs is de novo. From the data of SV database in this area, the CNV is rare

Table 9.2 The clinical features of patients

Patient (F/M)	CNV loci	Diagnosis	Seizure	ID/DD	EEG	MRI	Other clinical features
3940 (F)	2q24.1	Epilepsy	Onset 4 years, tonic seizure (always in sleep)	Moderate	Epileptiform wave	/	Hyperactivity, flat foot
2332 (M)	2q33.2-q34	EOEE	Onset 3 months, generalized seizure, 5–6 times per year	Severe	Normal	Dysplasia of corpus callosum, strabismus	Autism, atrial septal defect, microcephaly, ocular hypertelorism, micrognathia, big ear, blepharoptosis
1549 (F)	5q13.2, Xp22.31	Infantile spasm	Onset 4 years, infantile spasm, generalized seizure, 9 times per year	Severe	Spike slow wave	Normal	None
5332 (M)	5q13.2	Epilepsy	Onset 1 year 1 month, tonic seizure, 6–30 times per year	Severe	Generalized and multifocal spike wave and spike slow wave	Myelin dysplasia, dysplasia of corpus callosum	Hypotonia
5319 (M)	5q33.1-q34	Epilepsy	Onset 1.5 years, generalized seizure with fever, twice per year	Moderate	Normal	Dysplasia of cerebral white matter	Small ear, strabismus, high palatine arch, autism
1277 (M)	17p13.2	Infantile spasm	Onset 5 months, infantile spasm last 3.5 years then cured by hormonotherapy	Severe	High-amplitude sharp wave	Pachygyria, agyria, schizencephaly, hydrocephalus	None
1583 (M)	22q11	Epilepsy	Onset 5 months, tonic seizure usually with fever, 1.5 times per months	Severe	Low to medium amplitude spike wave and spike slow wave with interictal EEGs in sleep	Cerebral dysplasia, increased lateral ventricle especially on the left side, wide arachnoid at temporal lobe and frontal lobe	Anomalous face with short philtrum, small auricle, high palatal arch, and up-warped upper lip; high muscular tension of both lower limbs
3568 (F)	22q11	Epilepsy	Onset 2.5 years, complex focal seizure, 2 times per months	Moderate	/	Hippocampal sclerosis, abnormal signal at temporal lobe and parietal lobe (especially right side)	None

(Gogliotti et al. 2011) are related to ion channels or receptors, and *CASP8* (Ma et al. 2007), *NRP2* (Maden et al. 2012) and *KLF7* (Caiazzo et al. 2011) takes part in neurodevelopment.

There were two patients (1583 and 3568) sharing the identical 48 gene-deleted CNV in 22q11.21-q11.22. Besides epilepsy and ID/DD, they also suffered from encephalodysplasia. Patient 1583 also had an anomalous face. Patient 1583 was a boy born after uneventful term delivery. He was the product of healthy parents without family history of epilepsy and ID/DD. However, this boy had tonic seizures since 5 months. The seizures are intermittent with a frequency of 1.5 times per month. His intellectual and motive developments were delayed. The patient also had an anomalous face with short philtrum, small auricle, high palatal arch, and up-warped upper lip. A brain MRI showed cerebral dysplasia, increased lateral ventricle especially on his left side, wide arachnoid at temporal lobe and frontal lobe. Interictal EEGs found some low to medium amplitude spike wave and spike slow wave in sleep. Patient 3568 was a girl born after uneventful term delivery. There was no history of epilepsy and ID/DD in her family. The girl had complex focal seizures since 2.5 years. The seizures occurred intermittently with a frequency of 2 time per month. Her intellectual development was delayed. The MRI showed that two-side hippocampal sclerosis and abnormal signal at parietal lobe and temporal lobe (especially at the right sides). Besides epilepsy and ID/DD, they also suffered from encephalodysplasia. The reason may be this CNV have some genes of neurodevelopment, such as *TBX1*, *ARVCF*, *RTN4R*, *and CRKL* (Tables 9.2 and 9.4).

Based on the locus of the deletion and the clinic features, these patients would almost certainly suffer from 22q11.2 deletion syndrome. This syndrome involves a series of syndromes such as DiGeorge syndrome (DGS) (Kelley et al. 1982), velocardiofacial syndrome (VCFS) (Scambler et al. 1991; Driscoll et al. 1992), conotruncal anomaly face syndrome (CTAF) (Matsuoka et al. 1994), some cases of autosomal dominant Opitz G/BBB syndrome (McDonald-McGinn et al. 1995; Fryburg et al. 1996; Lacassie and Arriaza 1996), and Cayler cardiofacial syndrome (asymmetric crying facies) (Giannotti et al. 1994). Among candidate pathogenic genes, *COMT* is related to both epilepsy (Doyle and Sellinger 1980) and ID/DD (Zhang et al. 2007; Li et al. 2009). *SNAP29* (Elfving et al. 2008) and *TBX1* (Sedghi et al. 2012) have been proven to related to epilepsy.

In 2006, 5q34 was reported to be a susceptibility locus for idiopathic generalized epilepsy (Hempelmann et al. 2006). A 6.45 Mb deletion in 5q33-q34 and a 713 Kb deletion in 5q33.2 were reported by Mefford in 2010 and 2011 (Mefford et al. 2010, 2011) to be related to epilepsy and ID/DD. Our study found a patient was with a deletion in 5q33.1-q34. Among candidate pathogenic genes, *CYFIP2* is highly expressed in the brain and contributes to both epilepsy (Hideyama et al. 2010) and ID/DD (Hoeffer et al. 2012). The other ID/DD gene is *GLRA1* (Al-Futaisi et al. 2012), which encodes a subunit of glycine receptor.

In a research in 2013, Speriz reported that 17p13.2 may be an epilepsy and ID/DD related genetic region as a duplication (Spreiz et al. 2014). A deletion of 17p13.2 was also reported to be associated with Miller-Dieker lissencephaly syndrome (Chen et al. 2010). This report, together with our findings, indicates that 17p13.2

Table 9.3 Phenotype description of patients with long deletion

Phenotype	Frequency	Patients
Neurologic		
ID/DD	100% (8/8)	All
Epilepsy	100% (8/8)	All
Microcephaly	12.5% (1/8)	2332
Encephalodysplasia	85.7% (6/7)	1227, 1583, 2332, 3568, 3519, 5332
Psychiatric		
Autism	4% (2/5)	2332, 5319
Craniofacial		
Facial dysmorphism	37.5% (3/8)	1583, 2332, 5319
Cleft lip/palate	12.5% (1/8)	2332
Strabismus	37.5% (2/8)	5319, 2332
Dyskinesia	50% (4/8)	5332, 1583, 1549, 1277
Syndrome		
EOEE	37.5% (3/8)	2332,1277, 1549

Table 9.4 The CNV hotpots of idiopathic epilepsy

CNV locus	Candidate gene	Type of CNVs	Subtype	Reference
1q21.1	GJA5, GJA8, HYDIN2	Deletion	JAE, JME	de Kovel et al. (2010), Mefford et al. (2010)
15q11.2	CYFIP1, NIPA2	Deletion	JAE, JME, CAE, EGTCS only	Zhang et al. (2009), de Kovel et al. (2010), Mefford et al. (2010)
15q13.3	CHRNA7	Deletion	JAE, JME, CAE, EGTCS only	Dibbens et al. (2009), Helbig et al. (2009), de Kovel et al. (2010)
16p11.2	KCTD13, SEZ6L2	Deletion	JME	Mefford et al. (2010)
16p13.11	NDE1	Deletion	JME, CAE, EGTCS only	de Kovel et al. (2010), Mefford et al. (2010)
22q11.2	SLC25A18	Deletion	EGTCS only	de Kovel et al. (2010)

may be an important genetic region for gyrus development. Among pathogenic gene candidates, *SRR*, *PAFAH1B1* and *MRPL40* should be considered. *AFAH1B1* was reported to be associated with lissencephaly (Cardoso et al. 2000; Kerjan and Gleeson 2007). *R* encodes serine racemase, which catalyzes L-serine to D-serine. D-serine is an important transmitter in brain and may be related to epilepsy (Ryu et al. 2010) and ID/DD (Klatte et al. 2013).

There were two patients (1549 and 5332) sharing the similar gene-deleted CNV in 5q13.2. Patient 1549 was a girl born after uneventful term delivery. She was the product of healthy parents without family history of epilepsy and ID/DD. However, this girl had generalized seizure since 4 years. The seizures are with a most frequency of 9 times per day. Her intellectual and motive developments were delayed.

Table 9.5 The CNV hotpots of epileptic encephalopathy

CNV locus	Candidate gene	Type of CNVs	Subtype	Reference
1p36	KLHL17	Deletion	IS	Paciorkowski et al. (2011)
2q32.3		Deletion	IS	Tiwari et al. (2013)
2q24.3	SCN1A	Deletion	Dravet syndrome	Wang et al. (2012)
3q11	EPHA6, GABRR3	Duplication	IS	Mefford et al. (2011)
4q3.1-q3.2	EPHA5	Duplication	Dravet syndrome	Lin et al. (2013)
7q11.23	STX1A	Deletion	IS	Paciorkowski et al. (2011)
14q12	FOXG1	Duplication	IS	Paciorkowski et al. (2011)
15q11-13	GABRA5, GABRB3,GABRG3	Duplication	IS	Paciorkowski et al. (2011), Tiwari et al. (2013)
16p11.2		Duplication	IS	Mefford et al. (2011), Tiwari et al. (2013)
Xp22	CDKL5	Deletion	IS	Mefford et al. (2011), Tiwari et al. (2013)

Her brain MRI was normal. Interictal EEGs found some low to medium amplitude spike wave and spike slow wave in sleep. Patient 5332 was a boy born after uneventful term delivery. There was no history of epilepsy and ID/DD in her family. The boy got complex focal seizures since 1 year 4 month. The seizures occurred intermittently with a frequency of 6–30 time per day. His intellectual development was delayed. EEGs found Generalized and multifocal spike wave and spike slow wave. The MRI showed that brain dysplasia with defect of myelination of white matter.

Patient 1549 also has a long deletion in Xp22.31. The patient 5332 also suffered from aphasia and muscle hypotonia. The level of galactose in urine was a little higher than the normal standard. 5q13.2 was not reported to be related to epilepsy or ID/DD. In this CNV, no gene in this CNV have been reported to involve in epilepsy and ID/DD. The candidate pathogenic genes in this deletion were *SMN1*, *SMN2*, *OCLN*, and *NAIP*. These genes are involved in neurodevelopment. *SMN1* and *SMN2* are important factor for motor neuron development, and associate with spinal muscular atrophy (Prior 2007). Knocking out *SMN2* would increase seizure susceptibility (Gogliotti et al. 2011). *OCLN* encodes tight junction protein occludin, which is involved in the early stage of neurodevelopment (Virgintino et al. 2004). Occludin was reported to be overexpressed in Alzheimer's disease and vascular dementia (Romanitan et al. 2007), so it may be related to ID/DD. Neurodevelopment related gene *NAIP*, which encodes Neuronal Apoptosis Inhibitory Protein (Mercer et al. 2000), was reported to decrease in brains of patients suffering with Down syndrome or Alzheimer's disease (Seidl et al. 1999). It is indicated that *NAIP* may be related to ID/DD.

In an infantile spasms related deletion in Xp22 reported by Mefford, *CDKL5* was reported as the candidate pathogenic gene (Mefford et al. 2011). In our study, a deletion of Xp22.31 in patient 1594 (who also had a 5q13.2 deletion) contained 4

Table 9.6 Candidate pathogenic genes in the CNVs

CNV locus	All	Epilepsy (seizure)	ID/DD	Synapse	Ion channel/ receptor	Transmitter	Neurodevelopment	High expression in CNS
2q24.1	*NR4A2, GPD2*		*GPD2*				*NR4A2*	*GPD2*
2q33.1-q34	*SATB2, KCTD18, CASP8, TRAK2, NRP2, ADAM23, KLF7, CREB1, MAP2, UNC80*	*ADAM23 MAP2*	*SATB2, CREB1*		*ADAM23, KCTD18, TRAK2, UNC80*		*SATB2, CASP8, NRP2, ADAM23, KLF7, CREB1, MAP2*	
5q13.2	*OCLN, SMN1, SMN2, NAIP*						*OCLN, SMN1, SMN2, NAIP*	
5q33.1-q34	*GLRA1, HAND1, CYFIP2, ADRA1B*	*CYFIP2*	*GLRA1, CYFIP2*		*GLRA1, ADRA1B*		*HAND1*	*CYFIP2*
17p13.2	*SRR, PAFAH1B1, MRPL40*		*MRPL40*			*SRR*	*PAFAH1B1*	
22q11.21-q11.22	*SEPT5, TBX1, COMT, ARVCF, RTN4R, SNAP29, CRKL*	*COMT, SNAP29 TBX1*	*COMT*	*SEPT5, COMT, SNAP29*	*RTN4R*	*COMT*	*TBX1, ARVCF, RTN4R, CRKL*	
Xp22.31	*STS, VCX, PNPLA4*	*PNPLA4*	*STS, VCX*					

genes, *HDHD1A*, *STS*, *VCX*, and *PNPLA4*. *PNPLA4* may be involved to epilepsy and ID/DD (Carrascosa-Romero et al. 2012). *STS* and *VCX* was proven to take part in X-linked ID/DD (Ben Khelifa et al. 2013). As this patient is a girl, this heterozygous microdeletion in X chromosome may play a less role in the pathogenic mechanism.

9.4 Systems Biological Analysis

Recently, system biology has provided a series of powerful tools for biomedicine studies. As an important analysis method, network reconstruction was used in biomarker detection (Mitra et al. 2013), drug discovery (Zou et al. 2013), and for studying the synaptic plasticity (He et al. 2014) mechanism of learning and memory (Kandel et al. 2014). Network reconstruction is very suitable for studying the pathogenic mechanism of complex disease in CNS, such as autism (Corominas et al. 2014), schizophrenia (Sun et al. 2010), and tumor induced epilepsy (Mittal et al. 2013).

In this study, we also tried to use the analysis tools of systems biology to predict the common pathogenic mechanism for epilepsy and ID/DD as complex diseases. By the Cystoscope 3.1.0 (Shannon et al. 2003), the network of the known epilepsy genes (in Supplementary Table 9.1) and CNV genes was constructed based on genetic interaction, pathway, and physical interaction in GENEMANIA database (Montojo et al. 2010). Form the constructed network, we found 70.5% of the CNV genes (158/224) to be involved in a network, while only 3.5% CNV genes are known epilepsy-related genes (Fig. 9.5a). This result indicated that most of these genes are potential epilepsy related genes. All of the patients in our cohort have suffered from epilepsy and ID/DD, so we believe that there is some common pathogenic mechanism.

Interestingly, we found the BGNADP motif which was constructed by *BTD*, *GALNT10*, *NMUR2*, *AUTS2*, *DLG2* and *PTPRD* (Fig. 9.5b). This motif was connected with each of the CNVs in our patients. The BGNADP motif is a small epilepsy and ID/DD related gene network. *BTD* is the gene of biotinidase. Mutations in *BTD* caused a disease called biotinidase deficiency, which is characterized by seizures, hypotonia, skin rash, ataxia hearing loss and optic atrophy (Hymes et al. 2001). *AUTS2* (autism susceptibility candidate 2) is associated with a series of neurologic disorders, such as autism, attention deficit hyperactivity disorder, dyslexia, ID/DD and epilepsy (Poot et al. 2011; Jolley et al. 2013; Nagamani et al. 2013; Oksenberg et al. 2013). *DLG2* encodes a membrane-associated guanylate kinase called PSD-93, which interacts at postsynaptic sites of neurons and forms a scaffold for the clustering receptors and ion channel. *DLG2* expression was reported to increase in epilepsy, indicating the role of *DLG2* in epilepsy (Liu et al. 2007). *PTPRD* is a member of the protein tyrosine phosphatase gene family. Deficiency of *PTPRD* results in ID/DD (Choucair et al. 2015). *PTPRD* is also an epilepsy candidate gene according to a genome-wide association study (Speed et al. 2014). *GALNT10* and *NMUR2* are members of the CNVs of our cohort. They have not yet

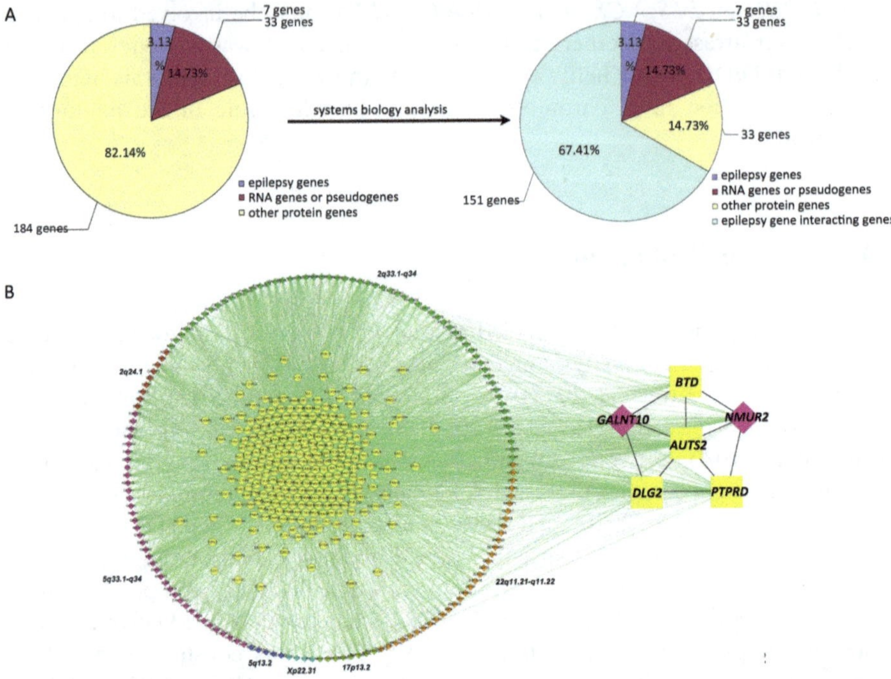

Fig. 9.5 The network of epilepsy genes and CNV genes. (**a**) Systems biology analysis showed 151 CNV genes (151/224, 67.41%) were interacting with the known epilepsy genes. (**b**) The network of epilepsy genes with CNV genes. The genes in different CNVs was labeled by different colors (2q33.1-q34 in green, 5q33.1-q34 in purple, 22q11.21-q11.22 in orange, 17p13.2 in deep green, 5q13.2 in blue, 2q24.1 in red and Xp22.31 in cyan)

been subjected to serious study, and should be targeted as candidate epilepsy and/or ID/DD genes in a further study. Our study has indicated that this BGNADP motif could be an important component in the common pathogenic mechanism. Further study should be needed to delineate the role of BGNADP in epilepsy and ID/DD.

References

Al-Futaisi AM, Al-Kindi MN, Al-Mawali AM, Koul RL, Al-Adawi S, Al-Yahyaee SA. Novel mutation of GLRA1 in Omani families with hyperekplexia and mild mental retardation. Pediatr Neurol. 2012;46(2):89–93.

Arai A, Saito T, Hanai S, Sukigara S, Nabatame S, Otsuki T, Nakagawa E, Takahashi A, Kaneko Y, Kaido T, Saito Y, Sugai K, Sasaki M, Goto Y, Itoh M. Abnormal maturation and differentiation of neocortical neurons in epileptogenic cortical malformation: unique distribution of layer-specific marker cells of focal cortical dysplasia and hemimegalencephaly. Brain Res. 2012;1470:89–97.

Barco A, Pittenger C, Kandel ER. CREB, memory enhancement and the treatment of memory disorders: promises, pitfalls and prospects. Expert Opin Ther Targets. 2003;7(1):101–14.

Barge-Schaapveld DQ, Ofman R, Knegt AC, Alders M, Hohne W, Kemp S, Hennekam RC. Intellectual disability and hemizygous GPD2 mutation. Am J Med Genet A. 2013; 161A(5):1044–50.

Bassuk AG, Geraghty E, Wu S, Mullen SA, Berkovic SF, Scheffer IE, Mefford HC. Deletions of 16p11.2 and 19p13.2 in a family with intellectual disability and generalized epilepsy. Am J Med Genet A. 2013;161A(7):1722–5.

Ben Khelifa H, Soyah N, Ben-Abdallah-Bouhjar I, Gritly R, Sanlaville D, Elghezal H, Saad A, Mougou-Zerelli S. Xp22.3 interstitial deletion: a recognizable chromosomal abnormality encompassing VCX3A and STS genes in a patient with X-linked ichthyosis and mental retardation. Gene. 2013;527(2):578–83.

Caciotti A, Tonin R, Rigoldi M, Ferri L, Catarzi S, Cavicchi C, Procopio E, Donati MA, Ficcadenti A, Fiumara A, Barone R, Garavelli L, Rocco MD, Filocamo M, Antuzzi D, Scarpa M, Mooney SD, Li B, Skouma A, Bianca S, Concolino D, Casalone R, Monti E, Pantaleo M, Giglio S, Guerrini R, Parini R, Morrone A. Optimizing the molecular diagnosis of GALNS: novel methods to define and characterize Morquio-A syndrome-associated mutations. Hum Mutat. 2015;36(3):357–68.

Caiazzo M, Colucci-D'Amato L, Volpicelli F, Speranza L, Petrone C, Pastore L, Stifani S, Ramirez F, Bellenchi GC, di Porzio U. Kruppel-like factor 7 is required for olfactory bulb dopaminergic neuron development. Exp Cell Res. 2011;317(4):464–73.

Cardoso C, Leventer RJ, Matsumoto N, Kuc JA, Ramocki MB, Mewborn SK, Dudlicek LL, May LF, Mills PL, Das S, Pilz DT, Dobyns WB, Ledbetter DH. The location and type of mutation predict malformation severity in isolated lissencephaly caused by abnormalities within the LIS1 gene. Hum Mol Genet. 2000;9(20):3019–28.

Carrascosa-Romero MC, Suela J, Alfaro-Ponce B, Cepillo-Boluda AJ. [X-chromosome-linked ichthyosis associated to epilepsy, hyperactivity, autism and mental retardation, due to the Xp22.31 microdeletion]. Rev Neurol. 2012;54(4): 241–48.

Chelly J, Khelfaoui M, Francis F, Cherif B, Bienvenu T. Genetics and pathophysiology of mental retardation. Eur J Hum Genet. 2006;14(6):701–13.

Chen CP, Liu YP, Lin SP, Chen M, Tsai FJ, Chen YT, Chen LF, Hwang JK, Wang W. Ventriculomegaly, intrauterine growth restriction, and congenital heart defects as salient prenatal sonographic findings of Miller-Dieker lissencephaly syndrome associated with monosomy 17p (17p13.2 --> pter) in a fetus. Taiwan J Obstet Gynecol. 2010;49(1):81–6.

Choucair N, Mignon-Ravix C, Cacciagli P, Abou Ghoch J, Fawaz A, Megarbane A, Villard L, Chouery E. Evidence that homozygous PTPRD gene microdeletion causes trigonocephaly, hearing loss, and intellectual disability. Mol Cytogenet. 2015;8:39.

Chulanova TA, Echikov SN, Sadovnikov VB, Shchipakina TG. Functional peculiarities of MAP2 in DBA/2J inbred mice as a component of genetic predisposition to seizures. Bull Exp Biol Med. 2001;132(5):1058–61.

Cooper GM, Coe BP, Girirajan S, Rosenfeld JA, Vu TH, Baker C, Williams C, Stalker H, Hamid R, Hannig V, Abdel-Hamid H, Bader P, McCracken E, Niyazov D, Leppig K, Thiese H, Hummel M, Alexander N, Gorski J, Kussmann J, Shashi V, Johnson K, Rehder C, Ballif BC, Shaffer LG, Eichler EE. A copy number variation morbidity map of developmental delay. Nat Genet. 2011;43(9):838–46.

Corominas R, Yang X, Lin GN, Kang S, Shen Y, Ghamsari L, Broly M, Rodriguez M, Tam S, Trigg SA, Fan C, Yi S, Tasan M, Lemmens I, Kuang X, Zhao N, Malhotra D, Michaelson JJ, Vacic V, Calderwood MA, Roth FP, Tavernier J, Horvath S, Salehi-Ashtiani K, Korkin D, Sebat J, Hill DE, Hao T, Vidal M, Iakoucheva LM. Protein interaction network of alternatively spliced isoforms from brain links genetic risk factors for autism. Nat Commun. 2014;5:3650.

Daoud H, Gruchy N, Constans JM, Moussaoui E, Saumureau S, Bayou N, Amy M, Vedrine S, Vu PY, Rotig A, Laumonnier F, Vourc'h P, Andres CR, Leporrier N, Briault S. Haploinsufficiency of the GPD2 gene in a patient with nonsyndromic mental retardation. Hum Genet. 2009;124(6): 649–58.

Derks EM, Ayub M, Chambert K, Del Favero J, Johnstone M, MacGregor S, Maclean A, McKechanie AG, McRae AF, Moran JL, Pickard BS, Purcell S, Sklar P, StClair DM, Wray NR,

Visscher PM, Blackwood DH. A genome wide survey supports the involvement of large copy number variants in schizophrenia with and without intellectual disability. Am J Med Genet B Neuropsychiatr Genet. 2013;162B(8):847–54.

Dibbens LM, Mullen S, Helbig I, Mefford HC, Bayly MA, Bellows S, Leu C, Trucks H, Obermeier T, Wittig M, Franke A, Caglayan H, Yapici Z, Sander T, Eichler EE, Scheffer IE, Mulley JC, Berkovic SF. Familial and sporadic 15q13.3 microdeletions in idiopathic generalized epilepsy: precedent for disorders with complex inheritance. Hum Mol Genet. 2009;18(19):3626–31.

Doyle RL, Sellinger OZ. Differences in activity in cerebral methyltransferases and monoamine oxidases between audiogenic seizure susceptible and resistant mice and deermice. Pharmacol Biochem Behav. 1980;13(4):589–91.

Driscoll DA, Budarf ML, Emanuel BS. A genetic etiology for DiGeorge syndrome: consistent deletions and microdeletions of 22q11. Am J Hum Genet. 1992;50(5):924–33.

Elfving B, Bonefeld BE, Rosenberg R, Wegener G. Differential expression of synaptic vesicle proteins after repeated electroconvulsive seizures in rat frontal cortex and hippocampus. Synapse. 2008;62(9):662–70.

Fryburg JS, Lin KY, Golden WL. Chromosome 22q11.2 deletion in a boy with Opitz (G/BBB) syndrome. Am J Med Genet. 1996;62(3):274–5.

Fukata Y, Lovero KL, Iwanaga T, Watanabe A, Yokoi N, Tabuchi K, Shigemoto R, Nicoll RA, Fukata M. Disruption of LGI1-linked synaptic complex causes abnormal synaptic transmission and epilepsy. Proc Natl Acad Sci U S A. 2010;107(8):3799–804.

Giannotti A, Digilio MC, Marino B, Mingarelli R, Dallapiccola B. Cayler cardiofacial syndrome and del 22q11: part of the CATCH22 phenotype. Am J Med Genet. 1994;53(3):303–4.

Glass IA, Swindlehurst CA, Aitken DA, McCrea W, Boyd E. Interstitial deletion of the long arm of chromosome 2 with normal levels of isocitrate dehydrogenase. J Med Genet. 1989; 26(2):127–30.

Gogliotti RG, Lutz C, Jorgensen M, Huebsch K, Koh S, Didonato CJ. Characterization of a commonly used mouse model of SMA reveals increased seizure susceptibility and heightened fear response in FVB/N mice. Neurobiol Dis. 2011;43(1):142–51.

Grayton HM, Fernandes C, Rujescu D, Collier DA. Copy number variations in neurodevelopmental disorders. Prog Neurobiol. 2012;99(1):81–91.

Grishin A, Li H, Levitan ES, Zaks-Makhina E. Identification of gamma-aminobutyric acid receptor-interacting factor 1 (TRAK2) as a trafficking factor for the K+ channel Kir2.1. J Biol Chem. 2006;281(40):30104–11.

Harvard C, Strong E, Mercier E, Colnaghi R, Alcantara D, Chow E, Martell S, Tyson C, Hrynchak M, McGillivray B, Hamilton S, Marles S, Mhanni A, Dawson AJ, Pavlidis P, Qiao Y, Holden JJ, Lewis SM, O'Driscoll M, Rajcan-Separovic E. Understanding the impact of 1q21.1 copy number variant. Orphanet J Rare Dis. 2011;6:54.

He Y, Kulasiri D, Samarasinghe S. Systems biology of synaptic plasticity: a review on N-methyl-D-aspartate receptor mediated biochemical pathways and related mathematical models. Biosystems. 2014;122:7–18.

Helbig I, Hartmann C, Mefford HC. Clarifying the role of the 22q11.2 microdeletion in juvenile myoclonic epilepsy. Epilepsy Behav. 2013;29(3):589–90.

Helbig I, Mefford HC, Sharp AJ, Guipponi M, Fichera M, Franke A, Muhle H, de Kovel C, Baker C, von Spiczak S, Kron KL, Steinich I, Kleefuss-Lie AA, Leu C, Gaus V, Schmitz B, Klein KM, Reif PS, Rosenow F, Weber Y, Lerche H, Zimprich F, Urak L, Fuchs K, Feucht M, Genton P, Thomas P, Visscher F, de Haan GJ, Moller RS, Hjalgrim H, Luciano D, Wittig M, Nothnagel M, Elger CE, Nurnberg P, Romano C, Malafosse A, Koeleman BP, Lindhout D, Stephani U, Schreiber S, Eichler EE, Sander T. 15q13.3 microdeletions increase risk of idiopathic generalized epilepsy. Nat Genet. 2009;41(2):160–2.

Hempelmann A, Taylor KP, Heils A, Lorenz S, Prud'homme JF, Nabbout R, Dulac O, Rudolf G, Zara F, Bianchi A, Robinson R, Gardiner RM, Covanis A, Lindhout D, Stephani U, Elger CE, Weber YG, Lerche H, Nurnberg P, Kron KL, Scheffer IE, Mulley JC, Berkovic SF, Sander

T. Exploration of the genetic architecture of idiopathic generalized epilepsies. Epilepsia. 2006;47(10):1682–90.

Hideyama T, Yamashita T, Nishimoto Y, Suzuki T, Kwak S. Novel etiological and therapeutic strategies for neurodiseases: RNA editing enzyme abnormality in sporadic amyotrophic lateral sclerosis. J Pharmacol Sci. 2010;113(1):9–13.

Hoeffer CA, Sanchez E, Hagerman RJ, Mu Y, Nguyen DV, Wong H, Whelan AM, Zukin RS, Klann E, Tassone F. Altered mTOR signaling and enhanced CYFIP2 expression levels in subjects with fragile X syndrome. Genes Brain Behav. 2012;11(3):332–41.

Hymes J, Stanley CM, Wolf B. Mutations in BTD causing biotinidase deficiency. Hum Mutat. 2001;18(5):375–81.

Jalava NS, Lopez-Picon FR, Kukko-Lukjanov TK, Holopainen IE. Changes in microtubule-associated protein-2 (MAP2) expression during development and after status epilepticus in the immature rat hippocampus. Int J Dev Neurosci. 2007;25(2):121–31.

Jolley A, Corbett M, McGregor L, Waters W, Brown S, Nicholl J, Yu S. De novo intragenic deletion of the autism susceptibility candidate 2 (AUTS2) gene in a patient with developmental delay: a case report and literature review. Am J Med Genet A. 2013;161A(6):1508–12.

Kandel ER, Dudai Y, Mayford MR. The molecular and systems biology of memory. Cell. 2014;157(1):163–86.

Kelley RI, Zackai EH, Emanuel BS, Kistenmacher M, Greenberg F, Punnett HH. The association of the DiGeorge anomalad with partial monosomy of chromosome 22. J Pediatr. 1982;101(2):197–200.

Kerjan G, Gleeson JG. Genetic mechanisms underlying abnormal neuronal migration in classical lissencephaly. Trends Genet. 2007;23(12):623–30.

Kim EH, Yum MS, Lee BH, Kim HW, Lee HJ, Kim GH, Lee YJ, Yoo HW, Ko TS. Epilepsy and other neuropsychiatric manifestations in children and adolescents with 22q11.2 deletion syndrome. J Clin Neurol. 2016;12(1):85–92.

Klatte K, Kirschstein T, Otte D, Pothmann L, Muller L, Tokay T, Kober M, Uebachs M, Zimmer A, Beck H. Impaired D-serine-mediated cotransmission mediates cognitive dysfunction in epilepsy. J Neurosci. 2013;33(32):13066–80.

Kogan JH, Gross AK, Featherstone RE, Shin R, Chen Q, Heusner CL, Adachi M, Lin A, Walton NM, Miyoshi S, Miyake S, Tajinda K, Ito H, Siegel SJ, Matsumoto M. Mouse model of chromosome 15q13.3 microdeletion syndrome demonstrates features related to autism spectrum disorder. J Neurosci. 2015;35(49):16282–94.

de Kovel CG, Trucks H, Helbig I, Mefford HC, Baker C, Leu C, Kluck C, Muhle H, von Spiczak S, Ostertag P, Obermeier T, Kleefuss-Lie AA, Hallmann K, Steffens M, Gaus V, Klein KM, Hamer HM, Rosenow F, Brilstra EH, Trenite DK, Swinkels ME, Weber YG, Unterberger I, Zimprich F, Urak L, Feucht M, Fuchs K, Moller RS, Hjalgrim H, De Jonghe P, Suls A, Ruckert IM, Wichmann HE, Franke A, Schreiber S, Nurnberg P, Elger CE, Lerche H, Stephani U, Koeleman BP, Lindhout D, Eichler EE, Sander T. Recurrent microdeletions at 15q11.2 and 16p13.11 predispose to idiopathic generalized epilepsies. Brain. 2010;133(Pt 1):23–32.

Lacassie Y, Arriaza MI. Opitz GBBB syndrome and the 22q11.2 deletion. Am J Med Genet. 1996;62(3):318.

Layouni S, Salzmann A, Guipponi M, Mouthon D, Chouchane L, Dogui M, Malafosse A. Genetic linkage study of an autosomal recessive form of juvenile myoclonic epilepsy in a consanguineous Tunisian family. Epilepsy Res. 2010;90(1-2):33–8.

Lee BI, Heo K. Epilepsy: new genes, new technologies, new insights. Lancet Neurol. 2014; 13(1):7–9.

Leoyklang P, Suphapeetiporn K, Siriwan P, Desudchit T, Chaowanapanja P, Gahl WA, Shotelersuk V. Heterozygous nonsense mutation SATB2 associated with cleft palate, osteoporosis, and cognitive defects. Hum Mutat. 2007;28(7):732–8.

Li J, Yu C, Li Y, Liu B, Liu Y, Shu N, Song M, Zhou Y, Zhu W, Li K, Jiang T. COMT val158met modulates association between brain white matter architecture and IQ. Am J Med Genet B Neuropsychiatr Genet. 2009;150B(3):375–80.

Lin WD, Chang KP, Wang CH, Chen SJ, Fan PC, Weng WC, Lin WC, Tsai Y, Tsai CH, Chou IC, Tsai FJ. Molecular aspects of Dravet syndrome patients in Taiwan. Clin Chim Acta. 2013; 421:34–40.

Liu FY, Wang XF, Li MW, Li JM, Xi ZQ, Luan GM, Zhang JG, Wang YP, Sun JJ, Li YL. Upregulated expression of postsynaptic density-93 and N-methyl-D-aspartate receptors subunits 2B mRNA in temporal lobe tissue of epilepsy. Biochem Biophys Res Commun. 2007;358(3):825–30.

Ma D, Williamson P, Januszewski A, Nogaro MC, Hossain M, Ong LP, Shu Y, Franks NP, Maze M. Xenon mitigates isoflurane-induced neuronal apoptosis in the developing rodent brain. Anesthesiology. 2007;106(4):746–53.

Maden CH, Gomes J, Schwarz Q, Davidson K, Tinker A, Ruhrberg C. NRP1 and NRP2 cooperate to regulate gangliogenesis, axon guidance and target innervation in the sympathetic nervous system. Dev Biol. 2012;369(2):277–85.

Matsuoka R, Takao A, Kimura M, Imamura S, Kondo C, Joh-o K, Ikeda K, Nishibatake M, Ando M, Momma K. Confirmation that the conotruncal anomaly face syndrome is associated with a deletion within 22q11.2. Am J Med Genet. 1994;53(3):285–9.

McDonald-McGinn DM, Driscoll DA, Bason L, Christensen K, Lynch D, Sullivan K, Canning D, Zavod W, Quinn N, Rome J. Autosomal dominant "Opitz" GBBB syndrome due to a 22q11.2 deletion. Am J Med Genet. 1995;59(1):103–13.

Mefford HC, Muhle H, Ostertag P, von Spiczak S, Buysse K, Baker C, Franke A, Malafosse A, Genton P, Thomas P, Gurnett CA, Schreiber S, Bassuk AG, Guipponi M, Stephani U, Helbig I, Eichler EE. Genome-wide copy number variation in epilepsy: novel susceptibility loci in idiopathic generalized and focal epilepsies. PLoS Genet. 2010;6(5):e1000962.

Mefford HC, Yendle SC, Hsu C, Cook J, Geraghty E, McMahon JM, Eeg-Olofsson O, Sadleir LG, Gill D, Ben-Zeev B, Lerman-Sagie T, Mackay M, Freeman JL, Andermann E, Pelakanos JT, Andrews I, Wallace G, Eichler EE, Berkovic SF, Scheffer IE. Rare copy number variants are an important cause of epileptic encephalopathies. Ann Neurol. 2011;70(6):974–85.

Mercer EA, Korhonen L, Skoglosa Y, Olsson PA, Kukkonen JP, Lindholm D. NAIP interacts with hippocalcin and protects neurons against calcium-induced cell death through caspase-3-dependent and -independent pathways. EMBO J. 2000;19(14):3597–607.

Mertz LG, Christensen R, Vogel I, Hertz JM, Nielsen KB, Gronskov K, Ostergaard JR. Angelman syndrome in Denmark. Birth incidence, genetic findings, and age at diagnosis. Am J Med Genet A. 2013;161A(9):2197–203.

Messmer K, Remington MP, Skidmore F, Fishman PS. Induction of tyrosine hydroxylase expression by the transcription factor Pitx3. Int J Dev Neurosci. 2007;25(1):29–37.

Mitra S, Das S, Chakrabarti J. Systems biology of cancer biomarker detection. Cancer Biomark. 2013;13(4):201–13.

Mittal S, Shah AK, Barkmeier DT, Loeb JA. Systems biology of human epilepsy applied to patients with brain tumors. Epilepsia. 2013;54(Suppl 9):35–9.

Montojo J, Zuberi K, Rodriguez H, Kazi F, Wright G, Donaldson SL, Morris Q, Bader GD. GeneMANIA Cytoscape plugin: fast gene function predictions on the desktop. Bioinformatics. 2010;26(22):2927–8.

Mulley JC, Scheffer IE, Desai T, Bayly MA, Grinton BE, Vears DF, Berkovic SF, Dibbens LM. Investigation of the 15q13.3 CNV as a genetic modifier for familial epilepsies with variable phenotypes. Epilepsia. 2011;52(10):e139–42.

Nagamani SC, Erez A, Ben-Zeev B, Frydman M, Winter S, Zeller R, El-Khechen D, Escobar L, Stankiewicz P, Patel A, Cheung SW. Detection of copy-number variation in AUTS2 gene by targeted exonic array CGH in patients with developmental delay and autistic spectrum disorders. Eur J Hum Genet. 2013;21(3):343–6.

Oksenberg N, Stevison L, Wall JD, Ahituv N. Function and regulation of AUTS2, a gene implicated in autism and human evolution. PLoS Genet. 2013;9(1):e1003221.

Olszewski AK, Radoeva PD, Fremont W, Kates WR, Antshel KM. Is child intelligence associated with parent and sibling intelligence in individuals with developmental disorders? An investigation in youth with 22q11.2 deletion (velo-cardio-facial) syndrome. Res Dev Disabil. 2014;35(12):3582–90.

Owuor K, Harel NY, Englot DJ, Hisama F, Blumenfeld H, Strittmatter SM. LGI1-associated epilepsy through altered ADAM23-dependent neuronal morphology. Mol Cell Neurosci. 2009;42(4):448–57.

Paciorkowski AR, Thio LL, Rosenfeld JA, Gajecka M, Gurnett CA, Kulkarni S, Chung WK, Marsh ED, Gentile M, Reggin JD, Wheless JW, Balasubramanian S, Kumar R, Christian SL, Marini C, Guerrini R, Maltsev N, Shaffer LG, Dobyns WB. Copy number variants and infantile spasms: evidence for abnormalities in ventral forebrain development and pathways of synaptic function. Eur J Hum Genet. 2011;19(12):1238–45.

Pichler I, Schwienbacher C, Zanon A, Fuchsberger C, Serafin A, Facheris MF, Marroni F, Pattaro C, Shen Y, Tellgren-Roth C, Gyllensten U, Gusella JF, Hicks AA, Pramstaller PP. Fine-mapping of restless legs locus 4 (RLS4) identifies a haplotype over the SPATS2L and KCTD18 genes. J Mol Neurosci. 2013;49(3):600–5.

Poot M, van der Smagt JJ, Brilstra EH, Bourgeron T. Disentangling the myriad genomics of complex disorders, specifically focusing on autism, epilepsy, and schizophrenia. Cytogenet Genome Res. 2011;135(3-4):228–40.

Prior TW. Spinal muscular atrophy diagnostics. J Child Neurol. 2007;22(8):952–6.

Riley KN, Catalano LM, Bernat JA, Adams SD, Martin DM, Lalani SR, Patel A, Burnside RD, Innis JW, Rudd MK. Recurrent deletions and duplications of chromosome 2q11.2 and 2q13 are associated with variable outcomes. Am J Med Genet A. 2015;167A(11):2664–73.

Romanitan MO, Popescu BO, Winblad B, Bajenaru OA, Bogdanovic N. Occludin is overexpressed in Alzheimer's disease and vascular dementia. J Cell Mol Med. 2007;11(3):569–79.

Rosenfeld JA, Ballif BC, Lucas A, Spence EJ, Powell C, Aylsworth AS, Torchia BA, Shaffer LG. Small deletions of SATB2 cause some of the clinical features of the 2q33.1 microdeletion syndrome. PLoS One. 2009;4(8):e6568.

Ryu HJ, Kim JE, Yeo SI, Kim DS, Kwon OS, Choi SY, Kang TC. Potential roles of D-serine and serine racemase in experimental temporal lobe epilepsy. J Neurosci Res. 2010;88(11):2469–82.

Scambler PJ, Carey AH, Wyse RK, Roach S, Dumanski JP, Nordenskjold M, Williamson R. Microdeletions within 22q11 associated with sporadic and familial DiGeorge syndrome. Genomics. 1991;10(1):201–6.

Sedghi M, Nouri N, Abdali H, Memarzadeh M, Nouri N. A case report of 22q11 deletion syndrome confirmed by array-CGH method. J Res Med Sci. 2012;17(3):310–2.

Seidl R, Bajo M, Bohm K, LaCasse EC, MacKenzie AE, Cairns N, Lubec G. Neuronal apoptosis inhibitory protein (NAIP)-like immunoreactivity in brains of adult patients with Down syndrome. J Neural Transm Suppl. 1999;57:283–91.

Shannon P, Markiel A, Ozier O, Baliga NS, Wang JT, Ramage D, Amin N, Schwikowski B, Ideker T. Cytoscape: a software environment for integrated models of biomolecular interaction networks. Genome Res. 2003;13(11):2498–504.

Sisodiya SM, Mefford HC. Genetic contribution to common epilepsies. Curr Opin Neurol. 2011;24(2):140–5.

Speed D, Hoggart C, Petrovski S, Tachmazidou I, Coffey A, Jorgensen A, Eleftherohorinou H, De Iorio M, Todaro M, De T, Smith D, Smith PE, Jackson M, Cooper P, Kellett M, Howell S, Newton M, Yerra R, Tan M, French C, Reuber M, Sills GE, Chadwick D, Pirmohamed M, Bentley D, Scheffer I, Berkovic S, Balding D, Palotie A, Marson A, O'Brien TJ, Johnson MR. A genome-wide association study and biological pathway analysis of epilepsy prognosis in a prospective cohort of newly treated epilepsy. Hum Mol Genet. 2014;23(1):247–58.

Spreiz A, Haberlandt E, Baumann M, Baumgartner Sigl S, Fauth C, Gautsch K, Karall D, Janetschek C, Rostasy K, Scholl-Burgi S, Zotter S, Utermann G, Zschocke J, Kotzot D. Chromosomal microaberrations in patients with epilepsy, intellectual disability, and congenital anomalies. Clin Genet. 2014;86(4):361–6.

Sun J, Jia P, Fanous AH, van den Oord E, Chen X, Riley BP, Amdur RL, Kendler KS, Zhao Z. Schizophrenia gene networks and pathways and their applications for novel candidate gene selection. PLoS One. 2010;5(6):e11351.

Tiwari VN, Sundaram SK, Chugani HT, Huq AH. Infantile spasms are associated with abnormal copy number variations. J Child Neurol. 2013;28(10):1191–6.

Tuchman R, Moshe SL, Rapin I. Convulsing toward the pathophysiology of autism. Brain Dev. 2009;31(2):95–103.

Usui D, Shimada S, Shimojima K, Sugawara M, Kawasaki H, Shigematu H, Takahashi Y, Inoue Y, Imai K, Yamamoto T. Interstitial duplication of 2q32.1-q33.3 in a patient with epilepsy, developmental delay, and autistic behavior. Am J Med Genet A. 2013;161A(5):1078–84.

Van Buggenhout G, Van Ravenswaaij-Arts C, Mc Maas N, Thoelen R, Vogels A, Smeets D, Salden I, Matthijs G, Fryns JP, Vermeesch JR. The del(2)(q32.2q33) deletion syndrome defined by clinical and molecular characterization of four patients. Eur J Med Genet. 2005;48(3):276–89.

Vanlerberghe C, Petit F, Malan V, Vincent-Delorme C, Bouquillon S, Boute O, Holder-Espinasse M, Delobel B, Duban B, Vallee L, Cuisset JM, Lemaitre MP, Vantyghem MC, Pigeyre M, Lanco-Dosen S, Plessis G, Gerard M, Decamp M, Mathieu M, Morin G, Jedraszak G, Bilan F, Gilbert-Dussardier B, Fauvert D, Roume J, Cormier-Daire V, Caumes R, Puechberty J, Genevieve D, Sarda P, Pinson L, Blanchet P, Lemeur N, Sheth F, Manouvrier-Hanu S, Andrieux J. 15q11.2 microdeletion (BP1-BP2) and developmental delay, behaviour issues, epilepsy and congenital heart disease: a series of 52 patients. Eur J Med Genet. 2015;58(3):140–7.

Virgintino D, Errede M, Robertson D, Capobianco C, Girolamo F, Vimercati A, Bertossi M, Roncali L. Immunolocalization of tight junction proteins in the adult and developing human brain. Histochem Cell Biol. 2004;122(1):51–9.

Wang JW, Shi XY, Kurahashi H, Hwang SK, Ishii A, Higurashi N, Kaneko S, Hirose S. Prevalence of SCN1A mutations in children with suspected Dravet syndrome and intractable childhood epilepsy. Epilepsy Res. 2012;102(3):195–200.

Williams HJ, Owen MJ, O'Donovan MC. Schizophrenia genetics: new insights from new approaches. Br Med Bull. 2009;91:61–74.

Yamada K, Iwayama Y, Toyota T, Ohnishi T, Ohba H, Maekawa M, Yoshikawa T. Association study of the KCNJ3 gene as a susceptibility candidate for schizophrenia in the Chinese population. Hum Genet. 2012;131(3):443–51.

Yu HE, Hawash K, Picker J, Stoler J, Urion D, Wu BL, Shen Y. A recurrent 1.71 Mb genomic imbalance at 2q13 increases the risk of developmental delay and dysmorphism. Clin Genet. 2012;81(3):257–64.

Zhang F, Gu W, Hurles ME, Lupski JR. Copy number variation in human health, disease, and evolution. Annu Rev Genomics Hum Genet. 2009;10:451–81.

Zhang K, Gao J, An C, Gao X, Zheng Z, Li R, Huang S, Zhang F. An association study between cathechol-O-methyltransferase gene and mental retardation in the Chinese Han population. Neurosci Lett. 2007;419(1):83–7.

Zhang Y, Kong W, Gao Y, Liu X, Gao K, Xie H, Wu Y, Zhang Y, Wang J, Gao F, Wu X, Jiang Y. Gene mutation analysis in 253 Chinese children with unexplained epilepsy and intellectual/developmental disabilities. PLoS One. 2015;10(11):e0141782.

Zou J, Zheng MW, Li G, Su ZG. Advanced systems biology methods in drug discovery and translational biomedicine. Biomed Res Int. 2013;2013:742835.

Chapter 10
Using Systems Biology and Mathematical Modeling Approaches in the Discovery of Therapeutic Targets for Spinal Muscular Atrophy

Matthew E. R. Butchbach

Abbreviations

AC	Adenylate cyclase
cAMP	Cyclic AMP
cnPDE	Cyclic nucleotide phosphodiesterase
CRE	cAMP-response element
CREB	CRE binding protein
dbcAMP	Dibutyryl cAMP
ELISA	Enzyme-linked immunosorbent assay
ESS	Exonic splicing enhancer
FL-SMN	Full length SMN
GPCR	G protein-coupled receptor
IGF1R	Insulin-like growth factor 1 receptor
ODE	Ordinary differential equation
PDE	Partial differential equation
PKA	cAMP-dependent protein kinase
PP2A	Protein phosphatase 2A
SMA	Spinal muscular atrophy
SMN1	Survival motor neuron 1
SMN2	Survival motor neuron 2

M. E. R. Butchbach (✉)
Center for Applied Clinical Genomics, Nemours Biomedical Research, Nemours/Alfred I. duPont Hospital for Children, Wilmington, DE, USA

Center for Pediatric Research, Nemours Biomedical Research, Nemours/Alfred I. duPont Hospital for Children, Wilmington, DE, USA

Department of Pediatrics, Thomas Jefferson University, Philadelphia, PA, USA

Department of Biological Sciences, University of Delaware, Newark, DE, USA
e-mail: Matthew.Butchbach@nemours.org

A. Cheung-Hoi Yu, L. Li (eds.), *Systems Neuroscience*, Advances in Neurobiology 21, https://doi.org/10.1007/978-3-319-94593-4_10

SMNΔ7	SMN lacking exon 7
SNV	Single nucleotide variant

10.1 Introduction

Systems biology integrates biochemical, genetic and cellular approaches to provide a more comprehensive understanding of higher level processes in living organisms. In this integrated approach, the components of a system are best understood by their relationships within the system as well as with other systems. These interconnected components are referred to as networks. Mathematical modeling of networks is an essential facet of systems biology (Dhurjati and Mahadevan 2008). When studying complex and dynamic interactions, experimental and/or mathematical approaches provide a means to explore and understand the system of networks in question (Dhurjati and Mahadevan 2008). These approaches can also highlight ways in which this system can be manipulated. Systems biology can be used to identify novel pathways implicated in different diseases and to determine the optimal ways of manipulating regulatory networks to treat these diseases.

Network analysis and pathway connectivity approaches in systems biology have provided important insights into the pathogenesis of neurodegenerative diseases (Villoslada et al. 2009). These approaches have led to the identification of novel molecular pathways affected by diseases like hereditary ataxias. Systems biology and mathematical modeling approaches can be used in the development of therapeutic strategies for neurological diseases. Neurological diseases having a clearly genetic etiology, like the pediatric-onset motor neuron disease spinal muscular atrophy (SMA), are particularly amenable to systems biology and mathematical modeling approaches. To illustrate this application, we describe below how mathematical modeling is being used to more thoroughly understand the regulation of *SMN2*, an endogenous modifier gene for SMA, expression and to develop optimal therapeutic targets for this disease.

10.2 Spinal Muscular Atrophy

Proximal SMA is an autosomal recessive early-onset neurodegenerative disease characterized by the loss of α-motor neurons in the anterior horn of the spinal cord which leads to muscle weakness and atrophy (Crawford and Pardo 1996; Kolb and Kissel 2015). Proximally innervated muscles are preferentially affected over distal muscles in SMA. It is a leading genetic cause of infant and early childhood mortality across the world with an incidence of 1 in ~10,000 live births (Pearn 1978; Cuscó et al. 2002; Sugarman et al. 2012). The carrier frequency for SMA ranges from 1:25 to 1:50 (Zaldívar et al. 2005; Labrum et al. 2007; Hendrickson et al. 2009; Ben-Shachar et al. 2011; Su et al. 2011; Sugarman et al. 2012; Lyahyai et al. 2012;

Sangaré et al. 2014). While SMA is primarily a disorder affecting motor neurons, other cells are affected by this disease (Shababi et al. 2014). Arrhythmias and other cardiac abnormalities have been observed in mouse models for SMA (Heier et al. 2010; Shababi et al. 2010; Biondi et al. 2012; Shababi et al. 2012). SMA mice have also demonstrated abnormalities in the autonomic and enteric nervous systems (Bevan et al. 2010; Gombash et al. 2015). Loss of insulin-producing β-cells has been observed in the pancreas (Bowerman et al. 2012, 2014). While peripheral organ dysfunction in SMA has been described, it is not yet clear whether or not this dysfunction is a direct result of the disease or a consequence of motor neuron loss and muscle atrophy.

There is a high degree of phenotypic variability within the SMA population. As such, SMA is divided into 5 clinical grades based on age of onset and severity of the disease (Munsat and Davies 1992; Russman 2007). The more severe SMA (types 0 and I) patients have a short lifespan and usually die because of respiratory complications arising from weakness in the intercostals muscles. Due in part to better supportive care (Wang et al. 2007), type II SMA patients generally have a life expectancy into early adulthood. Type III SMA patients usually have a normal lifespan but have difficulty walking. Adult-onset type IV SMA patients generally have a fairly benign disease progression.

Most cases of SMA, regardless of clinical grade, result from large-scale deletions within chromosome 5q13.2 and the loss of the *Survival Motor Neuron 1* (*SMN1*) gene (Lefebvre et al. 1995). The *SMN* gene is duplicated in humans to give rise to *SMN1* and *SMN2* (Rochette et al. 2001). This duplication event is not perfect in that there are single nucleotide differences between *SMN1* and *SMN2*. The major difference between these two *SMN* genes is a translationally silent, C-to-T transition in exon 7(*SMN2 c.850C > T*) (Lorson et al. 1999; Monani et al. 1999). This position on exon 7 lies within an exonic splicing enhancer (ESS) sequence that regulated the inclusion of exon 7 in *SMN1* mRNA transcripts (Fig. 10.1). This ESS is disrupted in *SMN2* so that most (about 90%) of the *SMN2*-derived mRNAs lack exon 7 (SMNΔ7) after splicing. The resultant SMNΔ7 protein is unstable and not fully functional (Lorson and Androphy 2000; Burnett et al. 2009; Cho and Dreyfuss 2010). Some *SMN2* mRNAs—roughly 10%—contain exon 7 which results in the production of some full-length SMN (FL-SMN) protein from *SMN2*.

10.3 *SMN2* as an Endogenous Genetic Modifier of SMA Phenotype

Since the region of chromosome 5 containing the SMA locus is subject to unequal segmental duplication, *SMN1* and *SMN2* copy numbers are quite variable in the genome. Numerous studies have demonstrated an inverse relationship between *SMN2* copy number and disease severity amongst patient with SMA (reviewed in Butchbach 2016). As a general rule, those patients with milder forms of SMA have

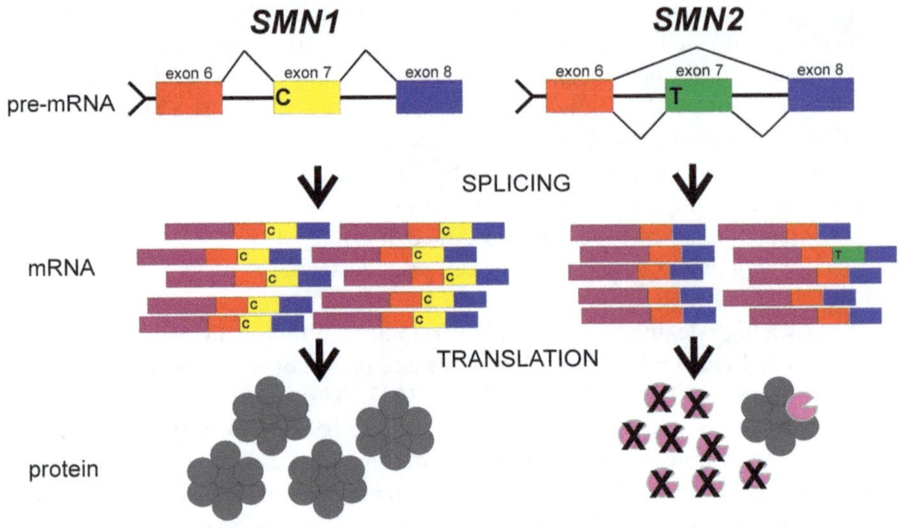

Fig. 10.1 Molecular difference between *SMN1* and *SMN2* and its effect on splicing. This figure is adapted from (Butchbach and Burghes 2004; Butchbach 2016)

higher *SMN2* copy numbers than severe SMA patients. There are some rare exceptions to this inverse relationship between *SMN2* copy number and disease severity in SMA. Some type II and III SMA patients have been shown to harbor only 2 copies of *SMN2* (Prior et al. 2009; Vezain et al. 2010; Bernal et al. 2010). These patients contain a rare single nucleotide variant (SNV) in exon 7 (*SMN2 c.859G > C*) that regulates exon 7 inclusion. *SMN2* is a genetic modifier of disease progression in SMA patients.

The modifier effect of *SMN2* is also observed in animal models for SMA. In *zSmn* (zebrafish orthologue to *SMN1*) mutant zebrafish, *SMN2* extends the survival of mutant larvae and rescues deficits in neuromuscular junction formation in these mutant fish (Hao Le et al. 2011). Transgenic insertion of *SMN2* into *mSmn* (murine orthologue to *SMN1*) nullizygous mice rescues embryonic lethality (Schrank et al. 1997; Monani et al. 2000; Hsieh-Li et al. 2000; Michaud et al. 2010). *SMN2* transgene copy number dictates the severity of the SMA phenotype in these mice. In other words, SMA mice with low *SMN2* copy numbers show a severe SMA phenotype (i.e. death within 8 days after birth) while high copy *SMN2* SMA mice have no phenotype (Monani et al. 2000; Hsieh-Li et al. 2000; Michaud et al. 2010). *SMN2* is, therefore, a major modifier of disease severity in humans as well as in animal models for SMA. These studies also show that *SMN2* is an ideal endogenous molecular target for the development of therapies for SMA.

10.4 Regulation of *SMN2* Expression by cAMP Signaling

Because of this phenotype modifying property, *SMN2* has been the target for numerous drug discovery strategies (Cherry et al. 2014). Targeting cyclic adenosine monophosphate (cAMP) signaling is of particular interest in developing inducers of *SMN2* expression. The cAMP signaling cascade (Fig. 10.2) regulates various cellular processes including gene expression, cell growth, metabolism and stress response (Kleppe et al. 2011). The *SMN2* promoter contains at least one cAMP-response element (CRE) that is able to bind to activated CRE-binding protein (phospho-CREB) (Majumder et al. 2004). The β2-adrenergic agonist salbutamol increases the amount of FL-SMN protein in SMA fibroblasts and leukocytes of SMA patients (Angelozzi et al. 2008; Tiziano et al. 2010). Forskolin, which stimulates adenylyl cyclase (AC) catalysis to produce cAMP from ATP, increases *SMN2* promoter activity (Majumder et al. 2004). The synthetic analogue dibutyryl cAMP (dbcAMP)—which activates cyclic AMP-dependent protein kinase (PKA)—also increases *SMN2* promoter activity (Majumder et al. 2004). We have recently shown that modulators of cAMP signaling significantly increase the number of gems—subnuclei foci of SMN protein (Liu and Dreyfuss 1996)—in fibroblasts derived

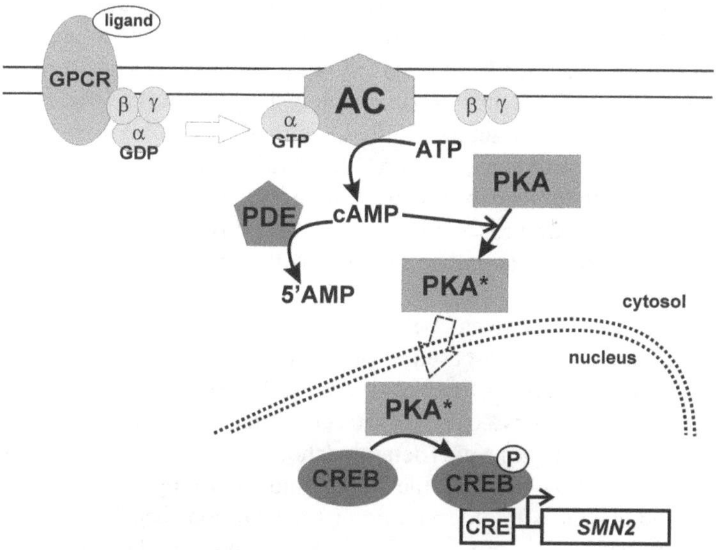

Fig. 10.2 Regulation of *SMN2* gene expression by the cAMP pathway. Ligand binds to and activates its membrane-bound G protein-coupled receptor (GPCR) leading to the dissociation of the $G_{\alpha,s}$ subunit from the GPCR. $G_{\alpha,s}$ then activates adenylyl cyclase (AC) which converts intracellular ATP into cAMP. cAMP then activates cAMP-dependent protein kinase—or protein kinase A (PKA). The catalytic PKA subunit, now freed of its regulatory subunits, acts on cAMP-response element-binding (CREB) protein. Phosphorylated CREB (phospho-CREB) binds to cAMP response elements (CREs) with the promoter regions of *SMN2*. Cyclic nucleotide phosphodiesterases (cnPDEs) diminish cAMP signaling by breaking down cAMP into AMP. This figure is adapted from (Mack et al. 2014)

from a type II SMA patient (Mack et al. 2014). Taken together, these studies show that modulation of cAMP signaling can increase SMN levels from *SMN2*.

10.5 Mathematical Modeling of Gene Expression

Mathematical models use mathematical concepts and terminology to describe a biological network using a set of variables and equations to define the relationships between these variables. Mathematical models are initially generated using available experimental data and domain knowledge. Through a process involving multiple iterations, the model assumptions are revised and refined in order to develop improved models that better fit the biological network (Dhurjati and Mahadevan 2008). This adaptability of mathematical models also makes it possible to integrate multiple pathways into a network model.

There are two types of mathematical models, quantitative and logic (Le Novère 2015). Quantitative models are linear representations of quantitative variables over time and can be used to compute concentrations of biomolecules and genes as well as durations of biomolecular interactions and processes. Quantitative models are precise and provide a direct comparison with experimentally-derived quantitative measurements but a priori knowledge of initial conditions and kinetic parameters is required to generate these models. Logic models, on the other hand, use qualitative activities and define phenotypes to compute transitions between two states and stable behaviors, known as attractors. While logic models are easy to generate and to use for simulation experiments, they are not useful for making quantitative predictions and it is difficult to select between multiple attractors. Historically, mathematical models have been designed to be either quantitative or logic; however, newer models which integrate quantitative modules with logic modules are being developed (Ryll et al. 2014).

Gene expression networks can be modeled mathematically using either thermodynamic, differential equation-based or Boolean (probabilistic) approaches (Ay and Arnosti 2011). The selection of modeling approach depends on the type of biological data available (qualitative or quantitative), the nature of the system to be modeled (static vs. dynamic), the level of detail and the scale of the model. Thermodynamic models are generated by factoring the quality and the arrangements of binding site for a biomolecule, for example, binding of transcription factors to their response elements within DNA. Thermodynamic models assume that the system is at a state of equilibrium and, hence, cannot describe the dynamic nature of the system. Differential equation models focus on regulatory interactions where time, state and space are viewed as continuous variables. As a result, differential equation models readily factor in the dynamic nature of the system in question. These models use ordinary differential equations (ODEs) if only one continuous variable, like time, is being factored or partial differential equations (PDEs) when multiple variables are being factored. Since ODEs and PDEs can be difficult to solve analytically, differential equation-based models can be hard to implement computationally, especially for larger biological networks. Boolean models represent relationships as one of two possible states, on or off, and can combine qualitative data into a logical structure. Instead of viewing variables as continuous, Boolean models consider time, state and space as discreet variables. While Boolean models are easy to analyze and implement computationally, they can be inaccurate if the system depends on fine details.

In most cases, there are unknown parameters within a mathematical model; as a result, these parameters need to be estimated so as to fit the proposed model with the experimental data. Parameter estimation begins with an initial estimate and new estimates are iteratively generated so as to minimize the error between simulated and experimental data. An objective function that measures model performance—such as the sign squared error or sum of squares of the residuals between model simulations and experimental data—is used in parameter estimation. More detailed information on parameter estimation approaches can be found elsewhere (Banga and Balsa-Canto 2008; Ay and Arnosti 2011). Parameter estimation is affected by both the structure of the model and the biological system for that model (Ay and Arnosti 2011). In order to assess how the model structure can affect parameter estimation, the effects of changes in parameter inputs on model outputs needs to be measured by process of sensitivity analysis. Local sensitivity analysis focuses on a specific set of parameter values at one point in time or space while global approaches examine the entire model over a range of parameter values. Further detailed information regarding the specifics of sensitivity analysis can be found elsewhere (Ingalls 2008). Sensitivity analysis is essential for the building and interpreting mathematical models.

10.6 Overall Strategy for Building Mathematical Models of Gene Expression

Mathematical modeling is an essential component of systems biology but it can appear daunting to biologists, especially those with limited experience in mathematical biology. Figure 10.3 provides a generic workflow for generating and testing a mathematical model for gene expression and cell signaling. This workflow is designed for differential equation-based models of the regulation of gene expression and cellular signaling. When beginning the process of model generation, one of the best sources of information for building mathematical models is the primary literature. There are also some recent reviews which describe the methodological details of generating a mathematical model for gene expression and cell signaling (Zi 2012; MacLean et al. 2016).

The first step in mathematical model generation is to create a comprehensive gene regulatory pathway map using existing biological information. CellDesigner is a convenient tool to graphically represent these pathway maps (Funahashi et al. 2003). CellDesigner uses standardized set of symbols known as Systems Biology

Fig. 10.3 Strategy for generating and testing mathematical models of gene expression and cell signaling

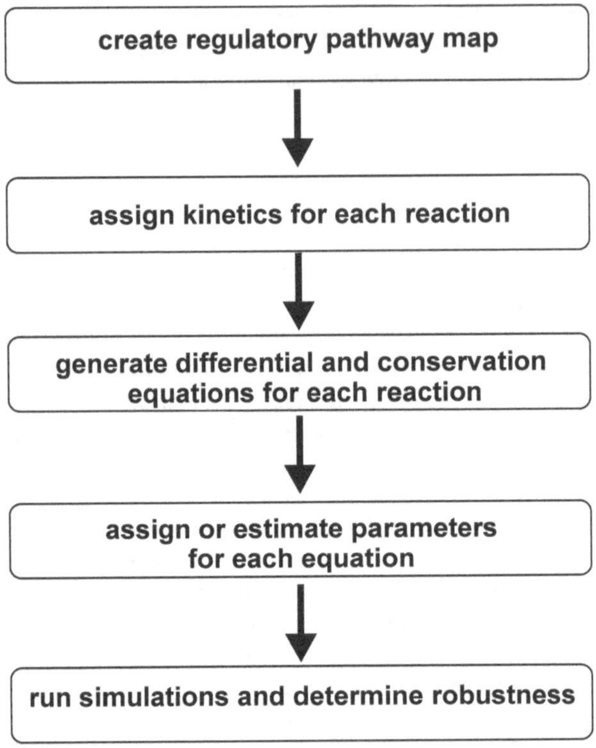

Graphical Notation (SBGN; (Hucka et al. 2003) to represent components of a biological network and their relationships (Le Novère et al. 2009). Complex Pathway Simulator (COPASI) is another platform-independent biological simulator program that can be used to generate mathematical models (Hoops et al. 2006; Mendes et al. 2009). The models can then imported into the mathematical software MATLAB using the Systems Biology Toolbox (Schmidt and Jirstrand 2006).

Once the pathway map has been generated, the kinetics for each reaction in the regulatory pathway must be assigned. The two primary components of a mathematical model, the differential equations and the conservation equations, can now be generated. The differential equations, which generally take the form of ODEs, represent changes in the components of a reaction in response to stimulation. The conservation equations are meant to show the balance between active and inactive forms of a signaling intermediate. The values of all of the parameters within each reaction kinetics equation must also be set from either a priori knowledge or be estimated using an objective function as described in the previous section (Banga and Balsa-Canto 2008; Ay and Arnosti 2011).

Simulations for the mathematical models can be completed once the parameters have been set and the initial concentrations of signaling components are estimated or determined from the literature. The robustness of a biological model can be assessed with sensitivity analysis as described in the previous section (Ingalls 2008;

MacLean et al. 2016). For a model to be considered robust, its outcomes must not be markedly affected by perturbations of the parameters or initial concentrations. With a robust model, the effects of altered expression of a component on the outcome, i.e. expression of the target gene, can be measured and future biological experiments can be designed with the assistance of mathematical models.

10.7 Mathematical Modeling of *SMN2* Regulation by cAMP Signaling

A systems biology approach can be used to investigate *SMN2* gene regulation. We recently developed mathematical models to characterize the regulation of *SMN2* expression by cAMP signaling (Mack et al. 2014). This approach is based on additive interactions between experimental data and mathematical models. We focused on the *SMN2* regulation by cAMP signaling because there is ample evidence in the literature showing that activation of cAMP signaling increases *SMN2* expression (Majumder et al. 2004; Angelozzi et al. 2008; Tiziano et al. 2010). The experimental data for these mathematical models were obtained from gem—a marker for SMN localization within the nucleus—assays in type II SMA fibroblasts because the reduction in gems correlates with SMN protein expression and SMA severity (Coovert et al. 1997) and this assay has been used in multiple studies identifying compounds which increase SMN expression (Andreassi et al. 2001, 2004; Sumner et al. 2003; Lunn et al. 2004; Grzeschik et al. 2005; Jarecki et al. 2005; Riessland et al. 2006; Mattis et al. 2006; Novoyatleva et al. 2008; Thurmond et al. 2008; Xiao et al. 2011). These gem inducing agents were validated by immunoblot or enzyme-linked immunosorbent assays (ELISAs).

The cAMP signaling treatment data were used to generate two distinct mathematical models, the full cAMP:*SMN2* and alternate cAMP:*SMN2* models (Fig. 10.4) (Mack et al. 2014). The full cAMP:*SMN2* model (Fig. 10.4a) factors in the effect of CREB activation on *SMN2* transcription. As some groups have suggested that cAMP signaling regulates *SMN2* expression post-transcriptionally by influencing FL-SMN protein stability (Burnett et al. 2009; Harahap et al. 2015), an alternate cAMP:*SMN2* model (Fig. 10.4b) was generated. Both models are extensions of a cAMP signaling mathematical model in yeast (Williamson et al. 2009). The models contain ODEs as well as conservation equations. The full cAMP:*SMN2* model contains seven ODEs and three conservation equations while the alternate cAMP:*SMN2* model contains five ODEs and two conservation equations (Mack et al. 2014). Simulated data from both models match with the experimental gem data showing that either model is valid. When these two models were combined, however, the resultant simulated data did not fit well with the experimental data suggesting that only one model correctly recapitulates the effect of cAMP signaling cascade on *SMN2* expression. Since the experimental data used to generate these models were fixed at one point in time, it is currently not possible to assess which mathematical

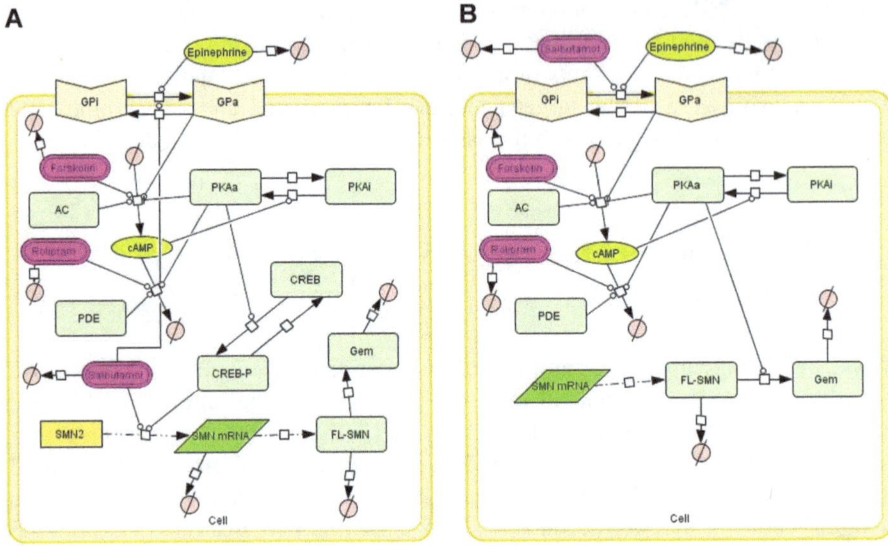

Fig. 10.4 Mathematical models for modulation of *SMN2* expression by cAMP signaling. Schematic representations of the full cAMP:*SMN2* (**a**) and alternate cAMP:*SMN2* (**b**) models for the regulation of *SMN2* expression by cAMP signaling. This figure is adapted from (Mack et al. 2014)

model more accurately simulates cAMP signaling-dependent regulation of *SMN2* expression. Future studies examining various facets of *SMN2* regulation including gem formation will allow better refinement and distinction between these two models.

10.8 Conclusions and Future Directions

Regulation of *SMN2* expression by cAMP signaling can be modeled mathematically. The regulation of *SMN2* by cAMP signaling is complex, multi-faceted and not completely understood. SMN is directly phosphorylated by PKA in vitro (Burnett et al. 2009; Wu et al. 2011). The interactions between SMN and other components of the core SMN macromolecular complex may be dependent upon PKA-dependent phosphorylation of SMN. PKA phosphorylation of SMN protein could not be factored into either mathematical model because the effects of PKA phosphorylation of SMN on its function and localization are not yet known. If PKA phosphorylation impacts SMN function, then this component of cAMP signaling can be factored into refined mathematical models of cAMP signaling and *SMN2* expression.

Another facet of gene regulation is the impact of other signaling pathways on the target pathway. For example, numerous extracellular stimuli including activation of ionotropic glutamate receptors, exercise and inhibition of insulin-like growth factor

I receptor (IGF1R) increases SMN expression in the spinal cord by AKT-mediated phosphorylation of CREB (Biondi et al. 2010, 2015; Branchu et al. 2013). CREB is regulated by the protein serine/threonine phosphatase 2A (PP2A) (Wadzinski et al. 1993). Protein serine/threonine phosphatase inhibitors like cantharidin and tautomycin have been shown to increase *SMN2* expression (Novoyatleva et al. 2008; Zhang et al. 2011). These natural product inhibitors may act through suppression of CREB dephosphorylation and, as a result, activation of the *SMN2* promoter. As new insights are gained as to how these other intracellular pathways affect CREB-mediated activation of *SMN2* expression, the intersection of AKT and PP2A with cAMP signaling can be integrated into current mathematical models of *SMN2* expression.

In addition to identifying the optimal component of the cAMP signaling pathway responsible for regulating *SMN2* expression, mathematical models can be used to predict the effects of drug combinations. For example, activation of AC by forskolin can act in concert with cyclic nucleotide phosphodiesterase (cnPDE) inhibition by rolipram to additively increase gem formation, as predicted mathematically (Mack et al. 2014). Once validated experimentally, mathematical modeling can used to design combination strategies that target different parts of a signaling cascade, in this case cAMP signaling. Furthermore, drug discovery efforts have identified numerous small molecule activators of *SMN2* expression that operate either by increasing *SMN2* transcription or alternative splicing to increase the proportion of *FL-SMN* transcripts (reviewed in Cherry et al. 2014). As the molecular targets and signaling pathways affected by these small molecules are identified, parallel mathematical models can be generated for each pathway as it relates to *SMN2* gene regulation. These pathways can then be integrated so as to create a comprehensive mathematical model for *SMN2* gene regulation. This comprehensive model can be used to predict which pathways could be modulated synergistically in order to maximize *SMN2* upregulation which will drive the development of combination therapeutic strategies for SMA.

In conclusion, mathematical modeling is a systems biology approach that can be used to understand how gene expression can be regulated by a signaling pathway. This approach has recently been applied to the regulation of *SMN2* expression by cAMP signaling. This systems-based mathematical modeling approach can ultimately aid in the development and optimization of cAMP signaling-based therapies for SMA. A similar approach could also be used for other molecular pathways that regulate *SMN2* expression. Mathematical models of these individual pathways regulating *SMN2* expression can then be integrated to create a more comprehensive model of *SMN2* gene regulation. Furthermore, mathematical modeling can be applied to other neurogenetic diseases wherein modifier genes, like *SMN2* for SMA, have been identified.

Acknowledgments This work was supported by an Institutional Development Award (IDeA) from the National Institute of General Medical Sciences of the National Institutes of Health (P30GM114736) and by the Nemours Foundation.

Conflict of Interest The author declares no conflict of interest.

References

Andreassi C, Angelozzi C, Tiziano FD, Vitali T, De Vincenzi E, Boninsegna A, et al. Phenylbutyrate increases SMN expression in vitro: relevance for treatment of spinal muscular atrophy. Eur J Hum Genet. 2004;12:59–65.

Andreassi C, Jarecki J, Zhou J, Coovert DD, Monani UR, Chen X, et al. Aclarubicin treatment restores SMN levels to cells derived from type I spinal muscular atrophy patients. Hum Mol Genet. 2001;10:2841–9.

Angelozzi C, Borgo F, Tiziano FD, Martella A, Neri G, Brahe C. Salbutamol increases SMN mRNA and protein levels in spinal muscular atrophy cells. J Med Genet. 2008;45:29–31.

Ay A, Arnosti DN. Mathematical modeling of gene expression: a guide for the perplexed biologist. Crit Rev Biochem Mol Biol. 2011;46:137–51.

Banga JR, Balsa-Canto E. Parameter estimation and optimal experimental design. Essays Biochem. 2008;45:195.

Ben-Shachar S, Orr-Urtreger A, Bardugo E, Shomrat R, Yaron Y. Large-scale population screening for spinal muscular atrophy: clinical implications. Genet Med. 2011;13:110–4.

Bernal S, Alías L, Barceló MJ, Also-Rallo E, Martínez-Hernández R, Gámez J, et al. The c.859G>C variant in the SMN2 gene is associated with types II and III SMA and originates from a common ancestor. J Med Genet. 2010;47:640–2.

Bevan AK, Hutchinson KR, Foust KD, Braun L, McGovern VL, Schmelzer L, et al. Early heart failure in the SMNΔ7 model of spinal muscular atrophy and correction by postnatal scAAV9-SMN delivery. Hum Mol Genet. 2010;19:3895–905.

Biondi O, Branchu J, Ben Salah A, Houdebine L, Bertin L, Chali F, et al. IGF-1R reduction triggers neuroprotective signaling pathways in spinal muscular atrophy mice. J Neurosci. 2015; 35:12063–79.

Biondi O, Branchu J, Sanchez G, Lancelin C, Deforges S, Lopes P, et al. In vivo NMDA receptor activation accelerates motor unit maturation, protects spinal motor neurons and enhances SMN2 gene expression in severe spinal muscular atrophy mice. J Neurosci. 2010;30:11288–99.

Biondi O, Lopes P, Desseille C, Branchu J, Chali F, Ben Salah A, et al. Physical exercise reduces cardiac defects in type 2 spinal muscular atrophy-like mice. J Physiol. 2012;590:5907–25.

Bowerman M, Michalski JP, Beauvais A, Murray LM, DeRepentigny Y, Kothary R. Defects in pancreatic development and glucose metabolism in SMN-depleted mice independent of canonical spinal muscular atrophy neuromuscular pathology. Hum Mol Genet. 2014;23:3432–44.

Bowerman M, Swoboda KJ, Michalski JP, Wang GS, Reeks C, Beauvais A, et al. Glucose metabolism and pancreatic defects in spinal muscular atrophy. Ann Neurol. 2012;72:256–68.

Branchu J, Biondi O, Chali F, Collin T, Leroy F, Mamchaoui K, et al. Shift from extracellular signal-related kinase to AKT/cAMP response element-binding protein pathway increases survival-motor-neuron expression in spinal-muscular-atrophy-like mice and patient cells. J Neurosci. 2013;33:4280–94.

Burnett BG, Muñoz E, Tandon A, Kwon DY, Sumner CJ, Fischbeck KH. Regulation of SMN protein stability. Mol Cell Biol. 2009;29:1107–15.

Butchbach MER. Copy number variations in the *survival motor neuron* genes: implications for spinal muscular atrophy and other neurodegenerative diseases. Front Mol Biosci. 2016;3:7.

Butchbach MER, Burghes AHM. Perspectives on models of spinal muscular atrophy for drug discovery. Drug Discov Today Dis Model. 2004;1:151–6.

Cherry JJ, Kobayashi DT, Lynes MM, Naryshkin NN, Tiziano FD, Zaworksi PG, et al. Assays for the identification and prioritization of drug candidates for spinal muscular atrophy. Assay Drug Dev Technol. 2014;12:315–41.

Cho S, Dreyfuss G. A degron created by SMN2 exon 7 skipping is a principal contributor to spinal muscular atrophy severity. Genes Dev. 2010;24:438–42.

Coovert DD, Le TT, McAndrew PE, Strasswimmer J, Crawford TO, Mendell JR, et al. The survival motor neuron protein in spinal muscular atrophy. Hum Mol Genet. 1997;6:1205–14.

Crawford TO, Pardo CA. The neurobiology of childhood spinal muscular atrophy. Neurobiol Dis. 1996;3:97–110.

Cuscó I, Barceló MJ, Soler C, Parra J, Baiget M, Tizzano E. Prenatal diagnosis for risk of spinal muscular atrophy. Br J Obstet Gynaecol. 2002;109:1244–9.

Dhurjati P, Mahadevan R. Systems biology: the synergistics interplay between biology and mathematics. Can J Chem Eng. 2008;86:127–41.

Funahashi A, Morohashi M, Kitano H, Tanimura N. CellDesigner: a process diagram editor for gene-regulatory and biochemical networks. Biosilico. 2003;1:159–62.

Gombash SE, Cowley CJ, Fitzgerald JA, Iyer CC, Fried D, McGovern VL, et al. SMN deficiency disrupts gastrointestinal and enteric nervous system function in mice. Hum Mol Genet. 2015;24:3847–60.

Grzeschik SM, Ganta M, Prior TW, Heavlin WD, Wang CH. Hydroxyurea enhances SMN2 gene expression in spinal muscular atrophy cells. Ann Neurol. 2005;58:194–202.

Hao Le T, Burghes AHM, Beattie CE. Generation and characterization of a genetic zebrafish model of SMA carrying the human SMN2 gene. Mol Neurodegener. 2011;6:24.

Harahap NIF, Nurputra DK, Rochmah MA, Shima A, Morisada N, Takarada T, et al. Salbutamol inhibits ubiquitin-mediated survival motor neuron protein degradation in spinal muscular atrophy cells. Biochem Biophys Rep. 2015;4:351–6.

Heier CR, Satta R, Lutz C, DiDonato CJ. Arrhythmia and cardiac defects are a feature of spinal muscular atrophy model mice. Hum Mol Genet. 2010;19:3906–18.

Hendrickson BC, Donohoe C, Akmaev VR, Sugarman EA, Labrousse P, Boguslavskiy L, et al. Differences in SMN1 allele frequencies among ethnic groups within North America. J Med Genet. 2009;46:641–4.

Hoops S, Sahle S, Gauges R, Lee C, Pahle J, Simus N, et al. COPASI—a complex pathway simulator. Bioinformatics. 2006;22:3067–74.

Hsieh-Li HM, Chang JG, Jong YJ, Wu MH, Wang NM, Tsai CH, et al. A mouse model for spinal muscular atrophy. Nat Genet. 2000;24:66–70.

Hucka M, Finney A, Sauro HM, Bolouri H, Doyle JC, Kitano H, et al. The systems biology markup language (SBML): a medium for representaion and exchange of biochemical network models. Bioinformatics. 2003;19:524–31.

Ingalls B. Sensitivity analysis: from model parameters to system behaviour. Essays Biochem. 2008;45:177–93.

Jarecki J, Chen X, Bernardino A, Coovert DD, Whitney M, Burghes AHM, et al. Diverse small-molecule modulators of SMN expression found by high-throughput compound screening: early leads towards a therapeutic for spinal muscular atrophy. Hum Mol Genet. 2005;14:2003–18.

Kleppe R, Krakstad C, Selheim F, Kopperud R, Døskeland SO. The cAMP-dependent protein kinase pathway as therapeutic target—possibilities and pitfalls. Curr Top Med Chem. 2011;11:1393–405.

Kolb SJ, Kissel JT. Spinal muscular atrophy. Neurol Clin. 2015;33:831–46.

Labrum R, Rodda J, Krause A. The molecular basis of spinal muscular atrophy (SMA) in South African black patients. Neuromuscul Disord. 2007;17:684–92.

Le Novère N. Quantitative and logic modelling of molecular and gene networks. Nat Rev Genet. 2015;16:146–58.

Le Novère N, Hucka M, Mi H, Moodie S, Schreiber F, Sorokin A, et al. The systems biology graphical notation. Nat Biotechnol. 2009;27:735–41.

Lefebvre S, Bürglen L, Reboullet S, Clermont O, Burlet P, Viollet L, et al. Identification and characterization of a spinal muscular atrophy-determining gene. Cell. 1995;80:155–65.

Liu Q, Dreyfuss G. A novel nuclear structure containing the survival of motor neurons protein. EMBO J. 1996;15:3555–65.

Lorson CL, Androphy EJ. An exonic enhancer is required for inclusion of an essential exon in the SMA-determining gene SMN. Hum Mol Genet. 2000;9:259–65.

Lorson CL, Hahnen E, Androphy EJ, Wirth B. A single nucleotide in the *SMN* gene regulates splicing and is responsible for spinal muscular atrophy. Proc Natl Acad Sci U S A. 1999;96:6307–11.

Lunn MR, Root DE, Martino AM, Flaherty SP, Kelley BP, Coovert DD, et al. Indoprofen upregulates the survival motor neuron protein through a cyclooxygenase-independent mechanism. Chem Biol. 2004;11:1489–93.

Lyahyai J, Sbiti A, Barkat A, Ratbi I, Sefiani A. Spinal muscular atrophy carrier frequency and estimated prevalence of the disease in Moroccan newborns. Genet Test Mol Biomarkers. 2012;16:215–8.

Mack SG, Cook DJ, Dhurjati P, Butchbach MER. Systems biology investigation of cAMP modulation to increase SMN levels for treatment of spinal muscular atrophy. PLoS One. 2014;9:e115473.

MacLean AL, Harrington HA, Stumpf MPH, Byrne HM. Mathematical and statistical techniques for systems medicine: the Wnt signaling pathway as a case study. Methods Mol Biol. 2016;1386:405–39.

Majumder S, Varadharaj S, Ghoshal K, Monani U, Burghes AHM, Jacob ST. Identification of a novel cyclic AMP response element (CRE-II) and the role of CREB-1 in the cAMP-induced expression of the survival motor neuron (SMN) gene. J Biol Chem. 2004;279:14803–11.

Mattis VB, Rai R, Wang J, Chang CWT, Coady T, Lorson CL. Novel aminoglycosides increase SMN levels in spinal muscular atrophy fibroblasts. Hum Genet. 2006;120:589–601.

Mendes P, Hoops S, Sahle S, Gauges R, Dada J, Kummer U. Computational models of biochemical networks using COPASI. Methods Mol Biol. 2009;500:17–59.

Michaud M, Arnoux T, Bielli S, Durand E, Rotrou Y, Jablonka S, et al. Neuromuscular defects and breathing disorders in a new mouse model of spinal muscular atrophy. Neurobiol Dis. 2010;38:125–35.

Monani UR, Lorson CL, Parsons DW, Prior TW, Androphy EJ, Burghes AHM, et al. A single nucleotide difference that alters splicing patterns distinguishes the SMA gene *SMN1* from the copy gene *SMN2*. Hum Mol Genet. 1999;8:1177–83.

Monani UR, Sendtner M, Coovert DD, Parsons DW, Andreassi C, Le TT, et al. The human centromeric survival motor neuron gene (*SMN2*) rescues embryonic lethality in *Smn-/-* mice and results in a mouse with spinal muscular atrophy. Hum Mol Genet. 2000;9:333–9.

Munsat TL, Davies KE. International SMA consortium meeting. Neuromuscul Disord. 1992;2:423–8.

Novoyatleva T, Heinrich B, Tang Y, Benderska N, Butchbach MER, Lorson CL, et al. Protein phosphatase 1 binds to the RNA recognition motif of several splicing factors and regulates alternative pre-mRNA processing. Hum Mol Genet. 2008;17:52–70.

Pearn J. Incidence, prevalence and gene frequency studies of chronic childhood spinal muscular atrophy. J Med Genet. 1978;15:409–13.

Prior TW, Krainer AR, Hua Y, Swoboda KJ, Snyder PC, Bridgeman SJ, et al. A positive modifier of spinal muscular atrophy in the *SMN2* gene. Am J Hum Genet. 2009;85:408–13.

Riessland M, Brichta L, Hahnen E, Wirth B. The benzamide M344, a novel histone deacetylase inhibitor, significantly increases SMN2 RNA/protein levels in spinal muscular atrophy cells. Hum Genet. 2006;120:101–10.

Rochette CF, Gilbert N, Simard LR. *SMN* gene duplication and emergence of the *SMN2* gene occured in distinct hominids: *SMN2* is unique to *Homo sapiens*. Hum Genet. 2001;108:255–66.

Russman BS. Spinal muscular atrophy: clinical classification and disease heterogeneity. J Child Neurol. 2007;22:946–51.

Ryll A, Bucher J, Bonin A, Bongard S, Gonçalves E, Saez-Rodriguez J, et al. A model integration approach linking signalling and gene-regulatory logic with kinetic metabolic models. Biosystems. 2014;124:26–38.

Sangaré M, Hendrickson B, Sango HA, Chen K, Nofziger J, Amara A, et al. Genetics of low spinal muscular atrophy carrier frequency in sub-Saharan Africa. Ann Neurol. 2014;75:525–32.

Schmidt H, Jirstrand M. Systems Biology Toolbox for MATLAB: a computational platform for research in systems biology. Bioinformatics. 2006;22:514–5.

Schrank B, Götz R, Gunnersen JM, Ure JM, Toyka KV, Smith AG, et al. Inactivation of the survival motor neuron gene, a candidate gene for human spinal muscular atrophy, leads to massive cell death in early mouse embryos. Proc Natl Acad Sci U S A. 1997;94:9920–5.

Shababi M, Habibi J, Ma L, Glascock JJ, Sowers JR, Lorson CL. Partial restoration of cardio-vascular defects in a rescued severe model of spinal muscular atrophy. J Mol Cell Cardiol. 2012;52:1074–82.

Shababi M, Habibi J, Yang HT, Vale SM, Sewell WA, Lorson CL. Cardiac defects contribute to the pathology of spinal muscular atrophy models. Hum Mol Genet. 2010;19:4059–71.

Shababi M, Lorson CL, Rudnik-Schöneborn S. Spinal muscular atrophy: a motor neuron disorder or a multi-organ disease? J Anat. 2014;224:15–28.

Su YN, Hung CC, Lin SY, Chen FY, Chern JPS, Tsai C, et al. Carrier screening for spinal muscular atrophy (SMA) in 107,611 pregnant women during the period 2005-2009: a prospective population-based cohort study. PLoS One. 2011;6:e17067.

Sugarman EA, Nagan N, Zhu H, Akmaev VR, Zhou Z, Rohlfs AM, et al. Pan-ethnic carrier screening and prenatal diagnosis for spinal muscular atrophy: clinical laboratory analysis of > 72400 specimens. Eur J Hum Genet. 2012;20:27–32.

Sumner CJ, Huynh TN, Markowitz JA, Perhac JS, Hill B, Coovert DD, et al. Valproic acid increases SMN levels in spinal muscular atrophy patient cells. Ann Neurol. 2003;54:647–54.

Thurmond J, Butchbach MER, Palomo M, Pease B, Rao M, Bedell L, et al. Synthesis and biological evaluation of novel 2,4-diaminoquinazoline derivatives as SMN2 promoter activators for the potential treatment of spinal muscular atrophy. J Med Chem. 2008;51:449–69.

Tiziano FD, Lomastro R, Pinto AM, Messina S, D'Amico A, Fiori S, et al. Salbutamol increases survival motor neuron (SMN) transcript levels in leukocytes of spinal muscular atrophy (SMA) patients: relevance for clinical trial design. J Med Genet. 2010;47:856–8.

Vezain M, Saukkonen AM, Goina E, Touraine R, Manel V, Toutain A, et al. A rare SMN2 variant in a previously unrecognized composite splicing regulatory element induces exon 7 inclusion and reduces the clinical severity of spinal muscular atrophy. Hum Mutat. 2010;31:E1110–25.

Villoslada P, Steinman L, Baranzini SE. Systems biology and its application to the understanding of neurological diseases. Ann Neurol. 2009;65:124–39.

Wadzinski BE, Wheat WH, Jaspers S, Peruski LF Jr, Lickteig RL, Johnson GL, et al. Nuclear protein phosphatase 2A dephosphorylates protein kinase A-phosphorylated CREB and regulates CREB transcriptional stimulation. Mol Cell Biol. 1993;13:2822–34.

Wang CH, Finkel RS, Bertini ES, Schroth M, Simonds A, Wong B, et al. Consensus statement for standard of care in spinal muscular atrophy. J Child Neurol. 2007;22:1027–49.

Williamson T, Schwartz JM, Kell DB, Stateva L. Deterministic mathematical models of the cAMP pathway in *Saccharomyces cerevisiae*. BMC Syst Biol. 2009;3:70.

Wu CY, Curtis A, Choi Y, Maeda M, Xu MJ, Berg A, et al. Identification of the phosphorylation sites in the survival motor neuron protein by protein kinase A. Biochim Biophys Acta. 2011;1814:1134–9.

Xiao J, Marugan JJ, Zheng W, Titus S, Southall N, Cherry JJ, et al. Discovery, synthesis and biological evaluation of novel SMN protein modulators. J Med Chem. 2011;54:6215–33.

Zaldívar T, Montejo Y, Acevedo AM, Guerra R, Vargas J, Garofalo N, et al. Evidence of reduced frequency of spinal muscular atrophy type I in the Cuban population. Neurology. 2005;65:636–8.

Zhang Z, Keleman O, Van Santen MA, Yelton SM, Wendlandt AE, Sviripa VM, et al. Synthesis and characterization of pseudocantharidins, novel phosphatase modulators that promote the inclusion of exon 7 into the SMN (survival of motoneuron) pre-mRNA. J Biol Chem. 2011;286:10126–36.

Zi Z. A tutorial on mathematical modeling of biological signaling pathways. Methods Mol Biol. 2012;880:41–51.

Chapter 11
Not Cure But Heal: Music and Medicine

Paulo E. Andrade and Joydeep Bhattacharya

Do you know that our soul is composed of harmony?
Leonardo Da Vinci

Despite evidence for music-specific mechanisms at the level of pitch-pattern representations, the most fascinating aspect of music is its transmodality. Recent psychological and neuroscientific evidence suggest that music is unique in the coupling of perception, cognition, action and emotion. This potentially explains why music has been since time immemorial almost inextricably linked to healing processes and should continue to be.

11.1 Introduction

Music captures our attention almost automatically, moves us emotionally, activates a wide range of brain regions, both at cortical and subcortical level, and engages a spectrum of processes pertaining to attentional, perceptual, memory, emotional, sensorimotor, mental simulation, perception-action, and communication (Koelsch 2012). Therefore, it is not surprising that music has therapeutic effects on the physiological and psychological well beings of individuals (Wheeler 2016). In fact, music therapy is defined as a systematic process in which carefully controlled music is used "in the treatment, rehabilitation, education and training of children and adults suffering from physical, mental or emotional disorder" (Alvin 1975, p. 4; see also Bunt and Stige 2014). The American Music Therapy Association defines it as "the clinical and evidence-based use of music interventions to accomplish individualized goals within a therapeutic relationship by a credentialed professional who has completed an approved music therapy program" (Juslin and Sloboda 2011).

P. E. Andrade · J. Bhattacharya (✉)
Department of Psychology, Goldsmiths, University of London, London, UK
e-mail: j.bhattacharya@gold.ac.uk

A. Cheung-Hoi Yu, L. Li (eds.), *Systems Neuroscience*, Advances in Neurobiology 21, https://doi.org/10.1007/978-3-319-94593-4_11

Despite anthropological and ethnomusicological evidence showing ancient belief on the healing powers of music (Gouk 2000; Merriam 1964), music therapy is a relatively new research field whose formalization of education and research started only a few decades ago (Bunt and Stige 2014). Music can have widespread beneficial effects, but the underlying mechanisms are not well understood or even adequately investigated. Therefore, music therapy as such was not considered amongst the mainstream medical intervention techniques till recently. In fact, medical practitioners in Western and westernized societies, differently from the non-Western medical and healing traditions (Gouk 2000), have been careful and even skeptical in considering music as an effective medium of healing and/or fitting into scientific procedures (Bonny 1986). The reason for this might reside on the fact that the final product of music is the "invisible" (sounds) with a great power of evoking emotions, making music the art that best lends itself to abstraction of our feelings. Indeed, for all pre-literate cultures, music had a sacred character, conveying cosmogonic and existentialist meanings, serving as an articulation point between the physical and the metaphysical (Andrade 2004). Although psychophysiological effects of music (e.g., on electrocardiogram and blood pressure) were documented almost a century ago (Hyde and Scalapino 1918), it is only from 1990s that an increasing amount of empirical studies investigating music-based intervention methods has taken place (Thaut 2005).

Investigations on music-therapy, however, have faced two major problems (Hillecke et al. 2005). The first is related to specificity, or a lack of it, i.e. the question whether observed outcomes are due to music-specific ingredients or to other factors common in the treatment of psychological disorders. In other words, the problem of specificity is a result of the adoption of the psychotherapy research tradition which did not control for unspecific factors influencing music-based intervention outcomes, such as extra-therapeutic aspects, therapeutic relationship, expectancy and placebo effects (Hillecke et al. 2005). The second problem is due to the theoretical heterogeneity among several music therapy approaches such as psychoanalytic music therapy, humanistic music therapy, behavioral music therapy, Nordoff-Robins music therapy, and music medicine. Such heterogeneity creates difficulties not only for an effective communication between music therapy centers but also in the search for working ingredients underlying successful music-based interventions, a crucial knowledge that could help both improving current interventions and guiding new hypotheses. Accordingly, the best way to handle these problems is to develop theories which are both coherent with available empirical knowledge and amenable to be tested and falsified.

Cognitive neuroscience has emerged as a promising scientific field which could give to music therapy its deserved scientific status. Particularly from 2000s, the development of advanced neuroimaging techniques has yielded important insights into the neural correlates of both listening to and engaging with music. It has been shown that music involves a multitude of brain areas dedicated to perception, cognition, i.e. short and long-term memory mechanisms, language, visuospatial processing, sequential processing and prediction, etc., motor skills, sensory-motor integration (Andrade and Bhattacharya 2014; Levitin and Tirovolas 2009; Zatorre

2005). Studies on the processing of musical emotions reveal activations of ancient structures in the limbic (e.g. hippocampus, amygdala) and paralimbic (e.g. caudal orbitofrontal cortex, insula, temporal pole, parahippocampal gyrus) brain regions (Koelsch 2010) as well as associated neurochemical changes related to reward, motivation, pleasure, stress and arousal (e.g. fear) (Chanda and Levitin 2013).

These findings from cognitive neuroscience of music have served as the basis for the development of music-based interventions to ameliorate memory, attention, language, spatial awareness, motor and executive functions (Koelsch 2009; Shannon 2010; Thaut 2005). Psychophysiological effects of music-evoked emotions are guiding interventions aimed at reducing anxiety, stress, pain, treatment of depression, etc. (Chanda and Levitin 2013; Koelsch 2015).

A neurological music therapy has been proposed and defined as "the therapeutic application of music to cognitive, sensory, and motor dysfunctions due to neurologic disease of the human nervous system" (Thaut 2005, p. 126). In the same line, other authors offer explanative models of the factors or working ingredients of music-therapy underlying the positive effect of music on the psychological and physiological health of individuals, such as modulation of attention, emotion, cognition, behavior, and communication (Hillecke et al. 2005; Koelsch 2009).

There has been an upsurge in the number of research publications on the main neurobiological and neurocognitive principles underlying evidence-based music therapy. Nonetheless, it is important to keep in mind that music therapy is a growing and multidisciplinary field that builds on knowledge from acoustics to neurobiology and biomedical research, from psychology to neurology, from sociology to musicology and ethnomusicology (Hillecke et al. 2005). In this chapter we intend to offer an integrated and convergent framework in which we directly address the connections of universal features and functions of musical behavior with the neuroscientific perspective of music therapy, linking anthropological/ethnomusicological and developmental data with the set of already known working ingredients underlying successful music therapy. Our main aim is to provide readers with the relevant references and sufficient information to enable them to form a theoretical and empirical framework with which the available research findings could be critically evaluated and testable hypotheses be formulated for future research.

11.2 Musical Behavior: Universal, Ancient and Precocious

The notion that music is a universal behavior which goes back to the origins of human species appears undisputable (Conard et al. 2009; Mithen 2006). Musical ubiquity across space and time has led many scholars, since Darwin's publication of *Descent of Man* in 1871, to propose that music might be a biological adaptation (Mithen 2006; Wallin and Merker 2001; but see also Justus and Hutsler 2005; Patel 2010; Pinker 1997). It is though debatable which musical traits were under selection pressure (Fitch 2005; Honing et al. 2015; McDermott 2008; Merker et al. 2015; Trainor 2015).

One perhaps wonders why a book chapter on the therapeutic use of music would explore some universals and evolutionary issues of music. Strange as it may seem, this link is of crucial relevance for understanding the profound impact of music on human beings and, hence, its relevance and potential use in clinical and therapeutic settings. Our idea on the importance of understanding typicality of musical behaviors for the cognitive neuroscience of music is well illustrated in Shepherd's words (1994, p. 9) "nothing in neurobiology makes sense except in the light of behavior". In other words, naturally occurring, ancient and universal behaviors, as is the case of music and language, are presumed to be mediated by neural circuits with a deep evolutionary history. Thus, deepening the understanding of what is relevant in terms of behaviors feeds the knowledge of what is relevant in terms of neural circuits and vice-versa.

Actually, several features of musical behaviors that are consistent with an evolutionary account of music represent, in our view, the substrate from which the most cogent arguments for its therapeutic use arises. Music is universal across extant and extinct human cultures. Despite cultural idiosyncrasies, there are important similarities (universalities) in pitch and rhythm structures and in the functions of music across cultures, and most importantly, perception of and attraction for music emerge very early in ontogenetic development. Processing of musical patterns by infants is similar to that of adults so that they respond better to melodies in diatonic scales as well as to consonant patterns and to complex metric rhythm, and further they possess absolute pitch early in life which changes to relative pitch later and have long-term musical memory as well. Finally, there is evidence that structural components such as pitch contour and pitch interval are encoded automatically, even by non-musicians. Taken together these evidence suggests that our auditory pathways are likely to be hard-wired to deal with music-related stimuli (Andrade and Bhattacharya 2003; Justus and Hutsler 2005; McDermott and Hauser 2005; Zatorre 2005).

Music is also a nonverbal form of communication. Historically it is a shared group activity, spontaneous and improvisational involving sound-movement synchronizations (Juslin and Sloboda 2011). Although music is usually defined in terms of an aural phenomenon based on patterned sound along pitch and time dimensions, there is, nevertheless, an emergent consensus that music is inseparable from movement, i.e. it is both sonic and embodied (Cross 2001; Dissanayake 2009; Merker et al. 2015). Moreover, the deep links between music, emotion and movement occurring from infancy to adulthood are also universal.

These profoundly organic features of music together along with its polysemic nature are perhaps the principal reason why serving as an important medium of healing is amongst the universal functions of music across societies, including courtship, praying, mourning and instructing (Blacking 1973; Cross 2001; Gouk 2000; Dissanayake 2009).

11.3 Music-and-Movement Therapy

A main feature of music across cultures is that it is a shared activity involving body movements, such as toe-tapping, head-nodding, hand-clapping (even in solitary appreciation recently allowed by the phonographic industry) and mainly dance (Blacking 1973; Cross 2001; Mithen 2006). In other words, when one moves, plays or sing along with music, the sensory experience of musical patterns is intimately coupled with action. Unsurprisingly some cultures employ terms to define music that are far more inclusive than the Western notion of music, like the word *nkwa* that for the Igbo people of Nigeria denotes "singing, playing instruments and dancing" (Cross 2001, p. 29). Or take *sangeet*, the Sanskrit word for music, which literally means singing together.

Moreover, developmental precursors of music in infancy through early childhood musical behaviors reveal an intense interest in music in the form of universal proto-musical behaviors which are exploratory and kinesthetically embedded and closely bound to vocal play and whole body movement. This scenario displaying an inextricable link between music, emotion and movement actually continues to unfold through adulthood giving origin to some new forms of shared (or not) activities such as singing and dance (Dissanayake 2000; Trehub 2003; Trevarthen 2000).

Indeed, in a way consistent with the view that music and dance are intertwined, some researchers propose that entrainment, i.e. coordination between internal and external rhythms (Merker et al. 2015), also referred to as sensorimotor synchronization (Repp and Su 2013), constitutes the most distinctive musical behavior. It is further argued that entrainment is at the heart of protomusical behaviors and music evolution, being vital to organism's adaptation for conferring survival benefits originated from primary selection pressures, such as better perceptual and predictive capacities, and from secondary selection pressures as well, such as facilitation of social interactions both at the level of mother-baby connections and group cohesion (Merker et al. 2015). Furthermore, similar to primary reward, like food and sex, with high adaptation value, music engages mesolimbic reward network (Koelsch 2015; Zatorre and Salimpoor 2013).

Consistent with the notion that music is both sound and action, and spontaneously elicits movements in the listeners, neuroimaging studies have frequently reported activations of motor areas in the brain, even during simple listening to music (Koelsch 2009). Particularly bilateral frontal and inferior frontal activations, such as premotor frontal areas BA6, dorsolateral prefrontal areas (BA8/9), inferior frontal areas as Broca (BA44/45), insula, and more anterior middle and inferior frontal cortices (46/47), are frequently observed in non-musicians (Platel et al. 1997; Zatorre et al. 1994) and musicians (Zatorre et al. 1998) even during passive listening to pitch sequences or Bach's music (Ohnishi et al. 2001), and during a task of musical imagery (Halpern and Zatorre 1999; Meister et al. 2004). Premotor cortex and supplementary motor areas and cerebellum are also activated during both reproduction (Sakai et al. 1999) and passive listening to rhythms (Chen et al. 2008).

More recent neurological findings have provided additional support for the existence of a natural link between sound and movement, with different sub-regions in the premotor cortex mediating distinct auditory-motor transformations. The ventral part of premotor cortex is particularly involved in anticipation and tap along with rhythms whereas the dorsal part is also recruited during movement synchronization and metrical organization. Even those sounds not clearly connected with a given action elicited activations in premotor sub-regions falling in between ventral and dorsal parts, supplementary motor areas and cerebellum as well (Chen et al. 2008).

Taken together, anthropological/ethnomusicological, developmental and neurological findings are consistent with the hypothesis that the capacity of music to spontaneously and unconsciously modulate motoric behavior is a relevant working ingredient of music therapy. This has been confirmed by the therapeutic potential of music in gait rehabilitation of stroke patients and other motoric problems such as those found in Parkinsons disease, autism, etc. For example, Thaut et al. (1997) found that combination of conventional physical therapy and rhythmic auditory stimulation where patients listen to a metronome or music tapes played over headsets while training their walking, has significantly ameliorated the walking patterns measured at post-test were compared to physical therapy alone.

According to Jankovic (2008) the four cardinal features of Parkinson's are: tremor at rest, rigidity, bradykinesia (the most important clinical feature characterized by slowness of movement) and postural instability, referred together as TRAP. The neurobiochemical signature of Parkinson's patients is a marked deficit in dopamine concentrations in the striatum (part of the subcortical structure in the brain known as basal ganglia).

There is evidence that the auditory-motor integration also occurs at the level of basal ganglia, i.e. amygdala, striatum (dorsal: caudate nucleus and putamen, ventral: nucleus accumbens) and globus pallidus via auditory association areas' projections to these sub-cortical structures. The basal ganglia are particularly involved in the voluntary control of complex movements (Pinel 2011), such as auditory sequencing and timing, musical rhythms (Janata and Grafton 2003; Thaut and Abiru 2010; Zatorre et al. 2007) and speech and language as well (Enard 2011). As pointed out by Thaut and Abiru (2010) these pathways may play a critical role in the facilitative effect of music and auditory rhythm on motor output in Parkinson's disease.

Further, the basal ganglia is also a part of a timing network particularly involved in the extraction of durations using regular beats as a reference (e.g. perception of metrical rhythms), whereas the cerebellum is involved in the perception of absolute duration regardless of the presence of regular beats (Teki et al. 2011). Teki et al. (2011) have shown that a striato-thalamo-cortical timing network, in which striatum and supplementary motor areas (also known for its involvement in timing) are interconnected, is crucial for the beat-based duration perception. Of particular relevance here is the close neurobiological link between rhythmic abilities and Parkinson's disease which is characterized by a primary loss of dopaminergic neurones in a structure of the mesencephalon named substantia nigra (Hughes et al. 1992). More specifically, the nigrostriatal neurons, i.e. neurons from the dopaminergic pathway connecting substantia nigra to the dorsal striatum (also known as mesostriatal dopamine pathway), are critical for interval timing so that Parkinson's patients are also impaired on time and rhythmic discrimination tasks (Teki et al. 2011). Actually, these connections between rhythmicity, motoric functions and dopamine have been confirmed by empirical findings that faster rhythmic auditory stimulation significantly enhanced gait velocity, cadence and walking pattern in Parkinson patients with and without medication (Thaut and Abiru 2010).

Since release of dopamine in the striatum is necessary for basal ganglia mediated motor behaviors, the evidence suggests that positive effects of music on impaired motoric abilities of Parkinson's patients include effects of rhythmicity on the mesostriatal dopamine pathway traditionally known for its motoric functions (planned movements) (Chanda and Levitin 2013; Wise 2009). There is an emerging agreement that the mesostriatal pathway also contributes to reward (Wise 2009). So it is likely that mesostriatal pathway also contributes to motivation beyond motoric functions.

In this motoric context, let us briefly discuss another neurological disorder, Autism Spectrum Disorders (ASD). ASD is mainly characterized by the social communication impairments (difficulties in acquire language and its idiosyncratic use, and impaired social interaction) but often accompanied by stereotyped repetitive behaviors, and increasing evidence indicates that perceptual-motor impairments may be common in this disorder (Srinivasan and Bhat 2013). For example, in the motor domain, patients suffering from ASD have problems with dual and multi-limb coordination, postural control, gait, and imitation and praxis (Srinivasan and Bhat 2013). As perception-action is intricately coupled for music, potentially it could be utilized to facilitate motoric processes in autistic patients (Overy and Molnar-Szakacs 2009; Srinivasan and Bhat 2013).

11.4 Music-and-Emotion Therapy

Possibly the most fascinating aspect of music is its power to induce emotions and influence mood. Music can leave people happy or sad, calm or anxious. Music's extraordinary ability to evoke powerful emotions is likely the main reason why we listen to music and why it is generally referred to as the "language of emotions" (Juslin and Sloboda 2011).

For example, Panksepp (1995) asked hundreds of young men and women why they felt music to be important in their lives, and 70% of both sexes responded it was "because it elicits emotions and feelings", and in the second place came "to alleviate boredom". Interestingly, the neural processes underlying aesthetic responses to music are much more clear and easily detectable scientifically than those elicited by the visual arts, probably because music has a more direct and powerful influence on subcortical emotional systems than the visual arts (Ramachandran and Hirstein 1999).

Indeed, of great relevance to the use of music in clinical and therapeutic settings on the basis of its putative emotional powers is the notion that music elicits emotions rather than merely expresses an emotion that the listener recognizes. Actually, most people experience a particularly intense, euphoric response to music, frequently accompanied by an autonomic or psychophysiological component, described as "shivers-down-the-spine" or "chills". Indeed, listening to music automatically elicits physiological changes in blood circulation, respiration, skin conductivity, body temperature, heart rate, etc. (Krumhansl 1997; Khalfa et al. 2002, 2008) which are autonomic responses of the sympathetic nervous system regulated by noradrenergic neurons in the brainstem and midbrain (Bernatzky et al. 2011; Chanda and Levitin 2013). Recent research provides direct evidence of dopamine release in the striatal dopaminergic region during the experience of chills (Salimpoor et al. 2011).

Consistent with the autonomic responses of the sympathetic nervous system elicited by music, the available literature provides cogent evidence that like biologically relevant visual stimuli, music activates primitive structures in the limbic and paralimbic areas of the brain involved in reward and fear (Blood and Zatorre 2001; Brown et al. 2004). Activation of limbic areas during listening to music was even observed in neonates (Perani et al. 2010). Therefore, music resonates with our basic emotional systems, bringing out many phylogenetically ancient affective emotions and it appears fair to assume that "our love of music reflects the ancestral ability of our mammalian brain to transmit and receive basic emotional sounds that can arouse affective feelings which are implicit indicators of evolutionary fitness" (Panksepp and Bernatzky 2002, p. 134).

Yes, music elicits real emotions! However, in our view, crucial for the study of therapeutic uses of music emotions is understanding what the main structural components of music are, how these components elicit emotions and at what extent these components are universal and/or culturally determined. This knowledge is important not only for the understanding of neurophysiological data but also for the formulation of hypotheses and experimental designs grounded on knowledge about musical parameters and related emotional effects.

Evidence indicates that different configurations of musical characteristics can induce different emotions, and some of these characteristics are both universal and shared with language. Overall, literature indicates that emotions in music can vary across many dimensions, such as mode (major-like or minor-like scales), consonance/dissonance, pitch register, tempo (i.e. number of beats per minute, can be fast or slow), loudness, and complexity (Juslin and Sloboda 2011; Laukka et al. 2013). Particularly tempo (fast or slow) and mode (major or minor), which are associated with listeners' arousal levels and moods, respectively, have been the most extensively examined central features underlying music emotions in comparison to other dimensions. Both major and minor scales are considered as mainly having consonant intervals, i.e. musical notes whose fundamental frequencies form small integer ratios with the first degree (Tonic), such as octave (2:1), perfect fifths (3:2) and fourths (4:3). The Western's major scale, however, contains more consonant intervals, such as a major third (5:4) and major sixth (5:3), compared to Western's minor scale which has a minor third (6:5) and minor sixth (8:3). Dissonant musical stimuli, in contrast, are those based on intervals whose fundamental frequencies stand in more complex ratios such as augmented fourth (45:32), minor second (16:15), and major seventh (15:8), this last being present in both major scale and harmonic minor scale commonly used in the Western music (Bidelman and Krishnan 2009; Tramo et al. 2001).

Music is capable of inducing strong emotions with both positive and negative valence consistently across subjects (Krumhansl 1997) and cultures (Laukka et al. 2013). Cross-culturally, individuals tend to readily associate melodies in major modes at fast tempos as happy and melodies in minor modes at slow tempos as sad, responses that are considered the most consistent emotional judgements in music (Fritz et al. 2009; Laukka et al. 2013). Whereas emotional responses to tempo are really precocious and appear to depend less on experience (Hannon and Trainor 2007), judgements based solely on mode are evident only from 6 years of age (Dalla Bella et al. 2001).

Expectation (hence prediction) constitutes another basic component of music perception (Huron 2006). It operates on a variety of levels, including melodic, harmonic, metrical, and rhythmic, and it addresses the question "what" and "when", that is, what tones or chords are expected to occur and when, in a given musical sequence. It is not only presumed to play an important role in how listeners group the sounded events into coherent patterns, but also to appreciate patterns of tension and relaxation contributing to music's emotional effects. Both cognitive and emotional responses to music depend on whether, when, and how the expectations are fulfilled (Juslin and Västfjäll 2008). For instance, there is a hierarchy of stability in the tones forming the Western major scale in which the most stable note is the first degree (and the octave) of the scale, i.e. the Tonic, that gives to the listener a sensation of resolution. The Tonic is followed in stability by the fifth and the third scale tones, respectively, for their harmonic frequencies/components being more closely related to the Tonic; the stability decreases from fourth to sixth, with the second and seventh degrees the most unstable in this order. In the C major key this continuum from stability to instability will be C, G, E, F, A, D and B, respectively.

The main method used in cross-cultural comparisons of musical expectations is the probe tone task, first developed by Krumhansl and Shepard (1979) for quantifying the perceived hierarchy of stability of tones. Here a melody is presented to the listeners many times, but followed on each occasion by a single probe-tone with varying degree of fitness. Using a rating scale, the listeners assess the degree to which the probe-tone fits their expectations about how the melody might continue (Krumhansl et al. 2000). Cross-cultural studies comparing the fitness ratings given by Indian and Western listeners to North Indian ragas (Castellano et al. 1984), by Western and Balinese listeners to Balinese music (Kessler et al. 1984), native Chinese and American listeners' responses to Chinese and British folk songs (see Thompson et al. 1997) all found strong agreement between listeners from these different musical cultures. Of course, there were effects of expertise depending on the listeners' familiarity with the particular musical culture.

Recent cross-cultural works on melodic expectancies, with Korean music (Nam 1998) and with music of indigenous people of the Scandinavian Peninsula (Krumhansl et al. 2000), have provided additional evidence for the universal reliance on the hierarchical arrangement of pitches, indicating that music draws on common psychological principles of expectation even if musical cultures have a distinct effect in these principles, although the exact way it is accomplished varies with culture. Given the universal presence of consonant intervals in musical scales across cultures, it is reasonable to assume that consonance influences judgment in these probe tone tasks (Laukka et al. 2013). However, another factor underlying probe tone tasks after controlling for consonance/dissonance is the statistical properties of the music, such as the number of times that different tones and tone combinations appear in the presented musical contexts (Krumhansl and Cuddy 2010).

Actually, an interaction between biology and culture appears to underlie the development of emotions. For instance, although evidence indicates that sad-happy judgments based on mode are the result of enculturation, being consistent around the age of six (Dalla Bella et al. 2001), this learning seems to depend on the sensitivity to consonance/dissonance (Cousineau et al. 2012; Hannon and Trainor 2007; Gosselin et al. 2015) which appears to be innate. Although findings on the innate preference for consonance in infants are mixed, sensitivity to consonance/dissonance and processing advantages for consonant stimuli are both present in listeners of all cultures, in young infants (Plantinga and Trehub 2014; Virtala and Tervaniemi 2017, but see also McDermott et al. 2016) and neonates as well (Perani et al. 2010). Indeed, just like adults, infants can detect minor interval changes in melodies of the same contour when melodies are based on consonant intervals, but not when based on dissonant intervals, and are also better in detecting subtle changes in consonant than in dissonant intervals (for a review see McDermott and Hauser 2005). Perani et al. (2010) reported not only differential patterns of brain activations for consonant and dissonant musical stimuli in neonates, but also neural emotional responses in the limbic system for dissonance.

Evidence suggests that consonance/dissonance sensitivity appears to be the universal guide of scale construction across musical cultures of either extant (see Justus and Hutsler 2005) or extinct human societies (Conard et al. 2009). Consonance/dissonance sensitivity, jointly with tempo and complexity (Laukka et al. 2013) is an important mechanism underlying valence-based judgments of music emotions cross-culturally (Fritz et al. 2009). A good evidence for the relevance of consonance/dissonance sensitivity to emotional evaluations based on mode comes from congenital amusics. Congenital amusia is a neurogenetic disorder apparently specific of musical abilities that affects, beyond singing in tune and dancing, pitch processing abilities such as melody discrimination and recognition, pitch direction, small pitch deviations, and sensitivity to consonance/dissonance (Cousineau et al. 2012, 2015; Gosselin et al. 2015). Congenital amusics' deficits in fine-grained pitch perception, harmonicity perception, high abnormal perception of consonance and dissonance and no preference for consonance is causally associated with their inability to make happy-sad judgments based uniquely on mode changes (Cousineau et al. 2015; Gosselin et al. 2015).

We can conclude that substantial evidence suggests that sensitivity to consonance/dissonance is the main innate ability that serves as a fundamental building block of musical enculturation, such as the implicit knowledge of the hierarchical organization of the musical notes in the tonal system (notes and chords that best fit to complete the musical expectations and, hence, to conclude the song), and of the valence-based judgements of music (Hannon and Trainor 2007; Laukka et al. 2013).

In short, musical excerpts played in major mode and fast tempos are frequently associated with happiness, whereas minor mode and slow tempos are considered sad. Considering tempo in isolation, fast-tempos is taken as happier than slow-tempos, although it is more correct to say that faster tempos reflect high-arousal emotions such as happiness, fear and anger, whereas slow tempos is used to express low-arousal emotions such as sadness, tenderness and love. In general, sounds that are loud, dissonant and fast induce high arousal with a negative valence of in the listeners and associated with negative and high arousal emotions such as anger, fear, whereas sounds that are smooth, consonant and at a slow or intermediate tempos induce low arousal and positive feelings such as peacefulness, love, etc., regardless of being music, speech or environmental sounds. Low-pitched sounds are less pleasant and associated with fear or anxiety and used to express negative valences in music. Unexpectedness and irregularity, jointly with dissonance, are consistently associated with negative valence of fear (Laukka et al. 2013; Vieillard et al. 2008).

11.4.1 Music Structure, Music Emotions and Music Therapy

Whereas the mesostriatal pathway is involved in both motor and reward-related functions, the mesolimbic and mesocortical dopamine pathways, often referred as to mesocorticolimbic pathways, appear to be specifically dedicated to process rewarding stimuli (e.g. pleasure) and rewarding aspects of reinforcement learning (Wise 2009). Mesocorticolimbic dopamine pathways are characterized by projections from the ventral tegmental area (mesencephalon) to nucleus accumbens (ventral striatum) and to the prefrontal cortex, respectively (Wise 2009).

Imaging and lesion studies reveal the subcortical foundations of emotional musical experiences in many brain areas that are homologous between humans and all of the other mammals (Blood and Zatorre 2001; Brown et al. 2004). Pleasant/consonant music stimuli are systematically associated with activations of both paralimbic areas (e.g. insula, orbitofrontal cortex and ventromedial prefrontal cortex), involved in reward/motivation, emotion, and arousal, and the mesocorticolimbic dopamine pathways, which is the most important reward pathway; in contrast, unpleasant/dissonant music are associated with activations parahippocampal gyrus (Blood and Zatorre 2001; Menon and Levitin 2005), a paralimbic structure which, jointly with amygdalae, is involved in unpleasant emotional states evoked by pictures with negative emotional valence (Lane et al. 1997); the amygdalae, in its turn, is a key structure in fear processing, and has strong reciprocal connections with parahippocampal gyrus (Mesulam 1998) which is deactivated with consonant music (Blood and Zatorre 2001). Recently it has been demonstrated that unilateral damage to amygdala selectively impairs the perception of emotional expression of fear in scary music (minor chords on the third and sixth degrees, implying the use of many out-of-key notes, and fast tempos), while recognition of happiness (major mode and fast tempo) was normal, and recognition of peacefulness (major mode and intermediate tempo played with pedal and arpeggio accompaniment) and sadness (minor mode at an average slow tempo) in music was less clearly affected by the medial temporal lobe resection (Gosselin et al. 2005; see also Khalfa et al. 2008). Further, patients with lesions in left amygdala show reduced hedonic pleasure in music listening (Griffiths et al. 2004).

We have seen that music can change heart rate variability (HRV), a measure of cardiac autonomic balance. For example, an early study (Umemura and Honda 1998) showed an increase in HRV, an indicator of less stress and greater resilience, during listening to classical music, i.e. Coeur Fragile by Richard Clayderman (minor mode, slow tempo) and Waltz of the Flowers by Tchaikovsky (major mode and slow tempo), which also induced comfort in the listeners. In contrast, decreases in HRV, an indicator of greater stress, was observed during listening to rock music which induced discomfort. However, it is important to note that these results reflect the effects of the musical excerpts specifically selected for this study and not a necessary difference between classical music and rock. In fact, both classical musical excerpts were consonant and in slow tempo. Although the authors did not inform about the rock music, it is likely that it was in fast tempo and had a greater degree of dissonance.

Actually, relaxing music (consonant and in slow tempos) can reduce autonomic responses such as heart rate, blood pressure, and respiratory rate, whereas music excerpts eliciting high arousal emotions such as happy and fear, are characterized by faster tempos and accentuated rhythms, and are often associated with increases in respiration rate, heart rate and blood pressure and skin conductance as well (considered a better measure of the autonomic nervous system because it is under strict control of the sympathetic branch of the nervous system) (Khalfa et al. 2002). Reciprocally, sad music (slow tempo) has smaller responses on these autonomic measures when compared to happy music (Khalfa et al. 2002, 2008; Krumhansl 1997). Some neurochemical findings are consistent with these results. In healthy subjects, stimulating music played at fast tempos, such as techno, increases plasma levels of stress hormones along the hypothalamic-pituitary-adrenal (HPA) axis, such as cortisol and adrenocorticotropic hormone, and other neurochemicals known to mediate stress response, such as norepinephrine (produced in the brainstem locus ceruleus and central and peripheral autonomic nervous system and known to regulate autonomic responses of heart rate, blood pressure, and respiration) and ß-endorphin. Consistently with the role of the amygdale in stress-related responses, this structure is "rich in cortisol receptors and interacts with norepinephrine input and hippocampal connections" (Chanda and Levitin 2013, p. 183). In contrast, relaxing music (characterized by slow tempos and consonance) has been found to decrease these stress-related neurochemicals such as cortisol, norepinephrine and ß-endorphin (for a review see Chanda and Levitin 2013). Interestingly, musical pleasure was shown to be associated with deactivation in the amygdala, supporting the anxiolytic effects of consonant music (Blood and Zatorre 2001).

It is important to remember that both relaxing and happy music are positively valenced and elicit activations of both mesostriatal and mesocortical dopamine pathways involved in reward. However, it is important to remind that dopamine is not the only neurochemical involved in reward (and could not be a 'pleasure' neurochemical per se), and the feeling of pleasure appears to depend on both the release of dopamine and endogenous opioids within the nucleus accumbens (Chanda and Levitin 2013, p. 180). In fact, subjective feelings of pleasure is the product of dynamic interactions of neurochemical concentrations including dopamine, opiods and norepinephrin. Nevertheless, the subjective feelings of pleasure associated with consonant music (happy or calming) and its potential use to soothe pain appear to be related, among other things, to opioid transmission in the nucleus accumbens associated with dopamine release in the mesocorticolimbic pathways (Chanda and Levitin 2013).

Despite several methodological limitations, an old study by Goldstein (1980) observed that by administering naloxone, a well-known opioid antagonist, responses of thrill and chills during music listening were attenuated, suggesting a causal link between positive-valenced music and release of endogenous opiods.

For its robust impact on emotions and socioemotional processes as well as for its associated psychophysiological effect, brain activations, and neurochemical effects (Bernatzky et al. 2011; Chanda and Levitin 2013; Juslin and Sloboda 2011; Koelsch 2015), the potential of music as an effective medium for reducing anxiety, pain, stress and depression has been investigated.

In a meta-analysis including 51 studies using randomized controlled trials, Cepeda et al. (2006) concluded that adding music therapy to standard care in patients with chronic pain or cancer significantly reduced pain and opioid requirements. In a randomized clinical trial, Bringman et al. (2009) found that the relaxing music was more efficient than preoperative administration of the benzodiazepine midazolam. i.e. an anxiolytic drug used worldwide for sedation during minor operations and intensive care, with relaxing music.

In a randomized controlled clinical trial, Siedliecki and Good (2006) assigned their 60 subjects with chronic non-malignant pain syndromes (with back, neck and/or joint pain for at least 6 months and receiving at least one form of traditional medical or surgical pain management) to a music group with music selected by researchers, a music group with music selected by patients or a control group (without music intervention). Although no statistically significant difference was found between the two music groups, both groups had diminished pain and depression symptoms as well as better motor power and abilities in comparison to the control group.

In the same line, cancer patients have a high level of physical and psychological distress. Although therapeutic effects of music have not been clearly demonstrated in the end-of-life care (Bradt and Dileo 2010), there is evidence that just listening to music can improve the psychological state of patients and promote their physical well-being in different oncological contexts (Richardson et al. 2008) including palliative care (Hilliard 2005), radiotherapy (Bradt et al. 2011) and chemotherapy (Lee et al. 2012). When combined with conventional cancer treatments, music therapy can alleviate anxiety and pain (Richardson et al. 2008) and also reduce analgesic requirements (Pyati and Gan 2007).

The most distressing form of cancer treatment is chemotherapy during which patients are in high need of alleviating anxiety, pain and ameliorating psychological state as well. In a pilot study, Lee et al. (2012) carried out the first investigation that systematically compared the EEG responses to relaxation treatment using either monochord or progressive muscle relaxation, thus pioneering for providing information on the possible neural mechanisms underlying the therapeutic effects by music in the oncological context and for testing music therapy against a proven psychological relaxation method. Both groups of patients showed significant improvement in their physical and psychological states and in state anxiety. Further, the EEG signals for both groups showed an increase of posterior theta band (3.5–7.5 Hz) and a decrease of midfrontal beta-2 band (20–29.5 Hz) oscillations during the latter phase of music therapy session. Interestingly, these combination of EEG markers reflect brain's response to relaxed states. These results are also consistent with the findings by Sammler et al. (2007) who reported an increase of frontal midline theta power during listening to pleasant music. Furthermore, only the music therapy group showed a change in the neuronal complexity in the theta band oscillations (Bhattacharya and Lee 2016).

Music-based interventions to reduce pain and anxiety, and promoting well-being also extend to children under different oncological (Barrera et al. 2002; Daveson 2001; Kain et al. 2004) and cardiac contexts (Hatem et al. 2006).

In general, the effectiveness of music as an additional medium to reduce pain in comparison to standard care has been demonstrated in a diversity of clinical populations since 1960s. However, most studies only compared treatments with and without music and did not inform about the action mechanisms underlying music-specific analgesic effects; a more parsimonious explanation could be that music just exerted a distraction effect. Roy et al. (2008) induced pain with thermal stimulations in the subjects to investigate the valence effects of pleasant (positive) and unpleasant (negative) musical excerpts which were matched in terms of arousal. Although valence did not change warmth perception and unpleasant music did not significantly affect pain, the pleasant excerpts, in contrast, significantly reduced pain intensity. This supports the notion that positive-valenced music contributes to analgesic effects.

One problem with the study by Roy et al. (2008) is that all pleasant musical stimuli were at fast tempos, i.e. high in arousal, to match with the high arousal levels of unpleasant excerpts. Therefore, further studies should be conducted to assess specific effects of high (fast tempos) and low (slow tempos) arousal in pleasant musical stimuli. Moreover, some of the pleasant excerpts in the study by Roy et al. (2008) were based on minor modes and some of those based on major modes had a certain degree of dissonance and modulations, thereby increasing the musical complexity. Music parameters like consonance/dissonance, mode and tempo, and complexity as well, are well known factors underlying music emotions.

11.4.2 Music, Emotional Communication and Socio-Cognitive Therapy

Basic emotions (e.g. happiness, sadness, anger, and fear), are more easily communicated through music, and emotional prosody virtually share with music the same patterns in tempo, mode (major/minor), harmony, tonality, pitch, intonation, contour, interval, rhythm, amplitude, timbre, etc., that are specifically involved in communicating emotions (Juslin and Sloboda 2011). Within the perspective that music and movement are inseparable we suggest that protomusical behaviors represent deeper links between music, language and social-cognition.

Human infants interact with their caregivers producing and responding to patterns of sound and action, a rhythmicity that is manifestation of a fundamental musical competence, a musicality that is part of a "natural drive in human sociocultural learning which begins in infancy" (Trevarthen 1999, p. 194). Thus, innate sensitivity to pitch and rhythm structure does not seem to be in vain since infant-directed speech (also known as baby talk or *motherese*), a sing-song-like way adults instinctively use to communicate with children. In comparison to normal speech motherese is slower, with higher average pitch and exaggerated pitch contours and these special features greatly facilitate speech perception and language acquisition (Kuhl 2004; Trehub 2003).

Beyond speaking melodiously, adults also sing play songs and lullabies to children, special genres of music whose common features among cultures are simple pitch contour, repetitions and narrow pitch range (Trehub 2003). Musicality also allows infants to follow and respond accordingly to temporal regularities in vocalization, movement, and time, allowing the initiation of temporally regular sets of vocalizations and movements (Trevarthen 1999).

These protomusical behaviors are so intertwined with protoverbal behaviors that "preverbal origins of musical skills cannot easily be differentiated from the prelinguistic stages of speech acquisition and from the basic alphabet of emotional communication" (Papousek 1996b, p. 92). It is even argued that the musical elements that participate in the process of early communicative development "pave the way to linguistic capacities earlier than phonetic elements" (Papousek 1996a, p. 43). In the pitch dimension, infant-caregiver interactions, cross-culturally, tend to exhibit the same range of characteristics such as exaggerated pitch contours on the caregiver's part ('motherese') and melodic modulation and primitive articulation on the infant's part, all in the context of the rhythmic and kinesthetic interactions. On the part of the infant, these activities develop into exploratory vocal play (between 4 and 6 months) which gives rise to repetitive babbling (from 7 to 11 months) from which emerges both variegated babbling and early words (between 9 and 13 months) (Kuhl 2004; Papousek 1996a, b).

These temporally-controlled interactions involving synchrony and turn-taking are employed in the modulation and regulation of affective state (Dissanayake 2000), and in the achievement and control of joint attention also referred as to 'primary inter-subjectivity' (Trevarthen 1999). Arguably, protomusical behaviors are often reciprocally imitative and also clearly emotionally charged and linked to social and emotion regulations in infancy. The turn-taking aspect of these games is the "rhythmic dance" between mother and child and adults across cultures play reciprocal imitative games with their children that embody the temporal turn-taking. As the infant develops, protomusic becomes music and continues to play this social role throughout life as an indispensable component of most diverse kinds of gatherings from occasional to magnificent ones, rituals and ceremonies (religious, social, healing, etc.). We know that imitation games with music and dance are universal, and the tribal dances itself can be seen as one of the most frequent forms of imitation game, used to develop the in-group sense, the feeling of both "being like the other" and "the other is like me", and thus pertaining to a group (Dissanayake 2000; Trevarthen 1999).

Finally, there is preliminary evidence that motherese (Seltzer et al. 2010) and relaxing music (Nilsson 2009) are associated with increases in oxytocin a neuropeptide released by the posterior pituitary gland which is known to mediate social bonding and affiliation (for a review see Chanda and Levitin 2013).

It is, therefore, in the universal role of music in the modulation and regulation of affective state and inter-subjectivity that relies the potential of music as an intervention tool in clinical contexts related to social communication impairments such as autism (Overy and Molnar-Szakacs 2009). Autism Spectrum Disorders (ASDs) are a group of neurological disorders characterized by social communication impairments, presence of stereotyped and repetitive behaviors and interests, with frequently co-ocurring motor impairments (Srinivasan and Bhat 2013). It was also recently demonstrated that mirror neurons in the posterior inferior frontal gyrus of high-functioning autistic children with autism showed no activity compared to matched controls, during tasks involving imitation and observation of emotion expressions. Subserved by a fronto-parietal network in the human brain the mirror neuron system is a neural network involved in both other's action observation and execution thus allowing imitation processes and grasping of other's intention (theory of mind). The findings by Dapretto et al. (2006) indicate that social-cognitive deficits in autism could be due to a dysfunction of the mirror neuron system.

Since proto-musical behaviors are inextricably linked to imitation, emotionally charged and linked to social and emotion regulations in infancy, features that continue to characterize music behaviors throughout life, it is proposed that the human mirror neuron system and the limbic system interacts in the understanding of and attribution of emotion to complex musical patterns (Overy and Molnar-Szakacs (2009). In the Shared Affective Motion Experience model, SAME, proposed by Overy and Molnar-Szakacs (2009), music is not only patterned sound sequences but above all their resulting intentionally and hierarchically organized sequences of expressive motor acts with emotional meaning, thus involving imitation, synchronization, emotion and social cognition. From this perspective, SAME model has important implications for music therapy and special education.

Improvisational music therapy, defined as the interactive use of live music for engaging clients to meet their therapeutic needs (Bruscia 1998), resembles musical behaviors as typically occurring in the natural social contexts of proto-musical behaviors and shared and improvisational group activities (Juslin and Sloboda 2011). Improvisational music therapy involves spontaneous self-expression, emotional communication and social engagement (Gold et al. 2006; Kim et al. 2009; Overy and Molnar-Szakacs 2009; Srinivasan and Bhat 2013). The use of improvisational music therapy with autistic children have been shown to enhance social skills by improving eye contact, social engagement, spontaneous initiation and emotional understanding, as well as verbal and gestural communication (Gold et al. 2006; Kim et al. 2009, see Srinivasan and Bhat 2013).

11.5 Music Cognition and Intervention in Learning Disabilities

Music is a highly structured sequential organization of sounds and, like language, an acoustically based form of communication with a set of rules for combining limited number of perceptual discrete acoustic elements (pitches in music, and phonemes in language) in an infinite number of ways. According to the shared syntactic integration resource hypothesis (SSIRH) music and language can represent distinct modular systems at the level of long-term perceptual representations (pitch classes and chords and their harmonic relations in music; words and their syntatic features in language) but share cognitive mechanisms underlying *online structural integration* of these representations (Patel 2010) which appears to be inextricably linked to a domain general working memory system (Fedorenko et al. 2007).

Evidence for shared cognitive mechanisms and neural resources involved in tracking auditory patterned sequences and underlying intrinsic rules (syntax) between both domains has been demonstrated in behavioral (Fedorenko et al. 2009), imaging and lesion studies (see Koelsch 2011). Similar evidence of likely sharing mechanisms has been made for children's linguistic abilities (phonology and literacy) and perception of musical sequences controlling for fine-grained pitch perception and rhythm (Zuk et al. 2013), as well as for the prosody of language and music (Patel et al. 2005; Zioga et al. 2016).

Music and language also seem to share some aspects of basic auditory perception. There is evidence that spectrotemporal auditory processing, particularly the processing of fast acoustic transitions, is essential for speech processing (and whose impairment can lead to reading disabilities) and that musical training can improve rapid temporal processing and reading as well (Tallal and Gaab 2006).

It is also argued that accurate detection of supra-segmental cues, i.e. non-phonetic cues such as words, phrases and prosody (stress, rhythm and intonation), are key mechanisms underlying phonological development. Particularly the rhythmic prosody, such as perception of slow amplitude modulation in speech, is considered a key mechanism for segregating syllable onsets and rhymes which are essential for the acquisition of phonological representation in child development (Goswami et al. 2002). It is also proposed that infants build grammatical knowledge of the ambient language by means of this 'prosodic bootstrapping', with rhythm playing one of the central roles (Goswami et al. 2002; Corriveau et al. 2007).

Overlaps between music and other domains, such as auditory perception mechanism involved in speech perception and sequencing processing involved in syntax, is the first necessary requirement for hypothesizing that music can have an impact on these abilities. In fact, there is much evidence that musical experience shapes the brain and is associated with increases in white and gray matter in the corpus callosum and in the auditory and motor cortices as well, and that musical training can have significant positive impact on academic abilities (Merrett et al. 2013; Moreno et al. 2011; Schellenberg 2004; Tierney and Kraus 2013). These results can be taken as evidence for the great potential of music as an intervention tool for learning disabilities.

Nowadays literature abound with evidence the musical abilities correlate positively with language (e.g. phonological abilities) and literacy skills (Anvari et al. 2002; Zuk et al. 2013) and that musical training can improve speech perception abilities and neural coding of speech sounds (Schön et al. 2004; Kraus et al. 2009; Strait et al. 2012), phonological awareness and literacy abilities (Degé and Schwarzer 2011; Register et al. 2007), working-memory (Ribeiro and Santos 2012) and executive functions (Moreno et al. 2011; Forgeard et al. 2008; Zuk et al. 2014), or even general intellectual abilities in the verbal and nonverbal domains (Schellenberg 2004).

Overy (2003) reported both a correlation between song-rhythm tapping and spelling and dyslexic children's improvement in spelling after a music intervention based in song-rhythm tapping, suggesting that segmentation processes are common to both of these skills.

Consistent with the notion of sharing cognitive mechanisms in the perception of rhythm and processing of patterned auditory sequences musicianship has been shown to improve or correlate positively with language skills in numerous areas such as reading ability, phonological awareness, pitch processing in speech, prosody perception, and other language related abilities (for a brief review see Zuk et al. 2013).

Zuk et al. (2013) aimed at investigating the relations between music and language at the level of "patterned sequence processing" in a novel music task called Musical Sequence Transcription Task (MSTT). To control for fine-grained pitch processing they asked forty-three 7 years old Brazilian students to listen to four-sound sequences based on the combination of only two different sound types: one in the low register (thick sound), corresponding to a perfect fifth with fundamental frequencies 110 Hz (A) and 165 Hz (E), and the other in the high register of (thin sound), corresponding to a perfect fourth, with 330 Hz (E) and 440 Hz (A). Children were required to mark the thin sound with a vertical line '|' and the thick sound with an 'O', but were never told that the sequences only consist of four sounds. Performances of second graders on MSTT task were positively correlated with phonological processing and literacy skills, and predicted their literacy abilities 3 years after in the fifth grade (Figuccio et al. 2015). The authors claim that this task can potentially be used as a collective tool for the early screening for children at risk for reading disability.

For its characteristics and for improving social competence music is also a promising medium of intervention with populations handicapped in social skills in general, including the populations with Attention Deficit Hyperactivity Disorder (ADHD) (Rickson 2006; Treurnicht Naylor et al. 2011). However, while music has been shown to be somewhat effective for autistic children, the results in children with ADHD are mixed (Treurnicht Naylor et al. 2011).

11.6 Conclusion

The field of neuroscience, especially the cognitive neuroscience, has been progressing rapidly. We now have myriad neuroimaging technologies available to reveal the intricate functioning of human brain operating across multiple spatiotemporal scales. This chapter presents an overview of the neuroscientific findings of music cognition and also attempts linking them with the working ingredients underlying music therapy. As an empirical research field, music therapy is in its infancy. Though cognitive neuroscience cannot answer sufficiently to all relevant issues in music therapy, we believe that with stronger dialogue between these two disciplines, our understanding about the underlying mechanisms of the healing power of music would be significantly improved (Magee and Stewart 2015; O'Kelly 2016).

References

Alvin J. Music therapy. London: Hutchinson; 1975.

Andrade PE. Uma abordagem evolucionária e neurocientífica da música. Neurociencias. 2004;1(1): 21–33.

Andrade PE, Bhattacharya J. Brain tuned to music. J R Soc Med. 2003;96(6):284–7.

Andrade PE, Bhattacharya J. Music: specialized to integrate? Empir Musicol Rev. 2014;9(3-4): 183–92.

Anvari SH, Trainor LJ, Woodside J, Levy BA. Relations among musical skills, phonological processing, and early reading ability in preschool children. J Exp Child Psychol. 2002;83(2):111–30.

Barrera ME, Rykov MH, Doyle SL. The effects of interactive music therapy on hospitalized children with cancer: a pilot study. Psychooncology. 2002;11(5):379–88.

Bernatzky G, Presch M, Anderson M, Panksepp J. Emotional foundations of music as a non-pharmacological pain management tool in modern medicine. Neurosci Biobehav Rev. 2011;35(9):1989–99. https://doi.org/10.1016/j.neubiorev.2011.

Bhattacharya J, Lee EJ. Modulation of EEG theta band signal complexity by music therapy. Int J Bifurcation Chaos. 2016;26:1650001. https://doi.org/10.1142/S0218127416500012.

Bidelman GM, Krishnan A. Neural correlates of consonance, dissonance, and the hierarchy of musical pitch in the human brainstem. J Neurosci. 2009;29(42):13165–71. https://doi.org/10.1523/JNEUROSCI.3900-09.2009.

Blacking J. How musical is man? Seattle: University of Washington Press; 1973.

Blood AJ, Zatorre RJ. Intensely pleasurable responses to music correlate with activity in brain regions implicated in reward and emotion. Proc Natl Acad Sci U S A. 2001;98(20):11818–23.

Bonny HL. Music and healing. Music Ther. 1986;6(1):3–12.

Bradt J, Dileo C. Music therapy for end-of-life care. Cochrane Database Syst Rev. 2010;(1):CD007169. https://doi.org/10.1002/14651858.CD007169.pub2. Review.

Bradt J, Dileo C, Grocke D, Magill L. Music interventions for improving psychological and physical outcomes in cancer patients. Cochrane Database Syst Rev. 2011;(8):CD006911. https://doi.org/10.1002/14651858.CD006911.pub2. Review.

Bringman H, Giesecke K, Thörne A, Bringman S. Relaxing music as pre-medication before surgery: a randomised controlled trial. Acta Anaesthesiol Scand. 2009;53:759–64.

Brown S, Martinez MJ, Parsons LM. Passive music listening spontaneously engages limbic and paralimbic systems. Neuroreport. 2004;15(13):2033–7.

Bruscia KE. Standards of integrity for qualitative music therapy research. J Music Ther. 1998;35(3):176–200.

Bunt L, Stige B. Music therapy: an art beyond words. London: Routledge; 2014.

Castellano MA, Bharucha JJ, Krumhansl CL. Tonal hierarchies in the music of North India. J Exp Psychol Gen. 1984;113(3):394–412.

Cepeda MS, Carr DB, Lau J, Alvarez H. Music for pain relief. Cochrane Database Syst Rev. 2006;(2):CD004843. https://doi.org/10.1002/14651858.CD004843.pub2.

Chanda ML, Levitin DJ. The neurochemistry of music. Trends Cogn Sci. 2013;17(4):179–93. https://doi.org/10.1016/j.tics.2013.02.007. Review.

Chen JL, Penhune VB, Zatorre RJ. Listening to musical rhythms recruits motor regions of the brain. Cereb Cortex. 2008;18(12):2844–54. https://doi.org/10.1093/cercor/bhn042.

Conard NJ, Malina M, Münzel SC. New flutes document the earliest musical tradition in southwestern Germany. Nature. 2009;460(7256):737–40. https://doi.org/10.1038/nature08169. Epub 2009 Jun 24.

Corriveau K, Pasquini E, Goswami U. Basic auditory processing skills and specific language impairment: a new look at an old hypothesis. J Speech Lang Hear Res. 2007;50(3):647–66.

Cousineau M, McDermott JH, Peretz I. The basis of musical consonance as revealed by congenital amusia. Proc Natl Acad Sci U S A. 2012;109(48):19858–63. https://doi.org/10.1073/pnas.1207989109.

Cousineau M, Oxenham AJ, Peretz I. Congenital amusia: a cognitive disorder limited to resolved harmonics and with no peripheral basis. Neuropsychologia. 2015;66:293–301. https://doi.org/10.1016/j.neuropsychologia.2014.11.031.

Cross I. Music, cognition, culture, and evolution. Ann N Y Acad Sci. 2001;930:28–42.

Dalla Bella S, Peretz I, Rousseau L, Gosselin N. A developmental study of the affective value of tempo and mode in music. Cognition. 2001;80(3):B1–10.

Dapretto M, Davies MS, Pfeifer JH, Scott AA, Sigman M, Bookheimer SY, Iacoboni M. Understanding emotions in others: mirror neuron dysfunction in children with autism spectrum disorders. Nat Neurosci. 2006;9(1):28–30.

Daveson BA. Music therapy and childhood cancer: goals, methods, patient choice and control during diagnosis, intensive treatment, transplant and palliative care. Music Ther Perspect. 2001;19(2):114–20.

Degé F, Schwarzer G. The effect of a music program on phonological awareness in preschoolers. Front Psychol. 2011;2:124. https://doi.org/10.3389/fpsyg.2011.00124.

Dissanayake E. Antecedents of the temporal arts in early mother–infant interaction. In: Wallin NL, Merker B, Brown S, editors. The origins of music. Cambridge: The MIT Press; 2000. p. 389–410.

Dissanayake E. Root, leaf, blossom, or bole: concerning the origin and adaptive function of music. In: Communicative musicality: exploring the basis of human companionship. New York: Oxford University Press; 2009. p. 17–30.

Enard W. FOXP2 and the role of cortico-basal ganglia circuits in speech and language evolution. Curr Opin Neurobiol. 2011;21(3):415–24. https://doi.org/10.1016/j.conb.2011.04.008.

Fedorenko E, Gibson E, Rohde D. The nature of working memory in linguistic, arithmetic and spatial integration processes. J Mem Lang. 2007;56:246–69.

Fedorenko E, Patel A, Casasanto D, Winawer J, Gibson E. Structural integration in language and music: evidence for a shared system. Mem Cogn. 2009;37(1):1–9. https://doi.org/10.3758/MC.37.1.1.

Figuccio M, Andrade P, Andrade O, Gaab N. Music perceptual abilities predict reading and writing skills in young readers: a longitudinal study. Poster accepted to the Massachusetts Neuropsychological Society's Annual Science Symposium, 2015.

Fitch WT. The evolution of music in comparative perspective. Ann N Y Acad Sci. 2005;1060:29–49. Review.

Forgeard M, Winner E, Norton A, Schlaug G. Practicing a musical instrument in childhood is associated with enhanced verbal ability and nonverbal reasoning. PLoS One. 2008;3(10):e3566. https://doi.org/10.1371/journal.pone.0003566.

Fritz T, Jentschke S, Gosselin N, Sammler D, Peretz I, Turner R, Friederici AD, Koelsch S. Universal recognition of three basic emotions in music. Curr Biol. 2009;19(7):573–6. https://doi.org/10.1016/j.cub.2009.02.058.

Gold C, Wigram T, Elefant C. Music therapy for autistic spectrum disorder. Cochrane Database Syst Rev. 2006;(2):CD004381. https://doi.org/10.1002/14651858.CD004381.pub2. Review.

Goldstein A. Thrills in response to music and other stimuli. Physiol Psychol. 1980;8:126–9.

Gosselin N, Paquette S, Peretz I. Sensitivity to musical emotions in congenital amusia. Cortex. 2015;71:171–82. https://doi.org/10.1016/j.cortex.2015.06.022.

Gosselin N, Peretz I, Noulhiane M, Hasboun D, Beckett C, Baulac M, Samson S. Impaired recognition of scary music following unilateral temporal lobe excision. Brain. 2005;128(Pt 3):628–40.

Goswami U, Thomson J, Richardson U, Stainthorp R, Hughes D, Rosen S, Scott SK. Amplitude envelope onsets and developmental dyslexia: a new hypothesis. Proc Natl Acad Sci U S A. 2002;99(16):10911–6.

Gouk P, editor. Musical healing in cultural contexts. Aldershot: Ashgate; 2000.

Griffiths TD, Warren JD, Dean JL, Howard D. "When the feelings gone": a selective loss of musical emotion. J Neurol Neurosurg Psychiatry. 2004;75:344–5.

Halpern AR, Zatorre RJ. When that tune runs through your head: a PET investigation of auditory imagery for familiar melodies. Cereb Cortex. 1999;9(7):697–704.

Hannon EE, Trainor LJ. Music acquisition: effects of enculturation and formal training on development. Trends Cogn Sci. 2007;11(11):466–72. Epub 2007 Nov 5. Review.

Hatem TP, Lira PI, Mattos SS. The therapeutic effects of music in children following cardiac surgery. J Pediatr. 2006;82(3):186–92.

Hillecke T, Nickel A, Bolay HV. Scientific perspectives on music therapy. Ann N Y Acad Sci. 2005;1060:271–82. Review.

Hilliard RE. Music therapy in hospice and palliative care: a review of the empirical data. Evid Based Complement Alternat Med. 2005;2(2):173–8.

Honing H, ten Cate C, Peretz I, Trehub SE. Without it no music: cognition, biology and evolution of musicality. Philos Trans R Soc Lond Ser B Biol Sci. 2015;370(1664):20140088. https://doi.org/10.1098/rstb.2014.0088.

Hughes AJ, Daniel SE, Kilford L, Lees AJ. Accuracy of clinical diagnosis of idiopathic Parkinson's disease: a clinico-pathological study of 100 cases. J Neurol Neurosurg Psychiatry. 1992;55(3):181–4.

Huron D. Sweet anticipation: music and the psychology of expectation. Cambridge: MIT Press; 2006.

Hyde IH, Scalapino W. The influence of music upon electrocardiograms and blood pressure. Am J Phys. 1918;46(1):35–8.

Janata P, Grafton ST. Swinging in the brain: shared neural substrates for behaviors related to sequencing and music. Nat Neurosci. 2003;6(7):682–7. Review.

Jankovic J. Parkinson's disease: clinical features and diagnosis. J Neurol Neurosurg Psychiatry. 2008;79(4):368–76. https://doi.org/10.1136/jnnp.2007.131045. Review.

Juslin PN, Sloboda J, editors. Handbook of music and emotion: theory, research, applications. Oxford: Oxford University Press; 2011.

Juslin PN, Västfjäll D. Emotional responses to music: the need to consider underlying mechanisms. Behav Brain Sci. 2008;31(5):559–75.

Justus T, Hutsler JJ. Fundamental issues in the evolutionary psychology of music: assessing innateness and domain specificity. Music Percept. 2005;23:1–27.

Kain ZN, Caldwell-Andrews AA, Krivutza DM, Weinberg ME, Gaal D, Wang SM, Mayes LC. Interactive music therapy as a treatment for preoperative anxiety in children: a randomized controlled trial. Anesth Analg. 2004;98(5):1260–6.

Kessler EJ, Hansen C, Shepard RN. Tonal schemata in the perception of music in Bali and in the West. Music Percept. 1984;1:276–95.

Khalfa S, Isabelle P, Jean-Pierre B, Manon R. Event-related skin conductance responses to musical emotions in humans. Neurosci Lett. 2002;328(2):145–9.

Khalfa S, Roy M, Rainville P, Dalla Bella S, Peretz I. Role of tempo entrainment in psychophysiological differentiation of happy and sad music? Int J Psychophysiol. 2008;68(1):17–26. https://doi.org/10.1016/j.ijpsycho.2007.12.001.

Kim J, Wigram T, Gold C. Emotional, motivational and interpersonal responsiveness of children with autism in improvisational music therapy. Autism. 2009;13(4):389–409. https://doi.org/10.1177/1362361309105660.

Koelsch S. A neuroscientific perspective on music therapy. Ann N Y Acad Sci. 2009;1169:374–84. https://doi.org/10.1111/j.1749-6632.2009.04592.x. Review.

Koelsch S. Towards a neural basis of music-evoked emotions. Trends Cogn Sci. 2010;14(3):131–7. https://doi.org/10.1016/j.tics.2010.01.002.

Koelsch S. Toward a neural basis of music perception—a review and updated model. Front Psychol. 2011;2:110. https://doi.org/10.3389/fpsyg.2011.00110.

Koelsch S. Brain and music. Chichester: Wiley-Blackwell; 2012.

Koelsch S. Music-evoked emotions: principles, brain correlates, and implications for therapy. Ann N Y Acad Sci. 2015;1337:193–201. https://doi.org/10.1111/nyas.12684.

Kraus N, Skoe E, Parbery-Clark A, Ashley R. Experience-induced malleability in neural encoding of pitch, timbre, and timing. Ann N Y Acad Sci. 2009;1169:543–57. https://doi.org/10.1111/j.1749-6632.2009.04549.x.Review.

Krumhansl CL. An exploratory study of musical emotions and psychophysiology. Can J Exp Psychol. 1997;51(4):336–53.

Krumhansl CL, Cuddy LL. A theory of tonal hierarchies in music. Music Percept. 2010;36:51–87.

Krumhansl CL, Shepard RN. Quantification of the hierarchy of tonal functions within a diatonic context. J Exp Psychol Hum Percept Perform. 1979;5(4):579–94.

Krumhansl CL, Toivanen P, Eerola T, Toiviainen P, Järvinen T, Louhivuori J. Cross-cultural music cognition: cognitive methodology applied to North Sami yoiks. Cognition. 2000;76(1):13–58.

Kuhl PK. Early language acquisition: cracking the speech code. Nat Rev Neurosci. 2004;5(11):831–43. Review.

Lane RD, Reiman EM, Bradley MM, Lang PJ, Ahern GL, Davidson RJ, Schwartz GE. Neuroanatomical correlates of pleasant and unpleasant emotion. Neuropsychologia. 1997;35(11):1437–44.

Laukka P, Eerola T, Thingujam NS, Yamasaki T, Beller G. Universal and culture-specific factors in the recognition and performance of musical affect expressions. Emotion. 2013;13(3):434–49. https://doi.org/10.1037/a0031388.

Lee EJ, Bhattacharya J, Sohn C, Verres R. Monochord sounds and progressive muscle relaxation reduce anxiety and improve relaxation during chemotherapy: a pilot EEG study. Complement Ther Med. 2012;20(6):409–16. https://doi.org/10.1016/j.ctim.2012.07.002.

Levitin DJ, Tirovolas AK. Current advances in the cognitive neuroscience of music. Ann N Y Acad Sci. 2009;1156:211–31. https://doi.org/10.1111/j.1749-6632.2009.04417.x.

Magee W, Stewart L. The challenges and benefits of a genuine partnership between music therapy and neuroscience: a dialog between scientist and therapist. Front Hum Neurosci. 2015;9:223.

McDermott JH. The evolution of music. Nature. 2008;453(7193):287–8. https://doi.org/10.1038/453287a.

McDermott JH, Hauser M. The origins of music: innateness, uniqueness, and evolution. Music Percept. 2005;23:29–59.

McDermott JH, Schultz AF, Undurraga EA, Godoy RA. Indifference to dissonance in native Amazonians reveals cultural variations in music perception. Nature. 2016. https://doi.org/10.1038/nature18635.

Meister IG, Krings T, Foltys H, Boroojerdi B, Müller M, Töpper R, Thron A. Playing piano in the mind—an fMRI study on music imagery and performance in pianists. Brain Res Cogn Brain Res. 2004;19(3):219–28.

Menon V, Levitin DJ. The rewards of music listening: response and physiological connectivity of the mesolimbic system. NeuroImage. 2005;28(1):175–84.

Merker B, Morley I, Zuidema W. Five fundamental constraints on theories of the origins of music. Philos Trans R Soc Lond Ser B Biol Sci. 2015;370(1664):20140095. https://doi.org/10.1098/rstb.2014.0095.Review.

Merrett DL, Peretz I, Wilson SJ. Moderating variables of music training-induced neuroplasticity: a review and discussion. Front Psychol. 2013;4:606. https://doi.org/10.3389/fpsyg.2013.00606. Review.

Merriam A. The anthropology of music. Evanston: Northwestern University Press; 1964.

Mesulam MM. From sensation to cognition. Brain. 1998;121(Pt 6):1013–52. Review.

Mithen S. The singing Neanderthals: the origins of music, language, mind, and body. Cambridge: Harvard University Press; 2006.

Moreno S, Bialystok E, Barac R, Schellenberg EG, Cepeda NJ, Chau T. Short-term music training enhances verbal intelligence and executive function. Psychol Sci. 2011;22(11):1425–33. https://doi.org/10.1177/0956797611416999.

Nam U. Pitch distribution in Korean court music. Evidence consistent with tonal hierarchies. Music Percept. 1998;16(2):243–7.

Nilsson U. Soothing music can increase oxytocin levels during bed rest after open-heart surgery: a randomised control trial. J Clin Nurs. 2009;18(15):2153–61. https://doi.org/10.1111/j.1365-2702.2008.02718.x.

O'Kelly J. Music therapy and neuroscience: opportunities and challenges. Voices. 2016;16(2):1–22.

Ohnishi T, Matsuda H, Asada T, Aruga M, Hirakata M, Nishikawa M, Katoh A, Imabayashi E. Functional anatomy of musical perception in musicians. Cereb Cortex. 2001;11(8):754–60.

Overy K. Dyslexia and Music. From timing deficits to musical intervention. Ann N Y Acad Sci. 2003; 999:497-505. Review

Overy K, Molnar-Szakacs I. Being together in time: musical experience and the mirror neuron system. Music Percept. 2009;26:489–504.

Panksepp J. The emotional source of "chills" induced by music. Music Percept. 1995;13:171–207.

Panksepp J, Bernatzky G. Emotional sounds and the brain: the neuro-affective foundations of musical appreciation. Behav Process. 2002;60(2):133–55.

Papousek H. Musicality in infancy research: biological and cultural origins of early musicality. In: Deliège I, Sloboda J, editors. Musical beginnings. Oxford: Oxford University Press; 1996a. p. 37–55.

Papousek M. Intuitive parenting: a hidden source of musical stimulation in infancy. In: Deliège I, Sloboda J, editors. Musical beginnings. Oxford: Oxford University Press; 1996b.

Patel AD. In: Bailar M, editor. Music, biological evolution, and the brain. Emerging disciplines. Houston: Rice University Press; 2010.

Patel AD, Foxton JM, Griffiths TD. Musically tone-deaf individuals have difficulty discriminating intonation contours extracted from speech. Brain Cogn. 2005;59(3):310–3.

Perani D, Saccuman MC, Scifo P, Spada D, Andreolli G, Rovelli R, Baldoli C, Koelsch S. Functional specializations for music processing in the human newborn brain. Proc Natl Acad Sci U S A. 2010;107(10):4758–63. https://doi.org/10.1073/pnas.0909074107.

Pinel JP. Biopsychology. 8th ed. Boston: Allyn & Bacon; 2011.

Pinker S. How the mind works. New York: Norton; 1997.

Plantinga J, Trehub SE. Revisiting the innate preference for consonance. J Exp Psychol Hum Percept Perform. 2014;40(1):40–9. https://doi.org/10.1037/a0033471.

Platel H, Price C, Baron JC, Wise R, Lambert J, Frackowiak RS, Lechevalier B, Eustache F. The structural components of music perception. A functional anatomical study. Brain. 1997;120(Pt 2):229–43.

Pyati S, Gan TJ. Perioperative pain management. CNS Drugs. 2007;21(3):185–211. Review.

Ramachandran VS, Hirstein W. The science of art: a neurological theory of aesthetic experience. J Conscious Stud. 1999;6(6-7):15–51.

Register D, Darrow AA, Standley J, Swedberg O. The use of music to enhance reading skills of second grade students and students with reading disabilities. J Music Ther. 2007;44(1):23–37.

Repp BH, Su YH. Sensorimotor synchronization: a review of recent research (2006-2012). Psychon Bull Rev. 2013;20(3):403–52. https://doi.org/10.3758/s13423-012-0371-2.Review.

Ribeiro FS, Santos FHD. Musical training and working memory span in beginners, veterans and with no musical knowledge children. Psicologia. 2012;25(3):559–67.

Richardson MM, Babiak-Vazquez AE, Frenkel MA. Music therapy in a comprehensive cancer center. J Soc Integr Oncol. 2008;6(2):76–81. Review.

Rickson DJ. Instructional and improvisational models of music therapy with adolescents who have attention deficit hyperactivity disorder (ADHD): a comparison of the effects on motor impulsivity. J Music Ther. 2006;43(1):39–62.

Roy M, Peretz I, Rainville P. Emotional valence contributes to music-induced analgesia. Pain. 2008;134(1-2):140–7.

Sakai K, Hikosaka O, Miyauchi S, Takino R, Tamada T, Iwata NK, Nielsen M. Neural representation of a rhythm depends on its interval ratio. J Neurosci. 1999;19(22):10074–81.

Salimpoor VN, Benovoy M, Larcher K, Dagher A, Zatorre RJ. Anatomically distinct dopamine release during anticipation and experience of peak emotion to music. Nat Neurosci. 2011;14:257–62.

Sammler D, Grigutsch M, Fritz T, Koelsch S. Music and emotion: electrophysiological correlates of the processing of pleasant and unpleasant music. Psychophysiology. 2007;44(2):293–304.

Schellenberg EG. Music lessons enhance IQ. Psychol Sci. 2004;15(8):511–4.

Schön D, Magne C, Besson M. The music of speech: music training facilitates pitch processing in both music and language. Psychophysiology. 2004;41(3):341–9.

Seltzer LJ, Ziegler TE, Pollak SD. Social vocalizations can release oxytocin in humans. Proc Biol Sci. 2010;277(1694):2661–6. https://doi.org/10.1098/rspb.2010.0567.

Shannon K. Neurologic music therapy: a scientific paradigm for clinical practice. Music Med. 2010;2(2):78–84.

Shepherd G. Neurobiology. Oxford: Oxford University Press; 1994.

Siedliecki SL, Good M. Effect of music on power, pain, depression and disability. J Adv Nurs. 2006;54(5):553–62.

Srinivasan SM, Bhat AN. A review of "music and movement" therapies for children with autism: embodied interventions for multisystem development. Front Integr Neurosci. 2013;7:22. https://doi.org/10.3389/fnint.2013.00022.

Strait DL, Parbery-Clark A, Hittner E, Kraus N. Musical training during early childhood enhances the neural encoding of speech in noise. Brain Lang. 2012;123(3):191–201. https://doi.org/10.1016/j.bandl.2012.09.001.

Tallal P, Gaab N. Dynamic auditory processing, musical experience and language development. Trends Neurosci. 2006;29(7):382–90. Epub 2006 Jun 27. Review.

Teki S, Grube M, Kumar S, Griffiths TD. Distinct neural substrates of duration-based and beat-based auditory timing. J Neurosci. 2011;31(10):3805–12. https://doi.org/10.1523/JNEUROSCI.5561-10.2011.

Thaut MH. Rhythm, music, and the brain: scientific foundations and clinical applications. New York: Taylor & Francis; 2005.

Thaut MH, Abiru M. Rhythmic auditory stimulation in rehabilitation of movement disorders: a review of current research. Music Percept. 2010;27:263–9.

Thaut MH, McIntosh GC, Rice RR. Rhythmic facilitation of gait training in hemiparetic stroke rehabilitation. J Neurol Sci. 1997;151(2):207–12.

Thompson WF, Cuddy LL, Plaus C. Expectancies generated by melodic intervals: evaluation of principles of melodic implication in a melody-completion task. Percept Psychophys. 1997;59(7):1069–76.

Tierney A, Kraus N. Music training for the development of reading skills. Prog Brain Res. 2013;207:209–41. https://doi.org/10.1016/B978-0-444-63327-9.00008-4.Review.

Trainor LJ. The origins of music in auditory scene analysis and the roles of evolution and culture in musical creation. Philos Trans R Soc Lond Ser B Biol Sci. 2015;370(1664):20140089. https://doi.org/10.1098/rstb.2014.0089.

Tramo MJ, Cariani PA, Delgutte B, Braida LD. Neurobiological foundations for the theory of harmony in western tonal music. Ann N Y Acad Sci. 2001;930:92–116. Review.

Trehub SE. The developmental origins of musicality. Nat Neurosci. 2003;6(7):669–73. Review.

Treurnicht Naylor K, Kingsnorth S, Lamont A, McKeever P, Macarthur C. The effectiveness of music in pediatric healthcare: a systematic review of randomized controlled trials. Evid Based Complement Alternat Med. 2011;2011:464759. https://doi.org/10.1155/2011/464759.

Trevarthen C. Musicality and the intrinsic motive pulse: evidence from human psychobiology and infant communication. Music Sci. 1999;3:155–215.

Trevarthen C. Autism as a neurodevelopmental disorder affecting communication and learning in early childhood: prenatal origins, post-natal course and effective educational support. Prostaglandins Leukot Essent Fatty Acids. 2000;63(1-2):41–6. Review.

Umemura M, Honda K. Influence of music on heart rate variability and comfort—a consideration through comparison of music and noise. J Hum Ergol (Tokyo). 1998;27(1-2):30–8.

Vieillard S, Peretz I, Gosselin N, Khalfa S, Gagnon L, Bouchard B. Happy, sad, scary and peaceful musical excerpts for research on emotions. Cognit Emot. 2008;22(4):720–52.

Virtala P, Tervaniemi M. Neurocognition of major-minor and consonance-dissonance. Music Percept. 2017;34(4):387–404.

Wallin NL, Merker B. The origins of music. Cambridge: MIT Press; 2001.

Wheeler BL, editor. Music therapy handbook. New York: Guilford Press; 2016.

Wise RA. Roles for nigrostriatal—not just mesocorticolimbic—dopamine in reward and addiction. Trends Neurosci. 2009;32(10):517–24. https://doi.org/10.1016/j.tins.2009.06.004.

Zatorre R. Music, the food of neuroscience? Nature. 2005;434(7031):312–5.

Zatorre RJ, Chen JL, Penhune VB. When the brain plays music: auditory-motor interactions in music perception and production. Nat Rev Neurosci. 2007;8(7):547–58. Review.

Zatorre RJ, Evans AC, Meyer E. Neural mechanisms underlying melodic perception and memory for pitch. J Neurosci. 1994;14(4):1908–19.

Zatorre RJ, Perry DW, Beckett CA, Westbury CF, Evans AC. Functional anatomy of musical processing in listeners with absolute pitch and relative pitch. Proc Natl Acad Sci U S A. 1998;95(6):3172–7.

Zatorre RJ, Salimpoor VN. From perception to pleasure: music and its neural substrates. Proc Natl Acad Sci U S A. 2013;110:10430–7.

Zioga I, Di Bernardi Luft C, Bhattacharya J. Musical training shapes neural responses to melodic and prosodic expectations. Brain Res. 2016;1650:267–82.

Zuk J, Andrade PE, Andrade OV, Gardiner M, Gaab N. Musical, language, and reading abilities in early Portuguese readers. Front Psychol. 2013;4:288. https://doi.org/10.3389/fpsyg.2013.00288.

Zuk J, Benjamin C, Kenyon A, Gaab N. Behavioral and neural correlates of executive functioning in musicians and non-musicians. PLoS One. 2014;9(6):e99868. https://doi.org/10.1371/journal.pone.0099868.

FSC
www.fsc.org
MIX
Papier | Fördert
gute Waldnutzung
FSC® C083411

Kolibri 360 GmbH
Ferdinand-Jühlke-Straße 7
99095 Erfurt, Deutschland
produktsicherheit@kolibri360.de